Adsorption Desalination and Cooling Systems: Advances in Design, Modeling and Performance

Adsorption Desalination and Cooling Systems: Advances in Design, Modeling and Performance

Editors

Jaroslaw Krzywanski
Norbert Skoczylas
Marcin Sosnowski

MDPI • Basel • Beijing • Wuhan • Barcelona • Belgrade • Manchester • Tokyo • Cluj • Tianjin

Editors
Jaroslaw Krzywanski
Jan Dlugosz University in Czestochowa
Poland

Norbert Skoczylas
The Strata Mechanics Research Institute of the Polish Academy of Sciences
Poland

Marcin Sosnowski
Jan Dlugosz University in Czestochowa
Poland

Editorial Office
MDPI
St. Alban-Anlage 66
4052 Basel, Switzerland

This is a reprint of articles from the Special Issue published online in the open access journal *Energies* (ISSN 1996-1073) (available at: https://www.mdpi.com/journal/energies/special_issues/adsorption_desalination).

For citation purposes, cite each article independently as indicated on the article page online and as indicated below:

LastName, A.A.; LastName, B.B.; LastName, C.C. Article Title. *Journal Name* **Year**, *Volume Number*, Page Range.

ISBN 978-3-0365-5913-1 (Hbk)
ISBN 978-3-0365-5914-8 (PDF)

Contents

About the Editors

Jaroslaw Krzywanski

Jaroslaw Krzywanski (1967) is a Professor at the Faculty of Science and Technology, Jan Dlugosz University in Czestochowa, Poland. He graduated from the Czestochowa University of Technology, Department of Mechanical Engineering, Institute of Thermal Machinery, Poland, and then earned his PhD. and D.Sc. Degrees (Doctor Habilitatus) from the Silesian University of Technology, Faculty of Energy and Environmental Engineering, Poland.

He has published more than 150 refereed works, including papers, two monographs, and conference proceedings, and serves as an editorial board member of several international journals.

His research interests are modelling energy processes and devices, including adsorption cooling and desalination systems, fluidization and circulating fluidized bed technology, chemical looping and calcium looping combustion, and gas emissions, using programmed and artificial intelligence methods.

Norbert Skoczylas

Assoc. Prof. Norbert Skoczylas (1979) is a graduate of the Faculty of Electrical and Computer Engineering, Cracow University of Technology, Poland. He wrote his doctoral thesis on the use of fuzzy logic for the assessment of methane and rock outburst risk. His habilitation was based on studies of the rock-gas system using original measurement methods. In his scientific work, he studies sorption and gas transport in porous media.

Marcin Sosnowski

Assoc. Prof. Marcin Sosnowski (1979), after graduating from the Faculty of Electrical Engineering at the Czestochowa University of Technology, Poland, in 2001, started working as an assistant at the Institute of Piston Engines and Control Techniques at the Faculty of Mechanical Engineering and Information Technology at the Czestochowa University of Technology. At that time, his main scientific interests were focused on the numerical modelling and simulation of flow processes with combustion. He analysed various techniques, prepared customized procedures, and implemented them into open-source codes. He also gained practical skills and engineering know-how by supervising and operating a combined heat and power generating unit at the WARTA S.A. Central Wastewater Treatment Plant in Czestochowa. After obtaining a Ph.D. degree in Mechanical Engineering, he started working in the position of analysis engineer at the Engineering Design Center, which is the R&D centre of General Electric, where he was responsible for the development and implementation of an innovative IT tool designed for the automatic discretization of computational domains describing combustion chambers of aircraft turbine engines and gas turbines. This allowed him to optimize the implementation process of the new product, for which he obtained the Green Belt in Six Sigma process quality management certificate. He conducted numerical analyses of the flow processes with combustion. He cooperated with engineers from the United States and India on projects directly related to numerical modelling. In 2009, he started working at the Faculty of Sciences and Technology of Jan Dlugosz University in Czestochowa, where he was a post-graduate studies manager, the deputy director and director of the Institute, vice dean for student and didactic affairs, and the dean of the Faculty of Science and Technology. He participated in several projects financed from external funds as a manager or coordinator. He has published approx. 100 scientific papers as an author or co-author, and he has reviewed 100+ scientific papers for journals with assigned Impact Factors. He was the member of scientific committee or organizing committee of 30+ conferences and served as topic editor, guest editor, or member of the editorial office of approx. 20 journals. He was the manager of a scientific project entitled "Modelling of thermal contact resistance in granular materials using computational fluid dynamics" financed by the Polish National Science Centre. He participated as a research worker/contractor in projects financed by the Polish National Centre for Research and Development and the Polish National Science Centre. His scientific activity was carried out in several scientific institutions, including international scientific internships and other forms of personal exchange. A measurable effect of his international cooperation are publications co-authored by representatives of recognized scientific centres from Japan, Saudi Arabia, China, Pakistan, Ukraine, and the Czech Republic.

energies MDPI

Editorial

Adsorption Desalination and Cooling Systems: Advances in Design, Modeling and Performance

Marcin Sosnowski [1,*], Jaroslaw Krzywanski [1] and Norbert Skoczylas [2]

[1] Faculty of Science and Technology, Jan Dlugosz University in Czestochowa, Armii Krajowej 13/15, 42-200 Czestochowa, Poland; j.krzywanski@ujd.edu.pl

[2] Strata Mechanics Research Institute of the Polish Academy of Sciences, Reymonta 27, 30-059 Kraków, Poland; skoczylas@imgpan.pl

* Correspondence: m.sosnowski@ujd.edu.pl

1. Introduction

The increase in energy efficiency, reducing energy demand, greenhouse gas emissions and the use of waste, renewable and recycled heat from low-temperature sources are significant challenges today and are key parts of the idea of the 4th Generation District Heating (4GDH). On the other hand, currently about one billion people around the world are suffering from water scarcity, and another three billion are approaching this situation. Only 2.5% of the total water globally is freshwater, of which around 70% is not available, and only 0.4% constitutes the most valuable part of freshwater. Adsorption cooling technology is one of the most effective ways of cooling and producing potable water from renewable and waste heat of the near ambient temperature, including sewage water, solar heat and underground resources.

The brief summary of the contributions accepted for the Special Issue of *Energies*, "Adsorption Desalination and Cooling Systems: Advances in Design, Modeling and Performance", is presented in this paper in order of the publication date.

Citation: Sosnowski, M.; Krzywanski, J.; Skoczylas, N. Adsorption Desalination and Cooling Systems: Advances in Design, Modeling and Performance. *Energies* **2022**, *15*, 4036. https://doi.org/10.3390/en15114036

Received: 22 May 2022
Accepted: 28 May 2022
Published: 31 May 2022

2. A Short Review of the Contributions in the Special Issue

Sztekler et al. [1] tested four newly developed silica-based porous materials and compared them with silica gel, an adsorber commonly paired with water. Extended sorption tests using mercury intrusion porosimetry, gas adsorption and dynamic vapor sorption were performed. The morphology of the samples was determined using a scanning electron microscope. The thermal properties were defined using simultaneous thermal analysis and a laser flash method. Metal organic silica (MOS) nanocomposites analysed in this study had thermal properties similar to those of commonly used silica gel. MOS samples have a thermal diffusivity coefficient in the range of 0.17–0.25 mm^2/s, whereas that of silica gel is about 0.2 mm^2/s. The highest water adsorption capacity was measured for AFSMo-Cu and was equal to 33–35%. For narrow porous silica gel, the mass uptake was equal to about 25%. In the case of water adsorption, it was observed that the pore size of the sorbent is essential, and adsorbents with pore sizes larger than 5 nm are the most recommended for working pairs with water.

Computational fluid dynamic (CFD) analysis of elements of an adsorption chiller with a desalination function were carried out in [2]. The authors presented the results of numerical tests on the elements of an adsorption chiller that comprises a sorption chamber with a bed, a condenser and an evaporator. The simulation is based on the data and geometry of a prototype refrigeration appliance. The simulation of this problem is unique and has not yet been performed and, so far, no simulation of the phenomena occurring in the systems on a real scale has been carried out. The presented results are part of the research covering the entire spectrum of designing an adsorption chiller. The full process of the numerical modeling of the thermal and flow phenomena taking place in the abovementioned components is presented. The computational mesh sensitivity analysis combined in the k-ε turbulence model was performed. To verify and validate the numerical results obtained, they were compared with the results of tests carried out on a laboratory stand at the AGH Center of Energy, Poland. The results of numerical calculations are in good agreement with the results of the experimental tests. The maximum deviation between the pressure obtained experimentally and by simulations is 1.8%, while the temperatures deviation is no more than 0.5%. The results allow the identification of problems and their sources, which allows for future structural modifications to optimize the operation of the device.

A verification of the possibility of increasing the cooling coefficient of performance (COP) and specific cooling power (SCP) of a laboratory adsorption chiller by optimising the length of cycle times and using a copper additive to silica gel with a mass fraction of 15% to increase heat transport in the bed was presented in [3]. The choice of copper among other considered additives was determined by the conclusions from the research on the sorption kinetics of various mixtures, price and availability and a high thermal conductivity. The device was operated in a two-bed mode aimed at producing cooling. The adsorbate was distilled water. The results were compared with those obtained under similar conditions when the beds were only filled with silica gel. As a result of the testing, it was found that the use of the copper additive with the sorbent increased both the COP and SCP. The tests were performed for different cycle times, of 100, 200, 300 and 600 s. With increasing cycle time, the COP also increased. In contrast, the specific cooling power increased only up to a certain point, whereafter its value decreased.

The effects of using steam to preheat the beds of an adsorption chiller with desalination function were investigated experimentally in [4]. The research was carried out on the adsorption chiller, working on a silica gel–water pair, installed in the AGH Center of Energy, Poland. The chiller was modified to preheat the sorbent with the use of steam. The results show that the use of steam instead of water for preheating the bed leads to higher temperatures in the heat exchanger and the bed. As a result, heat transfer from the heating medium to the bed is more intense, and a significant shortening of the desorption process was observed. In the case of using steam for preheating, the desorption time was about 30 s, while for water, it was 300 s. Thanks to this result, it is possible to reduce the size of the device and increase its efficiency. The proposed solution opens a new course of research on adsorption chillers and broadens the horizon of their applications, as steam is a by-product of many industrial processes.

Ahmad A. Alsarayreh et al. analyzed the performance of a variable mode adsorption chiller at different recooling water temperatures in [5]. Adsorption cooling can recover waste heat at low temperature levels, thereby saving energy and reducing greenhouse gas emissions. An air-cooled adsorption cooling system reduces water consumption and the technical problems associated with wet-cooling systems. However, it is difficult to maintain a constant recooling water temperature using such a system. To overcome this limitation, a variable mode adsorption chiller concept was introduced and investigated. A prototype adsorption chiller was designed and tested experimentally and numerically using the lumped model. Experimental and numerical results showed good agreement and a similar trend. The adsorbent pairs investigated in the chiller consisted of silicoaluminophosphate (SAPO-34)/water. The experimental isotherm data were fitted to the Dubinin–Astakhov (D–A), Freundlich, Hill, and Sun and Chakraborty (S–C) models. The fitted data exhibited satisfactory agreement with the experimental data except with the Freundlich model. In addition, the adsorption kinetics parameters were calculated using a linear driving force model that was fitted to the experimental data with high correlation coefficients. The results show that the kinetics of the adsorption parameters were dependent on the partial pressure ratio. Four cooling cycle modes were investigated: single stage mode and mass recovery modes with duration times of 25%, 50% and 75% of the cooling cycle time (denoted as short, medium and long mass recovery, respectively). The cycle time was optimised based on the maximum cooling capacity. The single stage, short mass recovery and medium mass recovery modes were found to be the optimum modes at lower (<35 °C), medium (35–44 °C) and high (>44 °C) recooling temperatures. Notably, the recooling water temperature profile is very important for assessing and optimising the suitable working mode.

Solovyeva et al. in [6] claims that the rapidly growing population, climate change and environment pollution puts heavy pressure on freshwater resources. However, the atmosphere is an immense worldwide and available water source. The adsorptive water harvesting from the atmosphere (AWHA) method is considered a promising alternative to desalination technologies for remote arid regions. The development of novel adsorbents with advanced water-adsorption properties is a prerequisite for practical realisation of this method. Metal–organic frameworks (MOFs) are a novel class of porous crystalline solids that bring a great potential for AWHA due to their extremely high specific surface area, porosity and tailored adsorption properties. The work addresses MIL-160 as a water adsorbent for AWHA. The water-adsorption equilibrium of MIL-160 was studied by volumetric method, the isosteric heat of adsorption was calculated, and finally, the potential of MIL-160 for AWHA was evaluated for climatic conditions of the deserts of Saudi Arabia, Mongolia, the Sahara, Atacama and Mojave as reference arid regions. MIL-160 was shown to ensure a maximum specific water productivity of 0.31–0.33 g_{H2O}/g_{ads} per cycle. High fractions of water extracted (0.90–0.98) and collected (0.48–0.97) could be achieved at a regeneration temperature of 80 °C with the natural cooling of the condenser by ambient air. The specific energy consumption for water production varied from 3.5 to 6.8 kJ/g, which is acceptable if solar heat is used to drive the desorption. The AWHA method employing MIL-160 is a promising way to achieve a freshwater supply in remote arid areas.

The optimisation of the operation of adsorption chillers with a desalination function was presented in [7]. The results of tests carried out on a three-bed adsorption chiller with desalination function are analysed in order to determine the effect of the cycle time on the COP and SCP. The working pair was silica gel and water. The results confirmed the effect of the duration of adsorption and desorption on the COP and SCP of the adsorption chiller. Increasing the duration of the cycle led to an increase in the COP.

The authors in [8] conducted energy and exergy analyses of an adsorption chiller to investigate the effect of recooling-water temperatures on the cooling capacity and COP with variable cycle modes. They investigated both the effect of the recooling-water temperature and the dead state temperature on the exergy destruction in the chiller components. Their results show that there is an optimum reheat cycle mode for each recooling-water temperature range. For the basic single stage cycle, the exergy destruction is mainly accrued in the desorber (49%), followed by the adsorber (27%), evaporator (13%), condenser (9%) and expansion valve (2%). The exergy destruction for the preheating process is approximately 35% of the total exergy destruction in the desorber. By contrast, the precooling process is almost 58% of the total exergy destruction in the adsorber. The exergy destruction decreases when increasing the recooling-water and the dead state temperatures, while the exergy efficiency increases. Nonetheless, the exergy efficiency decreases with an increase in the recooling-water temperature at fixed dead state temperatures. The effect of the mass recovery time in the reheat cycle on exergy destruction was also investigated, and the results show that the exergy destruction increases when the mass recovery time increases. The exergy destruction in the adsorbent beds was the most sensitive to the increase in mass recovery time.

The paper [9] presents experimental results of the metal-based and carbon nanotube additives influence on sorption kinetics of a silica-gel-based adsorption bed in an adsorption chiller. The purpose of the doping is to improve the efficiency of sorption processes within the bed by use of metallic and non-metallic additives characterised by higher thermal diffusivity than basic adsorption material. The higher the thermal conductivity of the bed, the faster the sorption processes take place, which directly translates into greater efficiency of the refrigerator. Sorption kinetics of pure silica gel sorbent doped with a given amount of aluminum (Al) and copper (Cu) powders and carbon nanotubes (CNT) were analyzed. The tests were performed on a DVS Dynamic Gravimetric Vapor Sorption System apparatus used for dynamic vapor sorption measurements. A decrease in the amount of adsorbed water was observed with an increase in the mass share of the additives in the performed studies. Experimental results show that CNTs seem to be the most promising additives as the sorption process time was reduced with the smallest decrease in water uptake. Any significant reduction of adsorption time was noted in case of the Al addition. Whereas in case of Cu doping, the delamination of the mixture was observed.

Skrobek et al. introduced the artificial intelligence (AI) approach for modeling fluidised adsorption beds in [10]. The idea of fluidised bed application allows a significantly increased heat transfer coefficient between adsorption bed and the surface of a heat exchanger, improving the performance of adsorption cooling and desalination systems [11,12]. The long short-term memory (LSTM) network algorithm was used, classified as a deep learning method, to predict the vapor mass quantity in the adsorption bed. The research used an LSTM network with two hidden layers. The network used in the study was composed of seven inputs and one output. The paper presents numerical research concerning mass prediction with the algorithm mentioned above for three sorbents in fixed ad fluidised beds. The results obtained by the developed algorithm of the LSTM network and the experimental tests are in good agreement of the matching the results above 0.95.

Experimental study of three-bed adsorption chiller with desalination function was carried out in [13]. The laboratory test stand included one evaporator, one condenser and three separate tanks for water, desalinated water and brine, respectively. The test stand's scheme and description were presented. All results were obtained during several test hours with stable temperature conditions in the range of 57–85 °C for the heating water. It was found that the COP increased from 0.20 to 0.58 when the heating water temperature increased from 57 to 85 °C. A similar finding is reported for SCP, which increased from 27 to 160 W/kg as the heating water temperature increased from 57 to 85 °C. It can be concluded that the heating water temperature strongly impacts the performance of the adsorption chiller.

The review concerning adsorbents, working pairs and coated beds for natural refrigerants in adsorption chillers is presented in [14]. The selected working pairs were thermodynamically characterised and ranked in terms of refrigerant evaporation temperature values. This was found to be a key parameter affecting the applicability of a given adsorbent/adsorbate pair and the value of SCP and COP, which are now commonly used comparison criteria of adsorption chillers. In the analysis of the coating studies, the focus was on the effect of individual parameters on the performance of the cooling system and the effect of using coated beds compared to packed beds. It was found that a fundamental problem in comparing the performance of different cooling systems is the use of different operating conditions during the tests. Therefore, the analysis compares the performance of the systems along with the most important thermodynamic cycle parameters for the latest studies.

3. Conclusions

The Special Issue of the journal *Energies* brings together research on the advances in design, modeling and performance of adsorption desalination and cooling systems and comprises eleven creative research articles and one review article. The authors of the papers are representatives of six countries, as depicted in Figure 1.

The papers published within the Special Issue are proof of the importance of design, modeling and performance of adsorption desalination and cooling systems. Further investigations concerning the application of adsorption technology for cooling and desalination can be found in [15–20]. Considering the large number of citations of the contributions to the Special Issue, the initiative of proposing topics or Special Issues dedicated to adsorption technology seems to be reasonable.

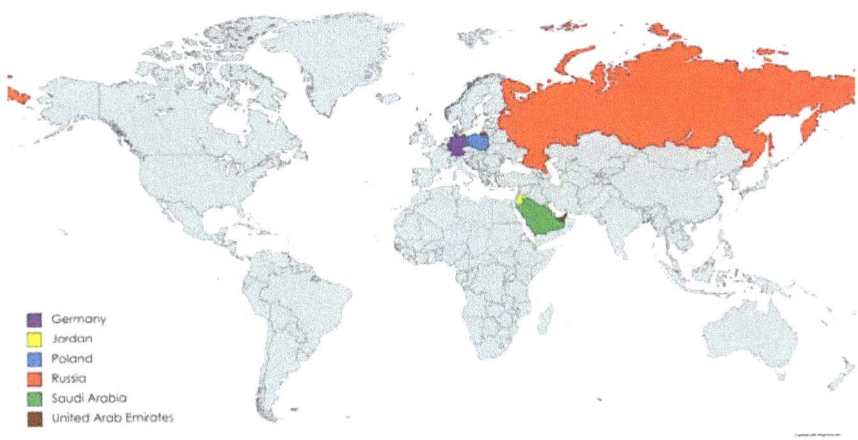

Figure 1. Authors' countries of origin.

Author Contributions: Conceptualization, M.S., J.K. and N.S.; methodology, M.S., J.K. and N.S.; software, M.S., J.K. and N.S.; validation, M.S., J.K. and N.S.; formal analysis, M.S., J.K. and N.S.; investigation, M.S., J.K. and N.S.; resources, M.S., J.K. and N.S.; data curation, M.S., J.K. and N.S.; writing—original draft preparation, M.S., J.K. and N.S.; writing—review and editing, M.S., J.K. and N.S.; visualization, M.S., J.K. and N.S.; supervision, M.S., J.K. and N.S.; project administration, M.S., J.K. and N.S.; funding acquisition, M.S., J.K. and N.S. All authors have read and agreed to the published version of the manuscript.

Funding: This research received no external funding.

Acknowledgments: We would like to express our gratitude to all the anonymous peer-reviewers who have read and evaluated the submissions to the Special Issue "Adsorption Desalination and Cooling Systems: Advances in Design, Modeling and Performance" of the *Energies* journal. Finally, we express our thanks to the authors of all the contributions published in the Special Issue.

Conflicts of Interest: The authors declare no conflict of interest.

References

1. Sztekler, K.; Mlonka-Mędrala, A.; Khdary, N.H.; Kalawa, W.; Nowak, W.; Mika, Ł. Possibility of Advanced Modified-Silica-Based Porous Materials Utilisation in Water Adsorption Processes—A Comparative Study. *Energies* **2022**, *15*, 368. [CrossRef]
2. Sztekler, K.; Siwek, T.; Kalawa, W.; Lis, L.; Mika, L.; Radomska, E.; Nowak, W. CFD Analysis of Elements of an Adsorption Chiller with Desalination Function. *Energies* **2021**, *14*, 7804. [CrossRef]
3. Sztekler, K.; Kalawa, W.; Mika, Ł.; Sowa, M. Effect of Metal Additives in the Bed on the Performance Parameters of an Adsorption Chiller with Desalination Function. *Energies* **2021**, *14*, 7226. [CrossRef]
4. Sztekler, K.; Kalawa, W.; Mika, L.; Lis, L.; Radomska, E.; Nowak, W. The Effects of Using Steam to Preheat the Beds of an Adsorption Chiller with Desalination Function. *Energies* **2021**, *14*, 6454. [CrossRef]
5. Alsarayreh, A.A.; Al-Maaitah, A.; Attarakih, M.; Bart, H.-J. Performance Analysis of Variable Mode Adsorption Chiller at Different Recooling Water Temperatures. *Energies* **2021**, *14*, 3871. [CrossRef]
6. Solovyeva, M.; Krivosheeva, I.; Gordeeva, L.; Aristov, Y. MIL-160 as an Adsorbent for Atmospheric Water Harvesting. *Energies* **2021**, *14*, 3586. [CrossRef]
7. Sztekler, K. Optimisation of Operation of Adsorption Chiller with Desalination Function. *Energies* **2021**, *14*, 2668. [CrossRef]
8. Alsarayreh, A.A.; Al-Maaitah, A.; Attarakih, M.; Bart, H.-J. Energy and Exergy Analyses of Adsorption Chiller at Various Recooling-Water and Dead-State Temperatures. *Energies* **2021**, *14*, 2172. [CrossRef]
9. Sztekler, K.; Kalawa, W.; Mika, Ł.; Mlonka-Medrala, A.; Sowa, M.; Nowak, W. Effect of Additives on the Sorption Kinetics of a Silica Gel Bed in Adsorption Chiller. *Energies* **2021**, *14*, 1083. [CrossRef]
10. Skrobek, D.; Krzywanski, J.; Sosnowski, M.; Kulakowska, A.; Zylka, A.; Grabowska, K.; Ciesielska, K.; Nowak, W. Prediction of Sorption Processes Using the Deep Learning Methods (Long Short-Term Memory). *Energies* **2020**, *13*, 6601. [CrossRef]

11. Krzywanski, J.; Grabowska, K.; Sosnowski, M.; Zylka, A.; Kulakowska, A.; Czakiert, T.; Sztekler, K.; Wesolowska, M.; Nowak, W. Heat transfer in adsorption chillers with fluidized beds of silica gel, zeolite, and carbon nanotubes. *Heat Transf. Eng.* **2021**, *43*, 172–182. [CrossRef]
12. Krzywanski, J.; Grabowska, K.; Sosnowski, M.; Zylka, A.; Kulakowska, A.; Czakiert, T.; Sztekler, K.; Wesolowska, M.; Nowak, W. Heat transfer in fluidized and fixed beds of adsorption chillers. In *E3S Web of Conferences*; EDP Sciences: Ulis, France, 2019; Volume 128, p. 01003.
13. Sztekler, K.; Kalawa, W.; Nowak, W.; Mika, L.; Gradziel, S.; Krzywanski, J.; Radomska, E. Experimental Study of Three-Bed Adsorption Chiller with Desalination Function. *Energies* **2020**, *13*, 5827. [CrossRef]
14. Boruta, P.; Bujok, T.; Mika, Ł.; Sztekler, K. Adsorbents, Working Pairs and Coated Beds for Natural Refrigerants in Adsorption Chillers—State of the Art. *Energies* **2021**, *14*, 4707. [CrossRef]
15. Grabowska, K.; Zylka, A.; Kulakowska, A.; Skrobek, D.; Krzywanski, J.; Sosnowski, M.; Ciesielska, K.; Nowak, W. Experimental Investigation of an Intensified Heat Transfer Adsorption Bed (IHTAB) Reactor Prototype. *Materials* **2021**, *14*, 3520. [CrossRef]
16. Kulakowska, A.; Pajdak, A.; Krzywanski, J.; Grabowska, K.; Zylka, A.; Sosnowski, M.; Wesolowska, M.; Sztekler, K.; Nowak, W. Effect of Metal and Carbon Nanotube Additives on the Thermal Diffusivity of a Silica Gel-Based Adsorption Bed. *Energies* **2020**, *13*, 1391. [CrossRef]
17. Sztekler, K.; Kalawa, W.; Mlonka-Medrala, A.; Nowak, W.; Mika, Ł.; Krzywanski, J.; Grabowska, K.; Sosnowski, M.; Debniak, M. The Effect of Adhesive Additives on Silica Gel Water Sorption Properties. *Entropy* **2020**, *22*, 327. [CrossRef] [PubMed]
18. Sosnowski, M. Evaluation of Heat Transfer Performance of a Multi-Disc Sorption Bed Dedicated for Adsorption Cooling Technology. *Energies* **2019**, *12*, 4660. [CrossRef]
19. Sztekler, K.; Kalawa, W.; Mika, L.; Krzywanski, J.; Grabowska, K.; Sosnowski, M.; Nowak, W.; Siwek, T.; Bieniek, A. Modeling of a Combined Cycle Gas Turbine Integrated with an Adsorption Chiller. *Energies* **2020**, *13*, 515. [CrossRef]
20. Krzywanski, J. A General Approach in Optimization of Heat Exchangers by Bio-Inspired Artificial Intelligence Methods. *Energies* **2019**, *12*, 4441. [CrossRef]

energies MDPI

Article

Experimental Study of Three-Bed Adsorption Chiller with Desalination Function

Karol Sztekler [1,*], Wojciech Kalawa [1], Wojciech Nowak [1], Lukasz Mika [1], Slawomir Gradziel [2], Jaroslaw Krzywanski [3] and Ewelina Radomska [1]

1 Department of Thermal and Fluid Flow Machines, Faculty of Energy and Fuels, AGH University of Science and Technology, Mickiewicza 30 St., 30-059 Krakow, Poland; kalawa@agh.edu.pl (W.K.); wnowak@agh.edu.pl (W.N.); lmika@agh.edu.pl (L.M.); radomska@agh.edu.pl (E.R.)
2 Department of Energy, Faculty of Environmental Engineering and Energy, Cracow University of Technology, Jana Pawla II 37 St., 31-864 Cracow, Poland; gradziel@mech.pk.edu.pl
3 Faculty of Science and Technology, Jan Dlugosz University in Czestochowa, A. Krajowej 13/15, 42-200 Czestochowa, Poland; j.krzywanski@ujd.edu.pl
* Correspondence: sztekler@agh.edu.pl

Received: 6 October 2020; Accepted: 6 November 2020; Published: 8 November 2020

Abstract: Energy efficiency is one of the most important topics nowadays. It is strictly related to energy demand, energy policy, environmental pollution, and economic issues. Energy efficiency can be increased and operating costs reduced by using waste heat from other processes. One of the possibilities is to use sorption chillers to produce chilled water and desalinated water. Low-temperature waste heat is not easy to utilize because of the low energy potential. Using adsorption chillers in low-temperature conditions allows utilizing waste heat and producing useful products in many regions of the world. The paper presents the results of an experimental study carried out on a three-bed adsorption chiller with desalination function, using silica gel and water as a working pair. The laboratory test stand included one evaporator, one condenser, and three separate tanks for water, desalinated water, and brine, respectively. The test stands scheme and description were presented. All results were obtained during several test hours with stable temperature conditions in the range of 57–85 °C for the heating water. It is found that the Coefficient of Performance (COP) increased from 0.20 to 0.58 when the heating water temperature increased from 57 to 85 °C. A similar finding is reported for Specific Cooling Power (SCP), which increased from 27 to 160 W/kg as the heating water temperature increased from 57 to 85 °C. It can be concluded that the heating water temperature strongly impacts the performance of the adsorption chiller.

Keywords: adsorption chiller; coefficient of performance; desalination; energy efficiency; low-temperature heat; silica gel; specific cooling power; waste heat recovery

1. Introduction

The dynamic development of the economy and industrialization in the nineteenth and twentieth century was associated with higher energy consumption, which resulted in an increasing demand for fossil fuels, especially bituminous coal and lignite. However, burning more fossil fuels has caused various ecological and environmental issues, which people have been trying to solve since recent years. One of the instruments to reduce the negative impact of burning fossil fuels on the environment are numerous European Union directives regarding climate protection. These directives concern, for example, the reduction of greenhouse gas emissions, supporting the development of renewable energy and low-carbon technologies, as well as the Emissions Trading System [1]. Similar efforts are also carried out at the global level, as exemplified by the Paris Agreement adopted at the Paris Climate Conference (COP21) in December 2015 [2], and the 24th Conference of the Parties to the United Nations

Framework Convention on Climate Change (COP24) at Katowice in 2018 [3]. The main aims of the agreements made during the conferences mentioned above are reducing energy and natural resources consumption and minimizing the level of pollution emitted to the environment.

According to the Energy Statistics in 2016 and 2017 [4], the global consumption of total energy in Poland in 2017 was 4410 PJ, while the energy consumption by households (including energy consumption by cars) was 1146 PJ, which accounted for approximately 26% of the global consumption of total energy in Poland. In Poland in 2016, the global consumption of total energy and the energy consumption by households (including energy consumption by cars) was 4216 and 1142 PJ, respectively. Furthermore, the energy consumption forecast prepared by Agencja Rynku Energii S.A. [5] indicates the growth of the total primary energy demand and the electricity demand in 2030 by 27% and 54%, respectively, compared to the demand in 2010. It is worth mentioning that electricity consumption by air conditioning and refrigeration systems is one of the factors that significantly contributes to the overall electricity consumption growth. Refrigeration has been developing dynamically since the nineteenth century due to its application in numerous fields, including production, transport and food storage, industry (the necessity of cooling various devices and machines), air conditioning of buildings, household refrigerators, etc. Additionally, more and more rigorous legal regulations concerning environmental protection, e.g., Regulation (European Union) No. 517/2014 of the European Parliament and of the Council of 16 April 2014, on fluorinated greenhouse gases and repealing Regulation (European Commission) No. 842/2006 [6], are being introduced. Therefore, searching for new possibilities for refrigeration systems development seems to be unavoidable.

The technology that meets the challenges of legal regulations and environment protection turned out to be adsorption refrigeration systems. The adsorption refrigerators are a more promising and attractive field for research than conventional electrically-driven vapor compression chillers, since the adsorption systems do not use undesirable chlorofluorocarbons that cause ozone depletion. Substances, e.g., water, used as refrigerants in the adsorption chillers, are environmentally friendly and do not cause ozone depletion [7].

The refrigeration cycles' working principle in adsorption chillers with water as a refrigerant is based on the sorbent's hygroscopic properties, which is a non-toxic substance. The sorbent adsorbs water vapor in the adsorption bed due to the electrostatic attraction and van der Waals forces. A silica-gel, aluminum fumarate, and ferro-aluminophospate can be used as sorbents [8]. Regardless of the sorbent used, heat must be supplied to the saturated sorbent to release the adsorbed water after the adsorption process. Thus, the adsorption chillers use heat to operate, unlike vapor compression chillers that use electricity to operate. The sources of the above mentioned heat include:

- Warm water from a district heating system [9];
- Direct burning of fossil fuels, e.g., gas [9];
- Waste heat from exhaust gases [9];
- Waste heat from power plants or combined heat and power systems [9];
- Renewable energy, especially solar energy [10].

Furthermore, adsorption chillers possess additional advantages over electrically-driven vapor compression chillers, such as the possibility of using water as a refrigerant [10,11], low level of noise, lack of vibrations, and lack of moving parts [9]. Adsorption chillers might be applied wherever the heat in the temperature range of 60–75 °C or a steam in the pressure range of 1-8 bar is available. Nevertheless, adsorption chillers possess a low Coefficient of Performance (COP), not exceeding approximately 0.6 [11], compared to vapor compressor chillers whose COP might be as high as 6 [12]. As a result, many methods of increasing the COP have been investigated. For example, Shabir et al. [13] investigated the COP of the adsorption chiller with different adsorbent/refrigerant pairs. It was found that among the investigated pairs (KOH6-PR/ethanol, WPT-AC/ethanol, Maxsorb-III/methanol, Maxsorb-III/CO_2, Maxsorb-III/n-butane, Maxsorb-III/R-134a, SAC-2/R32, and Maxsorb-III/R507a), the Maxsorb-III/methanol pair showed the highest COP. Uyun et al. [14] applied the internal heat recovery in the investigated adsorption chiller. The COP of the improved chiller increased by 13.4% compared to the COP of the chiller without heat recovery. Chorowski et al. [15] found that the chiller's COP can be raised by 7.5% by modifying only the control system, i.e., switching time and cycle allocation.

In addition to the low COP of adsorption chillers, a cyclic operation resulting in an irregular cold production is their significant drawback [16]. Thus, several (two, three, or more) sub-processes are working alternately, obtaining a quasi-continuous cooling capacity [17–19]. Two-bed and four-bed adsorption chillers have gained much attention in the scientific community in the last years. For example, Pan et al. [20] experimentally investigated the influence of the heating water temperature on the two-bed adsorption chiller's performance. Woo et al. [21] also examined the two-bed adsorption chiller's performance under different operating conditions, but their chiller possessed another water desalination function. Similar studies were conducted by Kim et al. [22], who investigated the water quality produced in the four-bed adsorption chiller. On the other hand, there are only a few studies concerning three-bed adsorption chillers in the literature. Saha et al. [23], who experimentally investigated the three-bed adsorption chiller's performance, reported that its Coefficient of Performance (COP) increased from approximately 0.28 to 0.38 when the heating water temperature increased from 60 to 90 °C. A slightly greater COP, i.e., 0.64, of the three-bed adsorption chiller was obtained by Chorowski and Pyrka [24]. Furthermore, it was concluded that the three-bed application allows chillers to work more effectively compared to two-bed chillers.

This paper presents the results of an experimental study on the influence of the heat source temperature on the Specific Cooling Power (SCP) and the COP of the three-bed adsorption chiller with desalination function, in which two beds work in parallel while the third bed works individually. The present work is the first in the literature dealing with the presented approach, i.e., two beds working in parallel and the third bed working individually, in the considered three-bed adsorption chiller. Water in the temperature range of 57–85 °C was used as a heat source for bed regeneration during the desorption process. Based on the literature review presented in the paragraphs above, it seems that the research on the three-bed adsorption chiller is necessary and justified due to the following reasons: Three-bed chillers are supposed to work more efficiently than two-bed chillers, but they are less complicated and require fewer elements than four-bed chillers. Furthermore, the three-bed adsorption chiller investigated in this work has the water desalination function, and to the best of the authors' knowledge, an experimental study on such a chiller has not been done before.

2. Materials and Methods

2.1. Experimental Setup

The study's base is the three-bed adsorption chiller, located in the Center of Energy at the AGH University of Science and Technology in Cracow, Poland. The simplified diagram and the pictures of the investigated three-bed adsorption chiller are shown in Figures 1 and 2, respectively, while Table 1 contains the chiller's main parameters. The investigated adsorption chiller can operate in two modes, i.e., cooling mode and water desalination mode. Additionally, the chiller can work in the two-bed as well as three-bed mode. The considered three-bed adsorption chiller comprises three beds as the efficiency, in terms of maximizing the cooling capacity and reducing the chilled water outlet temperature fluctuations, of such devices is higher than one with one or two adsorption beds [25]. For example, increasing the number of chiller's beds from two to four can improve efficiency by 70% [25]. Moreover, such an adsorption chiller provides additional functionality to conduct different operational strategies and divide adsorption and desorption processes between different beds. Independently from the operating modes, water and silica gel are used as refrigerant and adsorbent, respectively.

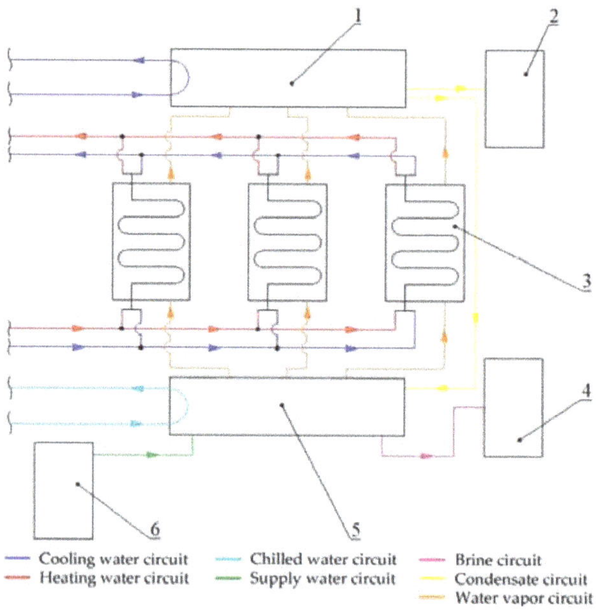

Figure 1. The simplified diagram of the investigated adsorption chiller with desalination function. 1—condenser; 2—distillate tank; 3—adsorbent bed; 4—brine tank; 5—evaporator; 6—deaerator.

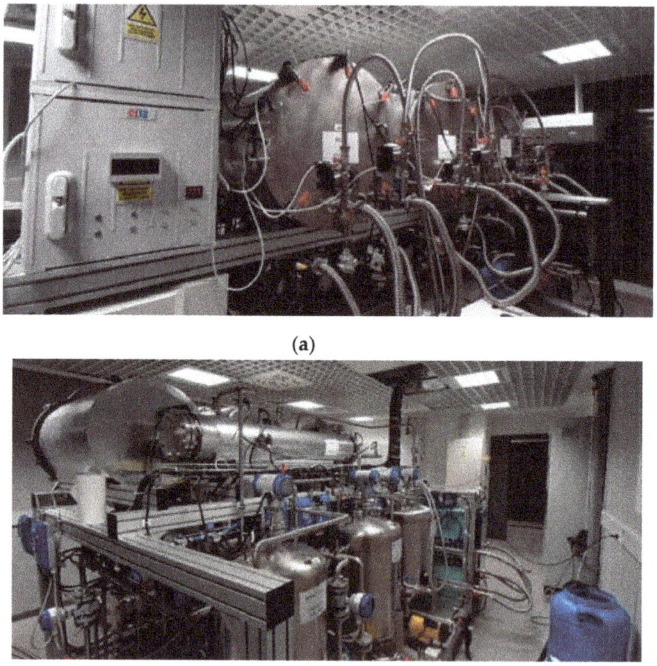

(**a**)

(**b**)

Figure 2. The picture of the investigated adsorption chiller with desalination function: (**a**) Front view, (**b**) back view.

Table 1. The nominal main parameters of the adsorption chiller.

	Parameter	Value	Unit
Evaporator	Cooling capacity	1.10	kW
	The chilled water inlet temperature	12	°C
	The chilled water outlet temperature	7	°C
	The chilled water mass flow	0.052	kg/s
Condenser	Capacity	2.00	kW
	The cooling water inlet temperature	20	°C
	The cooling water outlet temperature	22	°C
	The cooling water mass flow	0.250	kg/s
	Daily distillate production	40	kg

The investigated adsorption chiller consisted of six main elements: Condenser (marked with number 1 in Figure 1), distillate tank (2), adsorbent bed (3), brine tank (4), evaporator (5), and deaerator (6). Saline water (marked with green in Figure 1) from the deaerator flowed to the evaporator. Simultaneously, the chilled water flowed through the evaporator and transferred heat to the saline water. As a result, the saline water evaporated, and the chilled water temperature decreased, thereby obtaining a beneficial cooling effect. The saline water's evaporation at low temperature was possible due to a very low absolute pressure in the evaporator. Then, the water with high salinity, called brine (marked with purple in Figure 1), was pumped to the brine tank, while the evaporated water went to the adsorbent beds and the adsorption phase started. It should be noted that the operating mode of the investigated adsorption chiller for the needs of the experiment was set to three-bed. Hence, beds number one and two worked in parallel, while bed number three worked individually. It means that the adsorption took place simultaneously in beds number one and two, while at the same time, the desorption process was carried out in bed number three. Similarly, when the desorption was in beds number one and two, the adsorption was in bed number three. Nevertheless, the adsorption is an exothermic process, and the cooling water (marked with dark blue in Figure 1) dissipated the heat generated during the adsorption, as well as condensation. This phase ended when the cooling water reached the required temperature. After the adsorption/desorption phase, the three-way valves were adjusted so that the heating water (marked with red color in Figure 1) was flowing through the adsorber until the water reached a required temperature. That phase was called preheating, and the heating water was produced by using the electric water boiler with a power of 18 kW. After the preheating, the desorption phase started. During the desorption, the heating water flowed through the beds. As a result, the water collected earlier at the silica gel's surface was desorbed, its pressure increased, and in a gaseous state flowed to the condenser where it was condensed. The cooling water dissipated the heat of condensation. The part of the condensate (marked with yellow in Figure 1) collected in the bottom part of the condenser was transported to the evaporator through an expansion valve, while the other part was transported to the distillate tank, obtaining a useful distillate. The expansion valve function, which is not included in Figure 1, was to keep the pressure difference between the condenser and the evaporator. After the desorption phase, the heat recovery started. The three-way valves were adjusted so that the cooling water flowed through the adsorber, which was desorbed during the earlier phase. Heat accumulated in the adsorber was not dissipated immediately, but it was transported for some time to the heating water circuit.

As mentioned at the beginning of this section, the chiller can operate in the cooling mode and desalination mode, wherein these modes are simultaneous usually. The desalination mode is active when the saline water supplies the evaporator. The water evaporates but the salt does not, and the so-called brine that remained in the evaporator is pumped to the brine tank. The evaporated water goes to the condenser after the processes described in the previous paragraph. Finally, the water vapor condenses in the condenser, and the useful desalinated water, called distillate, is collected in the distillate tank. However, the experiments did not focus on the desalination mode, but on the cooling mode.

2.2. Measuring Devices and Data Acquisition System

The pressures, temperatures, and flow rates were measured during the experiment. The measuring devices and their main parameters are collected in Table 2. The PT-1000 sensors were used to measure the beds' temperatures, evaporator, condenser, inlet, and outlet temperatures of the cooling water, heating water, and chilled water. Pressure transducers measured the pressure in the condenser, evaporator, and beds. Electromagnetic flow meters measured the flow rates of the heating water, cooling water, and chilled water.

Table 2. Measuring devices and their main parameters.

Measured Value	Measuring Device	Range	Accuracy
Temperature	Pt-1000 sensor	−80 to 150 °C	±0.1 °C
Pressure	Pressure transducer	0 to 99 kPa	±0.5%
Flow rate	Electromagnetic flow meter	1–100 L/min	±0.5%

All of the measuring devices and measuring transducers were power-supplied from the switchgear fixed to the experimental setup supporting structure. The instrumentation, control, and automation section was equipped with a programmable logic controller. The appropriate software of the control system allowed controlling the pumps and valves, setting the duration time of each phase, as well as recording the measured values. The measured values were recorded every 5 s using the SCADA control system and a personal computer. The obtained data were exported in a tabular form in csv format, which allowed for analysis of the collected data. The analysis of the obtained data was done using Microsoft Excel.

All test stand's elements were thoroughly inspected. The measurements were made after the refrigerator was started and its working parameters stabilized. The stabilization time of the chiller's parameters was about 1 h.

2.3. Operating Conditions

Experimental studies were carried out to optimize the adsorption chiller's critical parameters, influencing its performance and efficiency. During the experiments, beds number one and two worked in parallel during both adsorption and desorption phases, while bed number three worked individually. Only one configuration of the sequences was used, and the duration time of each phase was determined on an individual basis. The duration time of each phase is presented in Table 3.

Table 3. The time of each phase.

Chiller Operation Phase	Time, (s)
Preheating	50
Adsorption/desorption	170
Heat recovery	30

As shown in Table 3, the total time of the device's operation cycle in the automatic mode consists of preheating time, desorption time, heat recovery time, and desorption time. Thus, the total time of the full working cycle is 420 s in the case of the investigated chiller. The desorption and adsorption of each bed are preceded by preheating and heat recovery, respectively, which is necessary for the proper operation of the device. However, the preheating and heat recovery extend the total time of the full cycle of each bed by 80 s, which is 23.5% of the one full cycle and may influence the coefficient of performance. As commonly known, the desorption is approximately 2 times faster than the adsorption, hence, the alternately-working beds are intentional. The parallel work of beds number one and two allows saving the preheating and heat recovery time in one bed since these processes are simultaneous in the investigated chiller. This should improve the three-bed adsorption chiller's coefficient of performance compared to the two-bed adsorption chiller within a certain heating water temperature range.

The experimental studies were carried out to investigate the influence of the heating water temperature used to regenerate the bed during the desorption on the Coefficient of Performance (COP), Specific Cooling Power (SCP), average evaporator temperature, average evaporator pressure, and average chilled water temperature. The beds' pressure and the average temperature of the evaporator, condenser, and beds, were also analyzed. The COP and SCP were calculated according to the formulas given in [26], which are presented below:

$$COP = \frac{CC}{HP}, \tag{1}$$

$$CC = \dot{m}_c \cdot c_{p,c} \cdot (T_{i,c} - T_{o,c}), \tag{2}$$

$$HP = \dot{m}_h \cdot c_{p,h} \cdot (T_{i,h} - T_{o,h}), \tag{3}$$

$$SCP = \frac{CC}{m_a}, \tag{4}$$

The symbols used in Equations (1)–(4) correspond to: CC—cooling capacity, (W); HP—heating power supplied to the system, (W); $\dot{m}$—flow rate, (kg/s); c_p—specific heat of water, which was assumed to be 4200 $Jkg^{-1}K^{-1}$; T—temperature, (°C); m_a—the mass of sorbent, which was 12 kg. Additionally, the subscripts c, h, i, and o correspond to chilled water, heating water, inlet, and outlet, respectively. The SCP determines the cooling capacity obtained from 1 kg of the sorbent. Additionally, the COP of the investigated three-bed adsorption chiller was compared with the COP of the two-bed adsorption chiller investigated by Sztekler et al. [26].

The heating water temperature was changed in the range of 57–85 °C. The cooling water was used to dissipate the heat from the condenser and beds, and its inlet temperature was 20 °C. The maximum temperature difference between the outlet and inlet cooling water from the condenser was approximately 2 °C. The adsorption chiller produced chilled water. The chilled water's outlet temperature varied from 8 to 16 °C, with a maximum temperature difference between the inlet and outlet temperature of 2.5 °C. The mass flow of each water circuit was constant during the experiments.

3. Results and Discussion

The experimental studies focused on the impact of heating water temperature, which varied in the range of 57–85 °C, on the three-bed adsorption chiller's main operating parameters. The results show that changing the temperature of heating water strongly influences the COP and SCP of the chiller with two beds working in parallel and one bed working individually, as presented in Figures 3 and 4, respectively. When the heating water temperature increased from 57 to 85 °C, the COP increased by 190% (from 0.20 to 0.58).

The COP of the investigated chiller, in this work the three-bed adsorption chiller, is difficult to be compared to the COP of the other chillers investigated by the other researchers, since the COP is influenced by many factors, including the beds' construction, capacity, and material. Therefore, the three-bed adsorption chiller's COP obtained in this work is compared only to the COP of the same chiller but working in a two-bed mode, as investigated in the authors' previous study [26]. It should be noted that the temperatures of chilled and cooling water in the case of the two-bed adsorption chiller were 13 and 18 °C, respectively, which is almost the same as in the case of the three-bed chiller.

As can be seen in Figure 3, the three-bed chiller's COP increased by about 190% when the heating water temperature increased from 57 to 85 °C. Similarly, the SCP, which determines the cooling capacity obtained from 1 kg of the sorbent, increased from 27 to 160 W/kg when the heating water temperature was increased from 57 to 85 °C. Additionally, for the heating water temperature above 75 °C, the COP of the considered three-bed adsorption chiller is higher by the maximum value of about 16% than the two-bed adsorption chiller's COP investigated in the previous work [26]. This confirms that the three-bed adsorption chiller performance in temperatures above 75 °C is higher than the chiller with two adsorption beds. On the other hand, the three-bed chiller's COP (two beds working in parallel and one bed working individually) is lower compared to the two-bed chiller's COP for the heating water temperature below 75 °C. A possible explanation of such a result is most likely due to a low heat transfer rate caused by a small temperature difference between the heating water and beds. As a result, a power shortage occurs in each chiller's element, especially in the evaporator, which is enhanced in the three-bed chiller due to an increased number of beds and larger sorbent mass.

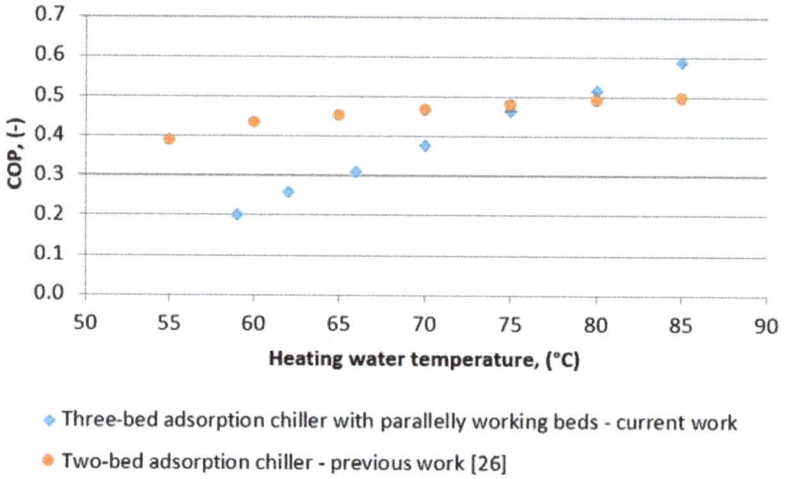

Figure 3. The influence of the heating water temperature on the coefficient of performance (COP).

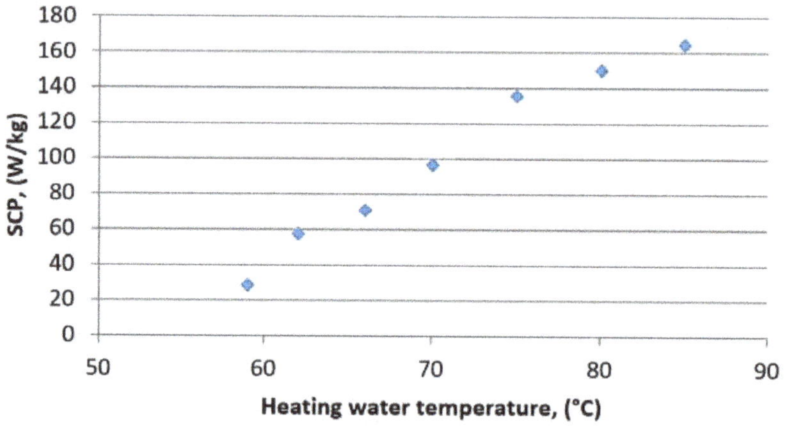

Figure 4. The influence of the heating water temperature on the specific cooling power (SCP).

The experiments' results indicate the strong influence of the heating water temperature on the COP and SCP. As the heating water temperature increased, both the COP and SCP rose due to a more intense water release from the silica gel and faster desorption process. The heating water temperature increment shortened the desorption time and accelerated the regeneration of the bed. As a result, more water from the evaporator was adsorbed by the sorbent. Consequently, the evaporator's cooling capacity increased, and the generated chilled water's temperature decreased, which is shown in Figure 5. The chilled water mass flow rate was constant during the experiments, and when the heating water temperature increased from 57 to 85 °C, the chilled water temperature decreased from approximately 16 to 10 °C. This means that the chilled water temperature decreased by 6 °C while maintaining a constant mass flow rate through the evaporator.

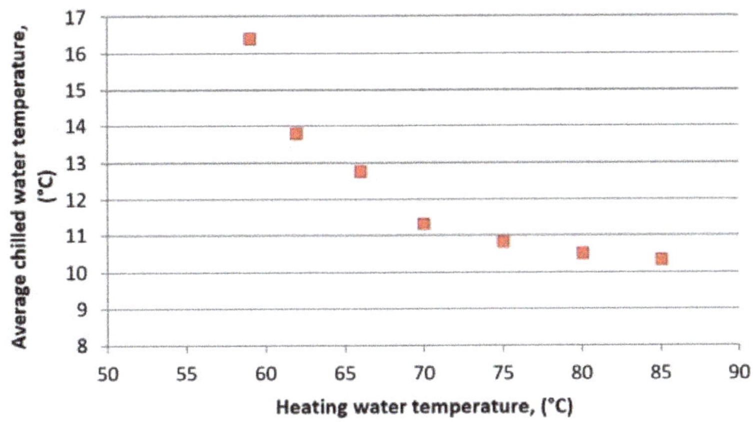

Figure 5. The influence of the heating water temperature on the average chilled water temperature.

The chilled water temperature drop caused by increasing the heating water temperature was related to the increment of the adsorbent sorption properties. As a result, the average evaporator pressure and the average evaporator temperature decreased, which is shown in Figure 6. As can be seen, the average evaporator pressure and the average evaporator temperature were 2 kPa and 16 °C, respectively, when the heating water temperature was 57 °C. As the heating water temperature increased to 85 °C, the average evaporator pressure and the average evaporator temperature decreased to 1.05 kPa and 7 °C, respectively. The evaporator temperature changes were a result of the processes in the chiller related to the adsorption/desorption of water vapor in the silica-gel in the beds.

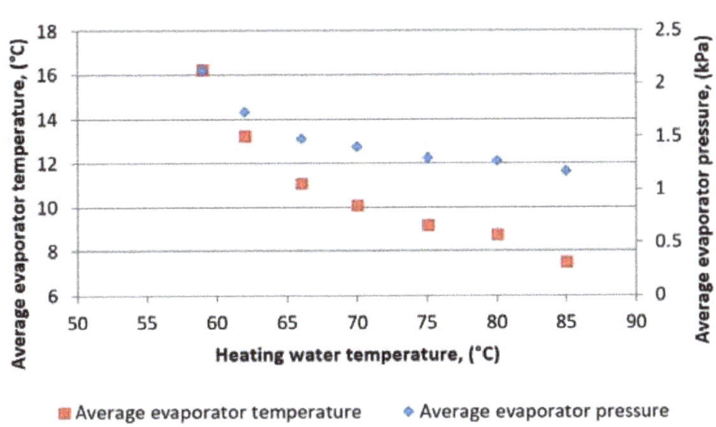

Figure 6. The influence of the heating water temperature on the average evaporator temperature and the average evaporator pressure.

As mentioned before, the heating water temperature was changed in the range of 57–85 °C. As shown in Figure 7, the temperature changes of beds are uniform, and the temperature changes resulted from the alternating adsorption/desorption processes. Additionally, when the heating water temperature decreased, the beds' temperatures decreased from 70 to 50 °C during the desorption. Moreover, for low heating water temperatures, the bed working individually heated up and cooled down faster than the two beds working in parallel. Moreover, as the heating water temperature decreased, the evaporator temperature increased, which indicates a low heat transfer rate from the evaporator to the water, and confirms a less intensive beds' operation. It is worth noting that the beds' temperatures were measured inside the heat exchanger, between the silica gel and the heat exchanger pipes, and manual settings set the time of the adsorption phases. The evaporator temperature increased from 7 to 16 °C when the heating water temperature decreased. On the other hand, the condenser temperature was constant, independently from the heating water temperature.

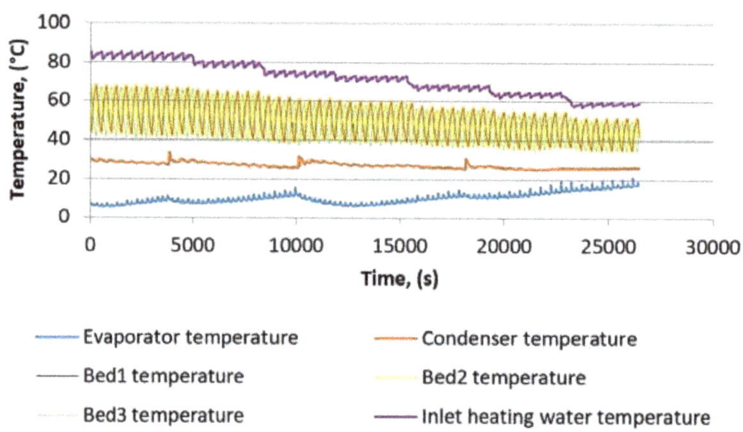

Figure 7. The changes in the evaporator, condenser, and bed temperatures in the function of time.

Figure 8 presents the beds' pressure in the function of time during the adsorption/desorption cycles. The results show that bed number one and bed number two worked in a parallel arrangement, which means they worked simultaneously during the adsorption/desorption phase, as explained in Section 2.1. Bed number three worked in a singular arrangement during the different working phases. The pressure changes in the beds reflect the operating cycle of the adsorption chiller. When the water vapor flows between the evaporator and the bed during the adsorption phase, the system is trying to equalize the pressure between the evaporator and bed. The pressure in the beds during the adsorption is approximately 1 kPa. During the desorption phase, the heating water is pumped through the beds. As a result, silica gel is heated, and the adsorbed water molecules evaporate. The released water vapor flows to the condenser, and the bed pressure increases to the condenser pressure. The bed pressure is 7.5 kPa during the desorption cycles. As can be seen, there is a dependency between the phases that is related to the pressure and temperature distribution. Furthermore, the pressure in bed numbers one and two is very similar, so the beds could be regarded as one big bed. However, the beds are divided into two beds to allow changing the beds' configuration into the two-bed mode or three-bed mode, as mentioned in Section 2.1. Additionally, the working configuration of the beds in the three-bed mode can be changed, i.e., beds one and two work in parallel, while bed three works singularly, or bed one and three work in parallel, while bed two works singularly, or bed two and three work in parallel, while bed one works singularly.

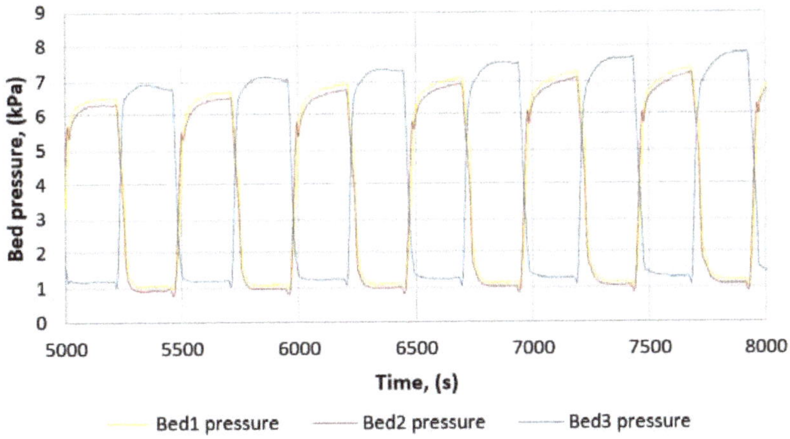

Figure 8. The changes in the pressure of beds in the function of time.

4. Conclusions

This paper deals with effective cooling and desalination technology, suitable for cleaner production and environmental conservation. It constitutes the first paper available in open literature dealing with the considered three-bed adsorption chiller with two beds working in parallel and one bed working individually, and the heating water's influence on its main operating parameters.

According to the obtained results, it can be concluded that the COP and SCP of the three-bed adsorption chiller are strongly dependent on the heating water temperature. When the heating water temperature increased from 57 to 85 °C, the COP increased from 0.20 to 0.58 (by 190%). Therefore, increasing the heating water temperature results in an increase in the COP. Furthermore, SCP, which determines the cooling capacity obtained from 1 kg of the adsorbent, increases. As the heating water temperature increased from 57 to 85 °C, the SCP also increased from 27 to 160 W/kg. An increase in the heating water temperature accelerates the desorption process; thus, the silica gel's water adsorbed is released faster, and the chiller's cooling capacity rises.

The experimental studies show that the COP of the three-bed adsorption chiller is higher than the two-bed adsorption chiller, which was investigated in the previous work [26], by about 16% for the heating water temperature of 85 °C. However, the three-bed chillers' COP was lower compared to the two-bed chiller for the heating water temperature below 75 °C. Such a decline in the COP was caused by the slow heat transfer rate, and thus power shortage, especially in the evaporator.

In further studies, the authors are planning to investigate experimentally the performance of the three-bed chiller with alternately working beds, which may improve the chiller's COP, especially at low heating water temperatures.

Author Contributions: Conceptualization, K.S. and W.N.; methodology, K.S. and S.G.; formal analysis, K.S. and L.M.; investigation, W.K.; resources, W.K.; data curation, K.S.; writing—original draft preparation, K.S. and E.R.; writing—review and editing, E.R. and J.K.; supervision, S.G., W.N., J.K., and L.M.; project administration, K.S.; funding acquisition, L.M. and W.N. All authors have read and agreed to the published version of the manuscript.

Funding: This research was funded by a subsidy from the Faculty of Energy and Fuels AGH University of Science and Technology, number 16.16.210.476.

Conflicts of Interest: The authors declare no conflict of interest.

References

1. Talus, K. European Union Energy. Available online: https://europa.eu/european-union/topics/energy_pl (accessed on 30 September 2020).
2. European Union, Paris Agreement. Available online: https://ec.europa.eu/clima/policies/international/negotiations/paris_en (accessed on 30 September 2020).
3. Ministry of the Environment of the Republic of Poland United Nations Climate Change Conference COP19. Available online: http://www.cop19.gov.pl/ (accessed on 10 January 2020).
4. Ministry of Energy. *Energy Statistics in 2016 and 2017*; Statistics Poland: Warsaw, Poland, 2018.
5. Agencja Rynku Energii, S.A. *Prognoza Zapotrzebowania Na Paliwa I Energię Do 2030 Roku*; Ministerstwo Gospodarki: Warsaw, Poland, 2009.
6. Regulation (EU) No 517/2014 of the European Parliament and of the Council of 16 April 2014 on Fluorinated Greenhouse Gases and Repealing Regulation (EC) No 842/2006. Available online: https://eur-lex.europa.eu/legal-content/EN/TXT/?uri=uriserv%3AOJ.L_.2014.150.01.0195.01.POL (accessed on 10 January 2020).
7. Shmroukh, A.N.; Ali, A.H.H.; Ookawara, S. Adsorption working pairs for adsorption cooling chillers: A review based on adsorption capacity and environmental impact. *Renew. Sustain. Energy Rev.* **2015**, *50*, 445–456. [CrossRef]
8. Lee, J.G.; Bae, K.J.; Kwon, O.K. Performance investigation of a two-bed type adsorption chiller with various adsorbents. *Energies* **2020**, *13*, 2553. [CrossRef]
9. Sztekler, K.; Kalawa, W.; Stefanski, S.; Krzywanski, J.; Grabowska, K.; Sosnowski, M.; Nowak, W.; Makowski, M. Using adsorption chillers for utilising waste heat from power plants. *Therm. Sci.* **2019**, *23*, S1143–S1151. [CrossRef]
10. Huang, L.; Zheng, R.; Piontek, U. Installation and operation of a solar cooling and heating system incorporated with air-source heat pumps. *Energies* **2019**, *12*. [CrossRef]
11. Sah, R.P.; Choudhury, B.; Das, R.K. A review on adsorption cooling systems with silica gel and carbon as adsorbents. *Renew. Sustain. Energy Rev.* **2015**, *45*, 123–134. [CrossRef]
12. Yu, F.W.; Chan, K.T.; Sit, R.K.Y.; Yang, J. Review of standards for energy performance of chiller systems serving commercial buildings. *Energy Procedia* **2014**, *61*, 2778–2782. [CrossRef]
13. Shabir, F.; Sultan, M.; Niaz, Y.; Usman, M.; Ibrahim, S.M.; Feng, Y.; Naik, B.K.; Nasir, A.; Ali, I. Steady-state investigation of carbon-based adsorbent-adsorbate pairs for heat transformation application. *Sustainability* **2020**, *12*. [CrossRef]
14. Uyun, A.S.; Miyazaki, T.; Ueda, Y.; Akisawa, A. High performance cascading adsorption refrigeration cycle with internal heat recovery driven by a low grade heat source temperature. *Energies* **2009**, *2*, 1170–1191. [CrossRef]
15. Chorowski, M.; Pyrka, P.; Rogala, Z.; Czuprynski, P. Experimental study of performance improvement of 3-bed and 2-evaporator adsorption chiller by control optimization. *Energies* **2019**, *12*. [CrossRef]
16. Rouf, R.A.; Jahan, N.; Alam, K.C.A.; Sultan, A.A.; Saha, B.B.; Saha, S.C. Improved cooling capacity of a solar heat driven adsorption chiller. *Case Stud. Therm. Eng.* **2020**, *17*, 100568. [CrossRef]
17. Krzywanski, J.; Grabowska, K.; Sosnowski, M.; Zylka, A.; Sztekler, K.; Kalawa, W.; Wojcik, T.; Nowak, W. An adaptive neuro-fuzzy model of a re-heat two-stage Adsorption Chiller. *Therm. Sci.* **2019**, *23*, S1053–S1063. [CrossRef]
18. Stefański, S.; Mika, Ł.; Sztekler, K.; Kalawa, W.; Lis, Ł.; Nowak, W. Adsorption bed configurations for adsorption cooling application. In Proceedings of the E3S Web of Conferences Energy and Fuels, Krakow, Poland, 19–21 September 2018.
19. Gwadera, M.; Kupiec, K. Adsorpcyjne układy chłodnicze. *Inżynieria Apar. Chem.* **2011**, *5*, 38–39.
20. Pan, Q.; Peng, J.; Wang, R. Experimental study of an adsorption chiller for extra low temperature waste heat utilization. *Appl. Therm. Eng.* **2019**, *163*, 114341. [CrossRef]
21. Woo, S.Y.; Lee, H.S.; Ji, H.; Moon, D.S.; Kim, Y.D. Silica gel-based adsorption cooling cum desalination system: Focus on brine salinity, operating pressure, and its effect on performance. *Desalination* **2019**, *467*, 136–146. [CrossRef]
22. Kim, Y.D.; Thu, K.; Masry, M.E.; Ng, K.C. Water quality assessment of solar-assisted adsorption desalination cycle. *Desalination* **2014**, *344*, 144–151. [CrossRef]

23. Saha, B.B.; Koyama, S.; Lee, J.B.; Kuwahara, K.; Alam, K.C.A.; Hamamoto, Y.; Akisawa, A.; Kashiwagi, T. Performance evaluation of a low-temperature waste heat driven multi-bed adsorption chiller. *Int. J. Multiph. Flow* **2003**, *29*, 1249–1263. [CrossRef]
24. Chorowski, M.; Pyrka, P. Modelling and experimental investigation of an adsorption chiller using low-temperature heat from cogeneration. *Energy* **2015**, *92*, 221–229. [CrossRef]
25. Chua, H.T.; Ng, K.C.; Malek, A.; Kashiwagi, T.; Akisawa, A.; Saha, B.B. Multi-bed regenerative adsorption chiller—Improving the utilization of waste heat and reducing the chilled water outlet temperature fluctuation. *Int. J. Refrig.* **2001**, *24*, 124–136. [CrossRef]
26. Sztekler, K.; Kalawa, W.; Nowak, W.; Mika, L.; Grabowska, K.; Krzywanski, J.; Sosnowski, M.; Al-Harbi, A.A. Performance evaluation of a single-stage two-bed adsorption chiller with desalination function. *J. Energy Resour. Technol.* **2020**, 1–22. [CrossRef]

Publisher's Note: MDPI stays neutral with regard to jurisdictional claims in published maps and institutional affiliations.

energies

MDPI

Article

Prediction of Sorption Processes Using the Deep Learning Methods (Long Short-Term Memory)

Dorian Skrobek [1,*], Jaroslaw Krzywanski [1], Marcin Sosnowski [1], Anna Kulakowska [1], Anna Zylka [1], Karolina Grabowska [1], Katarzyna Ciesielska [1] and Wojciech Nowak [2]

1 Faculty of Science and Technology, Jan Dlugosz University in Czestochowa, 13/15 Armii Krajowej Av., 42-200 Czestochowa, Poland; j.krzywanski@ujd.edu.pl (J.K.); m.sosnowski@ujd.edu.pl (M.S.); a.kulakowska@ujd.edu.pl (A.K.); a.zylka@ujd.edu.pl (A.Z.); k.grabowska@ujd.edu.pl (K.G.); k.ciesielska@ujd.edu.pl (K.C.)

2 Faculty of Energy and Fuel, AGH University of Science and Technology, A. Mickiewicza 30, 30-059 Cracow, Poland; wnowak@agh.edu.pl

* Correspondence: d.skrobek@ujd.edu.pl

Received: 17 November 2020; Accepted: 12 December 2020; Published: 14 December 2020

Abstract: The paper introduces the artificial intelligence (AI) approach for modeling fluidized adsorption beds. The idea of fluidized bed application allows a significantly increased heat transfer coefficient between adsorption bed and the surface of a heat exchanger, improving the performance of adsorption cooling and desalination systems. The Long Short-Term Memory (LSTM) network algorithm was used, classified as a deep learning method, to predict the vapor mass quantity in the adsorption bed. The research used an LSTM network with two hidden layers. The network used in the study is composed of seven inputs (absolute pressures in the adsorption chamber and evaporator, the temperatures in adsorption chamber and evaporator, relative pressure, the temperatures in the center of adsorption bed and 25 mm from the bed center, the kind of the solids mixture, the percentage value of the addition) and one output (mass of the sorption bed). The paper presents numerical research concerning mass prediction with the algorithm mentioned above for three sorbents in fixed ad fluidized beds. The results obtained by the developed algorithm of the LSTM network and the experimental tests are in good agreement of the matching the results above 0.95.

Keywords: sorption processes; deep learning; neural networks; Long Short-Term Memory (LSTM)

---

1. Introduction

The process when chemical compounds are bound to a solid phase is generally known as sorption. Adsorption occurs when adsorption of a substance takes place at the surface, while absorption occurs when the substance is absorbed in the entire volume of the solid phase. These processes can apply to volatile substances and particles dissolved in a liquid medium associated with the solid phase particles. Molecules and atoms can attach to surfaces in two ways. In the process of physical adsorption process between the adsorbate and the adsorbent, there are van der Waals interactions. In the process of chemical adsorption, molecules or atoms join with the surface to form chemical bonds.

The adsorption chillers [1] (Figure 1) are quiet, non-corrosive, reliable, environmentally friendly, and economical in operation appliances. They are consist of an evaporator, a condenser, separating valves, and a sorption bed. In some solutions, more than one sorption bed may be used. Adsorption chillers are capable of utilizing low-grade waste heat and renewable heat (e.g., solar energy) to produce cool and/or desalinated water. The adsorption chiller with silica gel-water, powered by a waste heat source, has been successfully commercialized in Japan [2]. Waste heat in the industry is rarely used and is currently usually discharged into the environment. The article [2] presents a three-stage adsorption chiller and computer program to simulate the cycle to predict its operation. Most often, in scientific studies, sorption processes are predicted using the nonlinear autoregressive network with exogenous inputs (NARX) [2–4] or feed forward neural network (FFNN) [3].

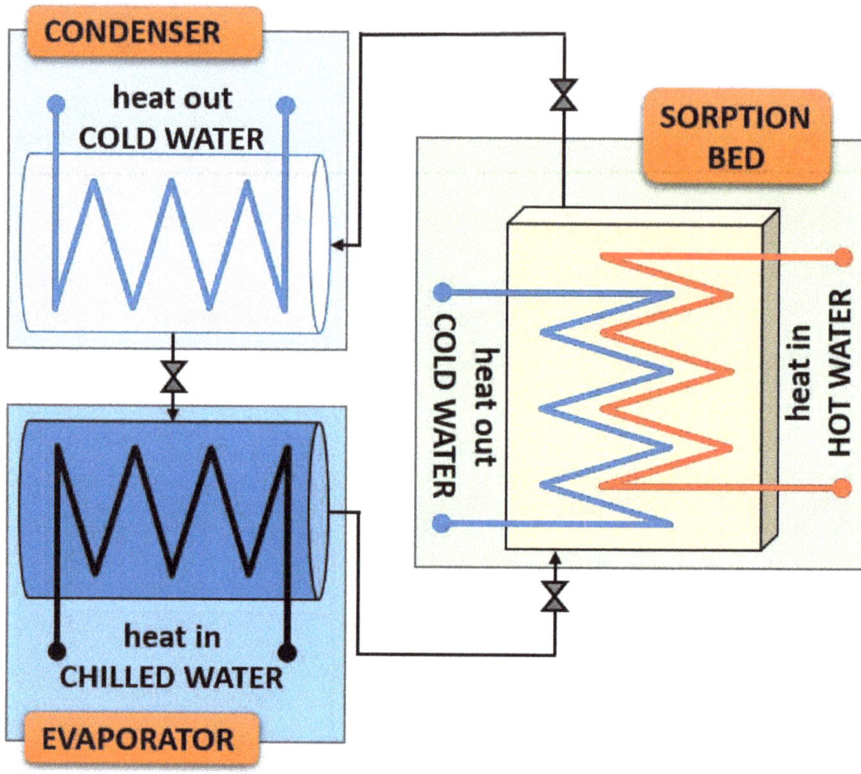

Figure 1. Scheme of adsorption chiller.

Neural networks (NNs) are used to predict various dependencies, among others, to predict the traffic volume [5], the efficiency and generator power of a supercritical coal-fired power plant [6,7], the hydrogen concentration in the syngas [8], in order to optimize a heat exchanger and adsorption chillers [8–10]. They come in many variants, feed-forward NN [7,11] fuzzy NN [10,12], recurrent NN (RNN) [13], and hybrid NN [14]. Recurrent Neural Networks (RNNs) by their chain-like structure and internal memory with loops are widely used. Recently, the deep learning model, such as RNNs, has been increasingly used [15]. The disadvantage of RNN is the vanishing gradient problem, which prevents them from modeling time series with long-term relationships such as wind speed and wind direction [16]. There have been several attempts to overcome the difficulty of training RNNs over the years. These difficulties were successfully addressed by the Long Short-Term Memory networks (LSTMs) [17], a type of RNN capable of learning long-term dependencies.

Long Short-Term Memory (LSTM) as a deep learning method can process sequential data [15] and is applied in many real-world problems, such as image captioning [18], music composition [19], predicting for COVID-19 [20], speech recognition [21], and human trajectory prediction in crowded places [22]. The papers [23,24] show algorithms by which there is the time at the input of neural networks and the data entered into the network are given in chronological order. In the presented article, no time variable was given at the input of the network. In the last few years, LSTM has gained popularity due to its ability to model long-term dependencies [25,26]. The long-term dependencies are typically learned from chronologically arranged input data, considering only forward dependencies, while dependencies learned from randomly fed inputs data have never been explored. NARX, FFNN, and LSTM are neural networks mainly dedicated to modeling time series cases. In this study, LSTM was used, which turned out to be one of the best and easy to interpret neural networks suitable for time-series problems.

The architecture of the LSTM-based model sought to be capable of describing the dynamics of sorption processes. Since most of the newly proposed LSTM-based prediction models are of one hidden-layer shallow architecture [27–29], their performance is poorer than those with several hidden layers models [30,31].

All time-series sets of data ought to be utilized during prediction by an LSTM model. Usually, the model's dataset is chronologically arranged from time epoch $t-1$ to t [32]. However, this may lead to filtering out, or ineffectively passing through the network structure of useful information. Therefore, it may be a good idea to consider randomizing data. Another reason for the sampling of data into our study is the periodicity of sorption cycles. Analyzing time-series data periodicity, especially in recurring patterns, will enhance the predictive performance from both forward and backward temporal perspectives [33]. However, based on our literature review, the dataset fed to LSTM is chronologically arranged, and the network itself uses forward and/or backward data prediction dependencies. The use of chronological data in the LSTM network may cause the network to start learning training data and incorrectly predict data, which is why it was decided that data would be entered into the network randomly within the research.

Since the literature review has already reported the advantages of the LSTM approach over other networks such as FFNN or NARX [34–36], the purpose of the paper is to use the LSTM network in the novel field of application, i.e., for adsorption processes in innovative fluidized adsorption beds. This work presents numerical research results related to predicting the adsorption bed mass using the Long Short-Term Memory. Therefore, the considered issue corresponds to the innovative concept of replacing the fixed adsorption beds in conventional adsorption chillers with fluidized beds described in detail in [37,38].

Adsorption chillers are promising appliances allowing to use of low-grade thermal energy [39–41], including renewable sources of energy such as solar heat, wastewater, underground resources, and waste heat, instead of high valued energy sources, e.g., electricity and fossil fuels-driven appliances [42–44].

The idea of fluidized bed application [45–47] significantly increases the heat transfer coefficient between the adsorption bed and the surface of a heat exchanger and the bed conductance of fluidized bed adsorption chillers, improving the performance of adsorption cooling and desalination systems [48–50]. Moreover, the set of experimental data used is unique because the advanced test stand was utilized, which allows for the fluidized state implementation into the adsorption bed under lowered pressure conditions, even up to 1500 Pa.

The present work is the first in the literature, dealing with the deep learning method, such as LSTM for modeling fluidized and fixed adsorption beds to the best of our knowledge. The data used in the deep learning network was recorded within the experimental research related to sorption processes. In the LSTM, the input dataset has been given in random order rather than in chronological order, and the network itself uses forward dependencies. This paper deals with an innovative approach consisting of a fluidized bed application. Such an idea allows for improving heat and mass transfer processes, with helps increase adsorption chiller' performance.

The second chapter contains a description of the test stand and research equipment, experimental research results, and the discussion on the algorithms used during the numerical research. The third section depicts the LSTM network hyperparameters and the structure of the LSTM network inputs and outputs as well as the results and their discussion. The work is finalized with a conclusion and proposal for further research.

2. Problem Formulation and Solving

2.1. Experimental Test

The data needed to predict the adsorption bed's mass comes from the previously conducted experimental studies carried out on the innovative test stand.

The test stand (Figure 2) consists of an evaporator, adsorption chamber, vacuum pump, three valves (V_1, V_2, V_3), and sensors: P_1—absolute pressure sensor in the adsorption chamber, P_2—absolute pressure sensor in the evaporator, P_3—relative pressure sensor, T_1—temperature sensor in the adsorption chamber, T_2—temperature sensor in the evaporator, T_3—temperature sensor in the adsorption bed (bed center), T_4—temperature sensor in the adsorption bed (25 mm from the bed center).

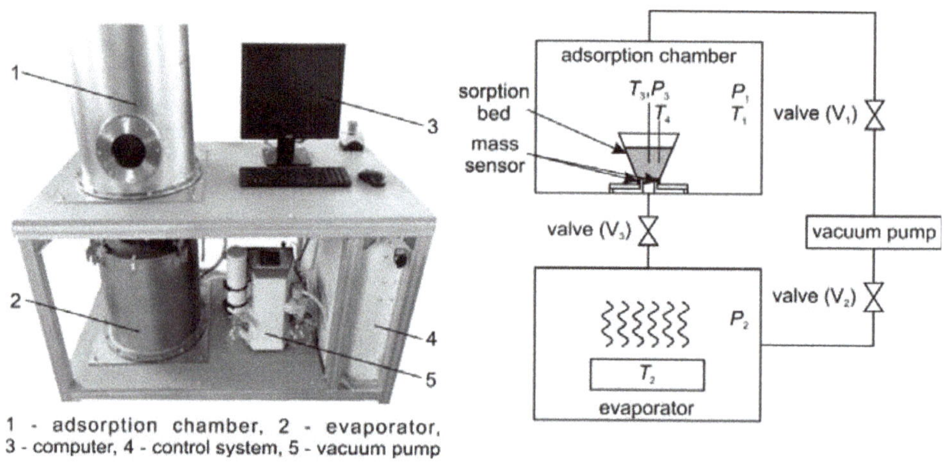

Figure 2. The testing stand and diagram of the test stand.

The first stage of work on the stand is to obtain the saturation pressure in the evaporator (P_2) in the temperature T_2. After obtaining the appropriate pressure for the evaporator (P_2) and the chamber (P_1), the water begins to boil, and the steam is released through the open valve (V_3) to the sorption bed, where the adsorption process takes place. The changes taking place in the bed are monitored using temperature sensors T_3, T_4, and the relative pressure sensor P_3 and mass sensors measuring the sorption bed's weight. In the test process, assumptions were made for the valve opening/closing time (V_3), according to Table 1. The table also shows the initial test conditions. Valves V_1 and V_2 are used to maintain the appropriate pressure difference in the evaporator and the adsorption chamber to keep the adsorbent's fluidized and fixed beds.

Table 1. Initial research conditions.

No.	Type of Material	Additive to the Mixture	t_0 [s] [1]	t_z [s] [2]	P_1 [bar] [3]	P_2 [bar] [4]	State	Mass of Sorbent in the Bed [g]
1	100%SG [5]	-	10	150	13	23	F [7]	55
2	100%SG [5]	-	10	150	21	23	S [8]	55
3	95%SG [5]	+5%Al [6]	10	150	13	23	F [7]	55
4	95%SG [5]	+5%Al [6]	10	150	21	23	S [8]	55
5	85%SG [5]	+15%Al [6]	10	150	13	23	F [7]	55
6	85%SG [5]	+15%Al [6]	10	150	21	23	S [8]	55

[1] the opening time of valve V_3, [2] stabilizing time of conditions in the chamber (valve V_3 state—closed), [3] pressure in the chamber, [4] pressure in the evaporator, [5] silica gel, [6] aluminum, [7] fluidized state, [8] stationary state.

Commercial silica gel from Fuji Silysia Chemical Ltd. (Greenville, USA) was employed for the research. Using the Analysette 3 Spartan shaker (FRITSCH GmbH, Idar-Oberstein, Germany), the material was separated to obtain granulation 250–300 µm. In the present study, aluminum (Al, granulation 45–450 µm) particles were used as an additive to improve the thermophysical properties of a silica gel (SG) adsorption bed, due to high thermal conductivity [51].

The exemplary results of the experiment are shown in Figure 3. They concern the tests of the 85% SG + 15%Al mixture for the stationary state. In this test and the other test variants (Table 1), the valve V_3 was open for 10 s. The figure below shows ten consecutive opening and closing cycles of the valve V_3.

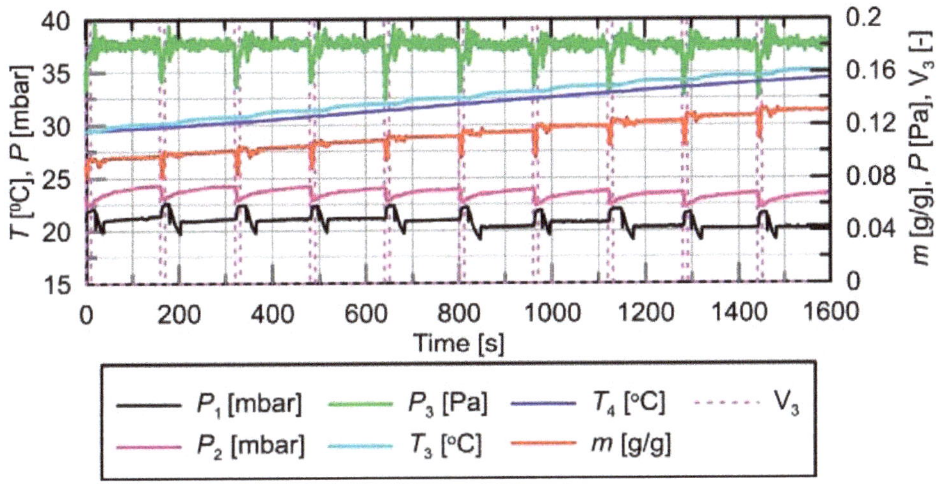

Figure 3. An exemplary result of the experimental test for the mixture 85%SG + 15%Al stationary state for ten consecutive cycles of valve V_3 opening and closing.

Based on Table 1, experimental studies were performed, and the data from these experiments were used as inputs (P_1, P_2, P_3, T_3, T_4, type of mixture, the percentage value of the additive) and outputs (sorption bed mass) of the LSTM network. Exemplary data that is entered into the LSTM network is shown in Figure 3; six such studies were performed as shown in Table 1. The test results from the six experiments were fed into the LSTM network as outlined above.

2.2. Recurrent Neural Network (RNN)

A recurrent Neural Network is a deep learning model consisting of neurons. It is mainly useful when considering sequence data, as each neuron can use its internal memory to store information about the previous input. This action resembles a loop (Figure 4) in which the output of a neuron at one specific stage is provided to the next neuron as an input. The RNN considers two inputs; the first is the current input, and the second is the previous computation [32]. RNNs contain an input layer, hidden layers, and; an output layer as with other neural networks.

All recurrent neural networks have the form of a chain of repeating modules of the neural network. In standard RNNs, this repeating module has a straightforward structure, such as a single *tanh* (hyperbolic tangent) layer (Figure 5).

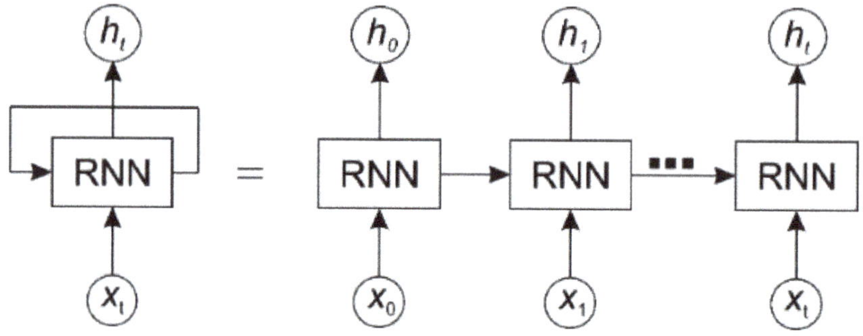

Figure 4. An unrolled Recurrent Neural Network.

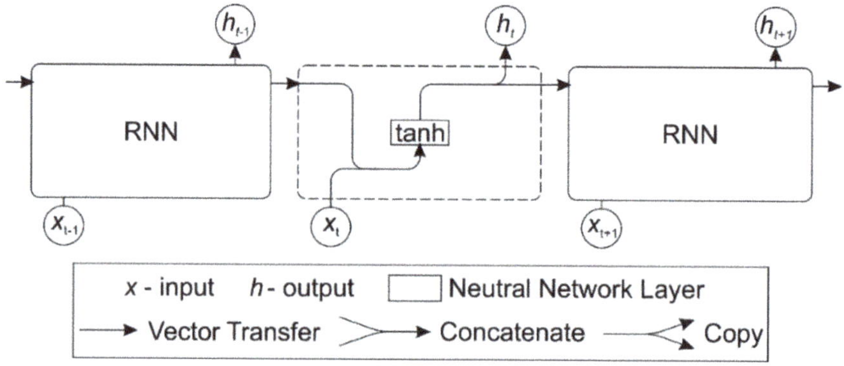

Figure 5. The repeating module in a standard RNN contains a single layer.

2.3. Long Short-Term Memory (LSTM)

Long Short-Term Memory networks (LSTMs) are a special kind of RNN, capable of learning long-term dependencies. They were introduced by Hochreiter and Schmidhuber in 1997 [17] and then refined and popularized by other researchers [52–54]. LSTMs have a chain-like structure show in Figure 6; the repeating module has a structure shown in Figure 7 (where *tanh*-hyperbolic tangent).

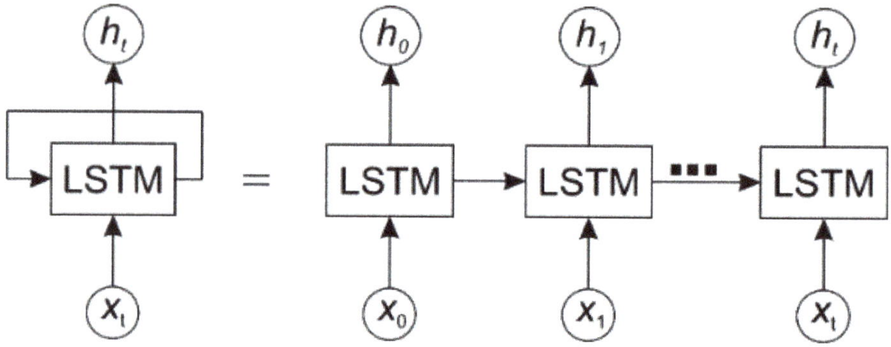

Figure 6. An unrolled LSTM.

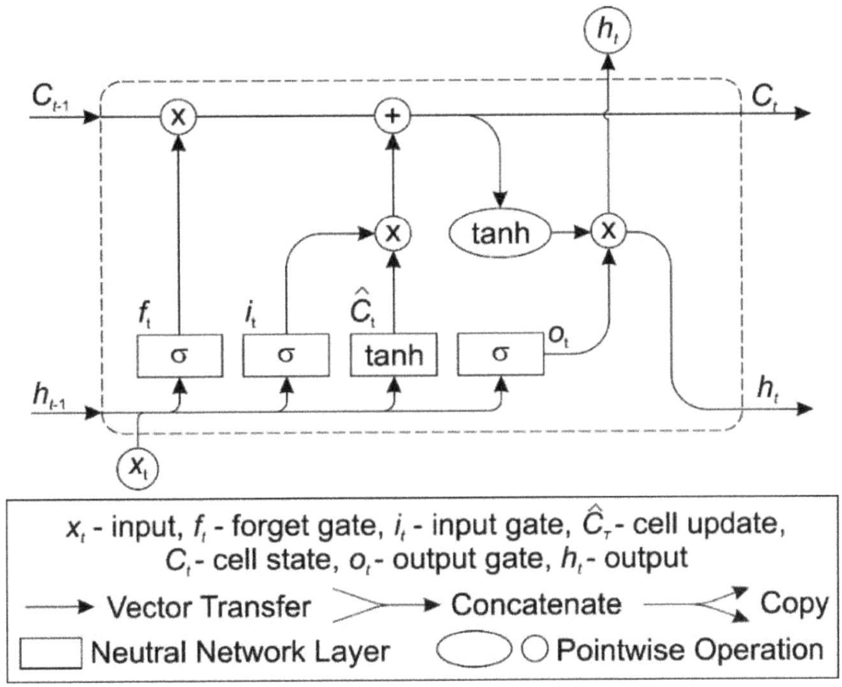

Figure 7. Graphical representation of the LSTM cell.

In order to implement the LSTM recurrent network, first, the LSTM cell should be implemented. The LSTM cell has three gates and two internal states, which should be determined to calculate the current output and current cell state. We distinguish the following LSTM cell gateways:

- forget gate f_t –filters information from the input and previous output and decides which information to remember or forget and discard,
- input gate i_t –controls the flow of activation to enter the cell,
- output gate o_t –controls the output flow of cell activation.

In addition to these three gates, the LSTM cell contains a cell update usually activated by the tanh function.

Three variables fall into each LSTM cell:

- input x_t,
- previous output h_{t-1}
- cell state C_{t-1}

Calculations for the LSTM cell in its individual layers can be described as follows.

- the forget gate ft (sigmoid layer):

$$f_t = \sigma\left(W_f \circ [h_{t-1}, x_t] + b_f\right) \tag{1}$$

- the input gate it (sigmoid layer):

$$i_t = \sigma(W_i \circ [h_{t-1}, x_t] + b_i) \tag{2}$$

- the cell state Ct:

$$\hat{c}_t = \tanh(W_c \circ [h_{t-1}, x_t] + b_c) \quad (3)$$

$$C_t = f_t \cdot C_{t-1} + i_t \cdot \hat{c}_t \quad (4)$$

- the output gate ot (sigmoid layer):

$$o_t = \sigma(W_o \circ [h_{t-1}, x_t] + b_o) \quad (5)$$

where: $\hat{c}_t$—the cell update; W_f, W_i, W_c, W_o—matrices of weights; b_f, b_i, b_c, b_o—bias vector.

The bias vector is specified as a numeric array. They are learnable parameters. When training the network, the biases vector in the first iteration is refilled with zeros.

The matrices of weights are specified as a numeric array, they are parameters that can be learned. The initial value of the weights in the algorithm is computed with the Glorot initializer [55] (also known as Xavier initializer). The Glorot initializer independently samples from a uniform distribution with zero mean and variance 2/(*numIn* + *numOut*) where *numIn*—number of inputs in *i*-th layer, *numOut*—number of outputs *i*-th layer.

The final stage of the calculations in the LSTM cell is defining the current output ht. The current output is calculated with the multiplication operation between the output gate layer and tanh layer of the current cell state C_t:

$$h_t = o_t \cdot \tanh(C_t) \quad (6)$$

The current output h_t passes through the network as the previous state for the next LSTM cell or as the input for the neural network output layer.

The structure of the LSTM network is shown in Figure 8. The same network settings were adopted in all studies. Hyperparameters (Table 2) were selected on the basis of a series of studies not presented in this article. From the LSTM network tests carried out earlier, the hyperparameters' values were selected based on the best fit. Every 30 epochs (the epoch is the full passage of the training algorithm through the entire training set) the learning coefficient changed its value according to the equation: *ilr* = 0.2 * *lr* (*lr*—current value of the learning coefficient).

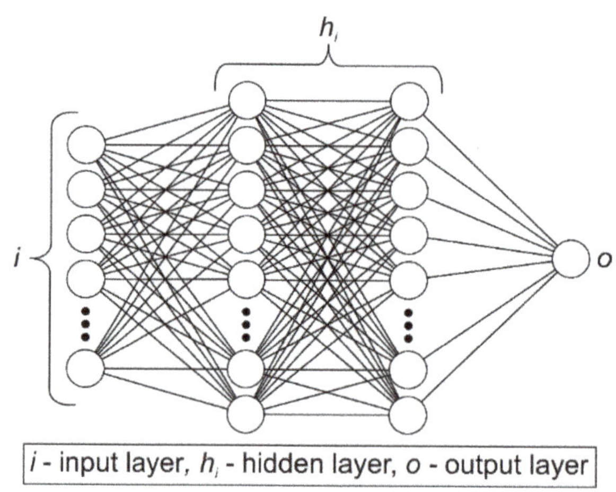

Figure 8. Diagram of the LSTM network.

The network input layer comprises the following inputs: P_1—absolute pressure in the adsorption chamber, P_2—absolute pressure in the evaporator, P_3—relative pressure, T_3—temperature in the adsorption bed (bed center), T_4—the temperature in the adsorption bed (25 mm from the bed center), type of the mixture, and the percentage value of the addition. The mass of the sorption bed constitutes the output of the neural network.

Table 2. The values of the hyperparameters for the LSTM.

Hyperparameter	Value
Number of epochs	200
Learning rate	0.005
Number of LSTM layers	2
Number of cells in layer 1	210
Number of cells in layer 2	190
Drop out layer	0.05

3. Results of Numerical Calculations

By adopting the assumptions, formulations, and experimental research results presented in the previous chapters, the LSTM network algorithm and a computer program were developed, which enabled predicting the sorption bed's mass during the sorption process. The experimental test results for the first ten valve V_3 opening cycles (6 tests, see Table 1) have been normalized to a range of 0 to 100 and divided into three parts. The training data is presented to the network during the training stage. Validation data is exploited to improve learning and possibly to stop training. Finally, the test data do not affect training and validation, and thus, provide an independent measurement of network performance after training. These data were randomized without duplication as follows:

(a) First numerical research (60-20-20):

- training data—60% of all data,
- validation data—20% of all data,
- test data—20% of all data,

(b) Second numerical research (70-15-15):

- training data—70% of all data,
- validation data—15% of all data,
- test data—15% of all data,

(c) Third numerical research (80-10-10):

- training data—80% of all data,
- validation data—10% of all data,
- test data—10% of all data.

Figures 9–11 show test data as the trend line (linear fit) for all studies and the 95% prediction interval of LSTM network results.

The first analysis of the prediction of mass in the sorption bed using the LSTM network concerned the division of data in the ratio of 60-20-20, the results of this study are shown in Figure 9 and Table 3.

Table 3. Coefficient of determination R^2 for the Linear Fit (60-20-20).

R^2						
All Data	**100%SG (F)**	**100%SG (S)**	**95%SG +5%Al (F)**	**95%SG +5%Al (S)**	**85%SG +15%Al (F)**	**85%SG +15%Al (S)**
0.9515	0.8934	0.9218	0.9891	0.9732	0.9848	0.9505

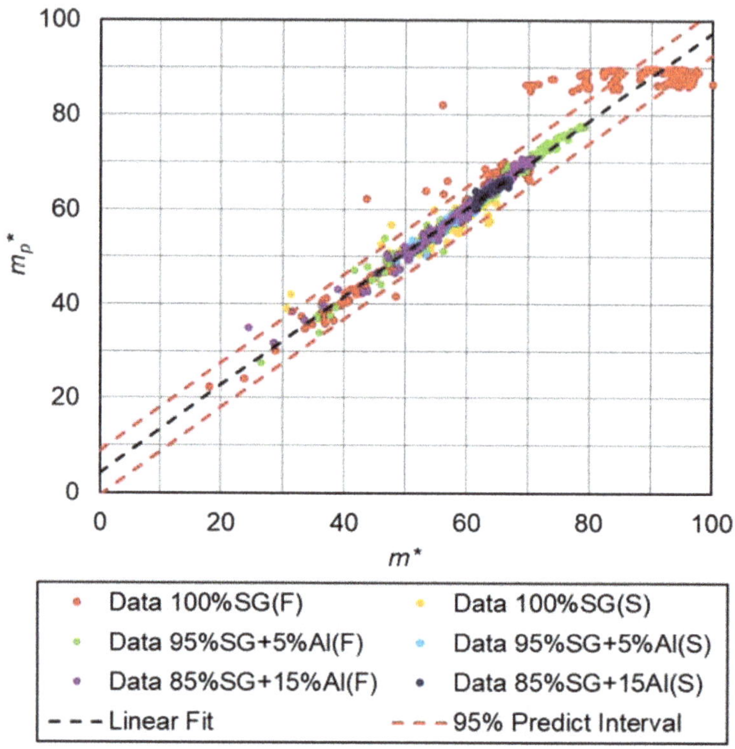

Figure 9. Comparison of the experimental results with the numerical results predicted by the LSTM (60-20-20).

Figure 9 shows the LSTM network operation results compared to the values obtained during the experiment. The LSTM network predicts the worst results for pure silica gel (100% SG) in a fluidized state.

Table 3 shows the fit for all data and individual mixture. The coefficient of determination for all data is 0.9515. The LSTM network predicts the worst values for pure silica gel (100% SG). In the case of fluidization, the coefficient of determination is 0.8934, and for the fixed bed, it is 0.9218, which may be the reason for the low repeatability of the cycles during the experiment. The network achieves the best match for the mixture 95% SG + 5%Al, where the coefficient of determination for fluidized and fixed bed was equal to 0.989, and 0.973, respectively.

The second analysis of mass prediction in the sorption bed using the LSTM network concerned the data division in the ratio of 70-15-15. The results of this study are presented in Figure 10 and Table 4.

Table 4. Coefficient of determination R^2 for the Linear Fit (70-15-15).

R^2						
All Data	**100%SG (F)**	**100%SG (S)**	**95%SG +5%Al (F)**	**95%SG +5%Al (S)**	**85%SG +15%Al (F)**	**85%SG +15%Al (S)**
0.9507	0.8404	0.9250	0.9788	0.9738	0.9800	0.9363

Figure 10 shows the result of the LSTM network in comparison with the values obtained during the experiment. As in the previous study, the LSTM network predicts the worst results for pure silica gel (100% SG) in a fluidized state.

All datasets achieved a fit of 0.9507, and Table 4 also shows the coefficient of determination of the individual mixture for the fit function. The coefficient of determination in this study is lower than in the previous study. The LSTM network predicts the worst values for pure silica gel (100% SG) in fluidized bed conditions. The coefficient of determination for fixed and fluidized bed was 0.9250 and 0.8404, respectively.

The model's best accuracy, in this case, was achieved for the fluidized bed of 85% SG + 15%Al mixture with R^2 equal to 0.98.

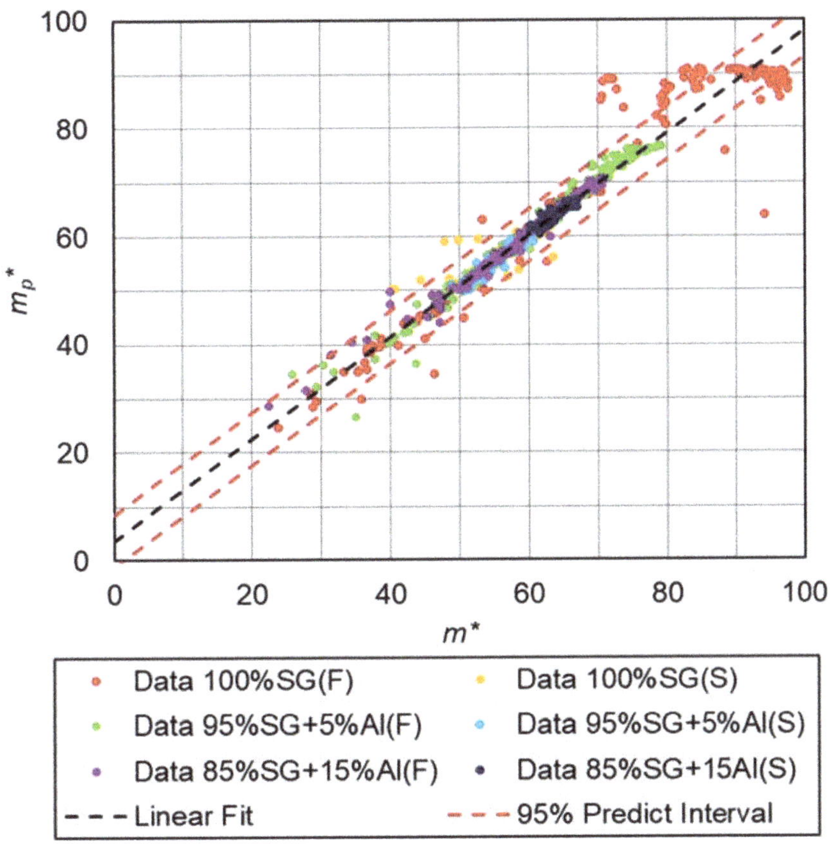

Figure 10. Comparison of the experimental results with the numerical results predicted by the LSTM (70-15-15).

The third analysis of the prediction of mass in the sorption bed using the LSTM network concerned the distribution of data in the ratio of 80-10-10. The results of this study are presented in Figure 11 and Table 5.

Table 5. Coefficient of determination R^2 for the Linear Fit (80-10-10).

	R^2					
All Data	**100%SG (F)**	**100%SG (S)**	**95%SG +5%Al (F)**	**95%SG +5%Al (S)**	**85%SG +15%Al (F)**	**85%SG +15%Al (S)**
0.9554	0.8670	0.9343	0.9915	0.9611	0.9874	0.9244

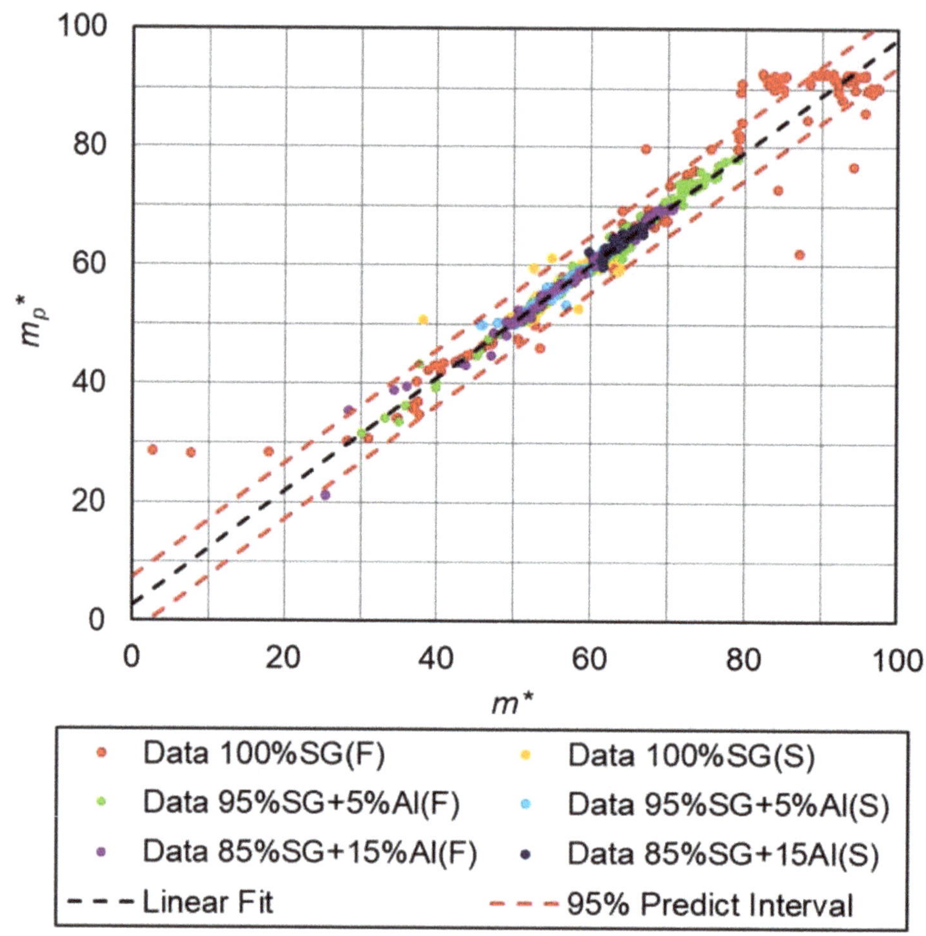

Figure 11. Comparison of the experimental results with the numerical results predicted by the LSTM (80-10-10).

Figure 11 shows the results of the LSTM in comparison with the values obtained during the experiment. As in previous studies, the LSTM network predicts the worst results for pure silica gel (100% SG), fluidized state. In this case, the network best predicts the results of the experimental research.

The coefficient of determination for all is equal to 0.9554. The accuracy of the developed model is the best of the two previous ones. Only a slight decrease in R^2 be seen for the 95% SG + 5%Al(S), and 95% SG + 15%Al(S) blends. The LSTM network prediction is still worst for the fluidized bed of pure silica gel (100% SG) with R^2 equal to 0.867. However, the best prediction was achieved for the fluidized bed of the mixture 95% SG + 5%Al with R^2 = 0.9915.

4. Conclusions

This paper deals with an innovative concept of a fluidized bed instead of a fixed adsorption beds application, currently employed in conventional adsorption chillers. The model, developed in the study, correctly predicts the vapor mass adsorbed in the adsorption chillers.

In this work, the Long Short-Term Memory networks, classified as a deep learning method, were also used to predict the sorption bed's mass. The LSTM network is one of the particular kinds of recursive networks that are capable of learning long-term dependencies.

The solution to predicting the results was based on the most accurate mapping of the experimental values by the LSTM network. In the mathematical model, all network inputs were normalized to the range <0–100> due to the different units of parameters used in the study.

The analysis was performed by splitting the input data set into three parts (training data, validation data, and test data), in three variants: 60-20-20, 75-15-15, 80-20-20. The LSTM network, while increasing the amount of data used for training, better reproduced the experimental results. By increasing the training data make it possible to increase the accuracy of LSTM. The division of data into training data, validation data, and test data in deep learning networks are problematic because increasing one of the above values reduces the other two. A better solution seems to be to increase the amount of data entered into the network, but in this case, it was impossible due to the number of sorption cycles that the adopted mixtures could perform. In order to increase the amount of data, the mass of the mixture should be increased, as well as the initial conditions under which the tests were performed, e.g., the absolute pressure in the adsorption chamber and evaporator.

The developed model using the LSTM network and the high accuracy of the obtained numerical results confirm that the LSTM network is suitable for predicting sorption processes.

The LSTM network predicted the worst experimental test for pure silica gel (100% SG) in the fluidized conditions where the coefficient of determination did not exceed the threshold of 0.9 since these experimental tests are the least repeatable. The test results for 100% SG are more difficult to predict because no additive in the mixture would stabilize the sorption processes during the experimental test, so the sorption cycles for 100% SG are not very repeatable. Due to its high thermal conductivity, aluminum's addition to the silica gel stabilizes the mixture, improving the sorption bed's thermophysical properties. The LSTM network achieved the best accuracy for the mixture of 95% silica gel with 5% aluminum of addition in the fluidized conditions. For data splitting of 80-10-10 the highest coefficient of determination was equal to 0.9915.

Future research is planned to conduct comparative studies of several deep learning methods.

Author Contributions: The contribution of co-authors in creating article is: conceptualization, D.S., J.K.; methodology, D.S., J.K., software, D.S., J.K.; validation, D.S., J.K., M.S., W.N.; formal analysis, J.K., M.S.; investigation, A.K., A.Z., K.G., K.C.; resources, A.K., A.Z., K.G., D.S., K.C., M.S., J.K.; data curation, A.K., A.Z., K.G., D.S., M.S., J.K.; writing—original draft preparation, D.S., J.K.; writing—review and editing, D.S., M.S., J.K.; visualization, D.S., J.K., M.S.; supervision, J.K.; project administration, J.K., K.G.; funding acquisition, J.K., M.S., K.G. All authors have read and agreed to the published version of the manuscript.

Funding: This research was funded by project No. 2018/29/B/ST8/00442, supported by the National Science Centre.

Conflicts of Interest: The authors declare no conflict of interest.

Nomenclature

Al	aluminum
F	fluidized state corresponding to fluidized bed conditions
LSTM	Long Short-Term Memory
m^*	normalized sorbent mass (experimental value), –
$m_p{}^*$	normalized sorbent mass predicted by the LSTM, –
NN	neural network
$n\%$	the percentage of the additive in the mixture, %
RNN	Recurrent Neural Network
S	stationary state corresponding to the fixed bed conditions
SG	silica gel

References

1. Sosnowski, M. Evaluation of Heat Transfer performance of a Multi-Disc Sorption Bed Dedicated for Adsorption Cooling Technology. *Energies* **2019**, *12*, 4660. [CrossRef]
2. Saha, B.B.; Koyama, S.; Lee, J.B.; Kuwahara, K.; Alam, K.C.A.; Hamamoto, Y.; Akisawa, A.; Kashiwagi, T. Performance evaluation of a low-temperature waste heat driven multi-bed adsorption chiller. *Int. J. Multiph. Flow* **2003**, *29*, 1249–1263. [CrossRef]
3. Scapino, L.; Zondag, H.A.; Diriken, J.; Rindt, C.C.M.; Van Bael, J.; Sciacovelli, A. Modeling the performance of a sorption thermal energy storage reactor using artificial neural networks. *Appl. Energy* **2019**, *253*, 113525. [CrossRef]
4. Argyropoulos, D.; Paraforos, D.S.; Alex, R.; Griepentrog, H.W.; Müller, J. NARX Neural Network Modelling of Mushroom Dynamic Vapour Sorption Kinetics. *IFAC-PapersOnLine* **2016**, *49*, 305–310. [CrossRef]
5. Hua, J.; Faghri, A. Applications of artificial neural networks to intelligent vehicle-highway systems. *Transp. Res. Rec.* **1994**, *1453*, 83.
6. Ashraf, W.M.; Uddin, G.M.; Arafat, S.M.; Afghan, S.; Kamal, A.H.; Asim, M.; Khan, M.H.; Rafique, M.W.; Naumann, U.; Niazi, S.G. Optimization of a 660 MWe Supercritical Power Plant Performance—A Case of Industry 4.0 in the Data-Driven Operational Management Part 1. Thermal Efficiency. *Energies* **2020**, *13*, 5592. [CrossRef]
7. Ashraf, W.M.; Uddin, G.M.; Kamal, A.H.; Khan, M.H.; Khan, A.A.; Ahmad, H.A.; Ahmed, F.; Hafeez, N.; Sami, R.M.Z.; Arafat, S.M. Optimization of a 660 MWe Supercritical Power Plant Performance—A Case of Industry 4.0 in the Data-Driven Operational Management. Part 2. Power Generation. *Energies* **2020**, *13*, 5619. [CrossRef]
8. Krzywanski, J.; Grabowska, K.; Herman, F.; Pyrka, P.; Sosnowski, M.; Prauzner, T.; Nowak, W. Optimization of a three-bed adsorption chiller by genetic algorithms and neural networks. *Energy Convers. Manag.* **2017**, *153*, 313–322. [CrossRef]
9. Krzywanski, J. A General Approach in Optimization of Heat Exchangers by Bio-Inspired Artificial Intelligence Methods. *Energies* **2019**, *12*, 4441. [CrossRef]
10. Krzywanski, J.; Grabowska, K.; Sosnowski, M.; Zylka, A.; Sztekler, K.; Kalawa, W.; Wojcik, T.; Nowak, W. An adaptive neuro-fuzzy model of a re-heat two-stage adsorption chiller. *Therm. Sci.* **2019**, *23*, 1053–1063. [CrossRef]
11. Park, D.; Rilett, L.R. Forecasting freeway link travel times with a multilayer feed-forward neural network. *Comput. Aided Civ. Infrastruct. Eng.* **1999**, *14*, 357–367. [CrossRef]
12. Yin, H.; Wong, S.; Xu, J.; Wong, C. Urban traffic flow prediction using a fuzzy-neural approach. *Transp. Res. Part C Emerg. Technol.* **2002**, *10*, 85–98. [CrossRef]
13. Van Lint, J.; Hoogendoorn, S.; Van Zuylen, H. Freeway travel time prediction with state-space neural networks: Modeling state-space dynamics with recurrent neural networks. *Transp. Res. Rec. J. Transp. Res. Board* **2002**, *1811*, 30–39. [CrossRef]
14. Yu, R.; Li, Y.; Shahabi, C.; Demiryurek, U.; Liu, Y. Deep learning: A generic approach for extreme condition traffic forecasting. In Proceedings of the 2017 SIAM International Conference on Data Mining, Houston, TX, USA, 27–29 April 2017; pp. 777–785.
15. Jozefowicz, R.; Zaremba, W.; Sutskever, I. An empirical exploration of recurrent network architectures. In Proceedings of the 32nd International Conference on Machine Learning (ICML-15), Lille, France, 6–11 July 2015; pp. 2342–2350.
16. Liu, H.; Chen, C. Data processing strategies in wind energy forecasting models and applications: A comprehensive review. *Appl. Energy* **2019**, *249*, 392–408. [CrossRef]
17. Hochreiter, S.; Schmidhuber, J. Long short-term memory. *Neural Comput.* **1997**, *9*, 1735–1780. [CrossRef]
18. Vinyals, O.; Toshev, A.; Bengio, S.; Erhan, D. Show and tell: A neural image caption generator. In Proceedings of the IEEE Conference on Computer Vision and Pattern Recognition, Boston, MA, USA, 7–12 June 2015; pp. 3156–3164.
19. Eck, D.; Schmidhuber, J. A first look at music composition using lstm recurrent neural networks. *Istituto Dalle Molle Di Studi Sull Intelligenza Artificiale* **2002**, *103*, 48.
20. Shahid, F.; Zameer, A.; Muneeb, M. Prediction for COVID-19 with deep learning models of LSTM, GRU, and Bi-LSTM. *Choas Solitions Fractals* **2020**, *140*, 110212. [CrossRef]

21. Graves, A.; Mohamed, A.-R.; Hinton, G. Speech recognition with deep recurrent neural networks. In Proceedings of the 2013 IEEE International Conference on Acoustics, Speech and Signal Processing, Vancouver, BC, Canada, 26–31 May 2013; pp. 6645–6649. [CrossRef]
22. Alahi, A.; Goel, K.; Ramanathan, V.; Robicquet, A.; Fei-Fei, L.; Savarese, S. Social lstm: Human trajectory prediction in crowded spaces. In Proceedings of the IEEE Conference on Computer Vision and Pattern Recognition, Las Vegas, NV, USA, 26 June–1 July 2016; pp. 961–971.
23. Martinez-Garcia, M.; Zhang, Y.; Suzuki, K.; Zhand, Y.-D. Deep Recurrent Entropy Adaptive Model for System Reliability Monitoring. *IEEE Trans. Ind. Inform.* **2020**, *17*, 839–848. [CrossRef]
24. Martinez-Garcia, M.; Zhang, Y.; Gordon, T. Memory pattern identification for feedback tracking control in human-machine system. *Hum. Factors* **2019**. [CrossRef]
25. Chen, Y.-Y.; Lv, Y.; Li, Z.; Wang, F.-Y. Long short-term memory model for traffic congestion prediction with online open data. In Proceedings of the 2016 IEEE 19th International Conference on Intelligent Transportation Systems (ITSC), Rio de Janeiro, Brazil, 1–4 November 2016; pp. 132–137. [CrossRef]
26. Wu, Y.; Tan, H. Short-term traffic flow forecasting with spatial temporal correlation in a hybrid deep learning framework. *arXiv* **2016**, arXiv:1612.01022.
27. Ma, X.; Tao, Z.; Wang, Y.; Yu, H.; Wang, Y. Long short-term memory neural network for traffic speed prediction using remote microwave sensor data. *Transp. Res. Part C Emerg. Technol.* **2015**, *54*, 187–197. [CrossRef]
28. Duan, Y.; Lv, Y.; Wang, F.-Y. Travel time prediction with lstm neural network. In Proceedings of the 2016 IEEE 19th International Conference on Intelligent Transportation Systems (ITSC), Rio de Janeiro, Brazil, 1–4 November 2016; pp. 1053–1058. [CrossRef]
29. Fu, R.; Zhang, Z.; Li, L. Using lstm and gru neural network methods for traffic flow prediction. In Proceedings of the 2016 31st Youth Academic Annual Conference of Chinese Association of Automation (YAC), Wuhan, China, 11–13 November 2016; pp. 324–328. [CrossRef]
30. Zhao, Z.; Chen, W.; Wu, X.; Chen, P.C.; Liu, J. Lstm network: A deep learning approach for short-term traffic forecast. *IET Intell Transp. Syst.* **2017**, *11*, 68–75. [CrossRef]
31. Graves, A.; Jaitly, N.; Mohamed, A.-R. Hybrid speech recognition with deep bidirectional LSTM. In Proceedings of the 2013 IEEE Workshop on Automatic Speech Recognition and Understanding, Olomouc, Czech Republic, 8–12 December 2013; pp. 273–278. [CrossRef]
32. LeCun, Y.; Bengio, Y.; Hinton, G. Deep learning. *Nature* **2015**, *521*, 436–444. [CrossRef] [PubMed]
33. Lipton, Z.C.; Berkowitz, J.; Elkan, C. A critical review of recurrent neural networks for sequence learning. *arXiv* **2015**, arXiv:1506.00019.
34. Kirbas, I.; Sozen, A.; Tuncer, A.D.; Kazancioglu, F.S. Comparative analysis and forecasting of Covid-19 cases in various European countries with ARIMA, NARN and LSM approaches. *Chaos Solitons Fractals* **2020**, *138*. [CrossRef]
35. Chitra, M.; Sutha, S.; Pappa, N. Application of deep neural techniques in predictive modelling for the estimation of Escherichia coli growth rate. *J. Appl. Microbiol.* **2020**. [CrossRef]
36. Thapa, S.; Zhao, Z.; Li, B.; Lu, L.; Fu, D.; Shi, X.; Tang, B.; Qi, H. Snowmelt-Driven Streamflow Prediction Using Machine Learning Techniques (LSTM, NARX, GPR, and SVR). *Water* **2020**, *12*, 1734. [CrossRef]
37. Box, G.E.; Jenkins, G.M.; Reinsel, G.C.; Ljung, G.M. *Time Series Analysis: Forecasting and Control*; John Wiley & Sons: Hoboken, NJ, USA, 2015.
38. Krzywanski, J.; Grabowska, K.; Sosnowski, M.; Zylka, A.; Kulakowska, A.; Czakiert, T.; Sztekler, K.; Wesolowska, M.; Nowak, W. Heat transfer in fluidized and fixed beds of adsorption chillers. *E3S Web Conf.* **2019**, *128*, 01003. [CrossRef]
39. Krzywanski, J.; Grabowska, K.; Sosnowski, M.; Zylka, A.; Kulakowska, A.; Czakiert, T.; Wesolowska, M.; Nowak, W. Heat transfer in adsorption chillers with fluidized beds of silica gel, zeolite, and carbon nanotubes. *Heat Transf. Eng.* **2019**, *128*, 01003.
40. Stanek, W.; Gazda, W.; Kostowski, W. Thermo-ecological assessment of CCHP (combined cold-heat-and-power) plant supported with renewable energy. *Energy* **2015**, *92*, 279–289. [CrossRef]
41. Stanek, W.; Gazda, W. Exergo-ecological evaluation of adsorption chiller system. *Energy* **2014**, *76*, 42–48. [CrossRef]
42. Aristov, Y.I. Review of adsorptive heat conversion/storage in cold climate countries. *Appl. Therm. Eng.* **2020**, *180*, 115848. [CrossRef]

43. Krzywanski, J.; Żyłka, A.; Czakiert, T.; Kulicki, K.; Jankowska, S.; Nowak, W. A 1.5D model of a complex geometry laboratory scale fuidized bed clc equipment. *Powder Technol.* **2017**, *316*, 592–598. [CrossRef]
44. Muskała, W.; Krzywański, J.; Czakiert, T.; Nowak, W. The research of CFB boiler operation for oxygen-enhanced dried lignite combustion. *Rynek Energii* **2011**, *1*, 172–176.
45. Krzywanski, J.; Fan, H.; Feng, Y.; Shaikh, A.R.; Fang, M.; Wang, Q. Genetic algorithms and neural networks in optimization of sorbent enhanced H2 production in FB and CFB gasifiers. *Energy Convers. Manag.* **2018**, *171*, 1651–1661. [CrossRef]
46. Chorowski, M.; Pyrka, P. Modelling and experimental investigation of an adsorption chiller using low-temperature heat from cogeneration. *Energy* **2015**, *92*, 221–229. [CrossRef]
47. Rogala, Z.; Kolasiński, P.; Gnutek, Z. Modelling and experimental analyzes on air-fluidised silica gel-water adsorption and desorption. *Appl. Therm. Eng.* **2017**, *127*, 950–962. [CrossRef]
48. Aristov, Y.I.; Glaznev, I.S.; Girnik, I.S. Optimization of adsorption dynamics in adsorptive chillers: Loose grains configuration. *Energy* **2012**, *46*, 484–492. [CrossRef]
49. Girnik, I.S.; Grekova, A.D.; Gordeeva, L.G.; Aristov, Y.I. Dynamic optimization of adsorptive chillers: Compact layer vs. bed of loose grains. *Appl. Therm. Eng.* **2017**, *125*, 823–829. [CrossRef]
50. Grabowska, K.; Krzywanski, J.; Nowak, W.; Wesolowska, M. Construction of an innovative adsorbent bed configuration in the adsorption chiller - Selection criteria for effective sorbent-glue pair. *Energy* **2018**, *151*, 317–323. [CrossRef]
51. Grabowska, K.; Sztekler, K.; Krzywanski, J.; Sosnowski, M.; Stefanski, S.; Nowak, W. Construction of an innovative adsorbent bed configuration in the adsorption chiller part 2. experimental research of coated bed samples. *Energy* **2021**, *215*, 119123. [CrossRef]
52. Kulakowska, A.; Pajdak, A.; Krzywanski, J.; Grabowska, K.; Zylka, A.; Sosnowski, M.; Wesolowska, M.; Sztekler, K.; Nowak, W. Effect of Metal and Carbon Nanotube Additives on the Thermal Diffusivity of a Silica Gel-Based Adsorption Bed. *Energies* **2020**, *13*, 1391. [CrossRef]
53. Song, X.; Kanasugi, H.; Shibasaki, R. Deep transport: Prediction and simulation of human mobility and transportation mode at a citywide level. In Proceedings of the Twenty-Fifth International Joint Conference on Artificial Intelligence, New York, NY, USA, 9–15 July 2016; pp. 2618–2624.
54. Yu, H.; Wu, Z.; Wang, S.; Wang, Y.; Ma, X. Spatiotemporal Recurrent Convolutional Networks for Traffic Prediction in Transportation Networks. *arXiv* **2017**, arXiv:1705.02699, 1501. [CrossRef] [PubMed]
55. Glorot, X.; Youshua, B. Understanding the difficult of training deep feedforward neural networks. In Proceedings of the Thirteen International Conference on Artificial Intelligence and Statistics, Sardinia, Italy, 13–15 May 2010; pp. 249–256.

Publisher's Note: MDPI stays neutral with regard to jurisdictional claims in published maps and institutional affiliations.

energies

MDPI

Article

Effect of Additives on the Sorption Kinetics of a Silica Gel Bed in Adsorption Chiller

Karol Sztekler, Wojciech Kalawa, Łukasz Mika, Agata Mlonka-Medrala, Marcin Sowa and Wojciech Nowak *

Faculty of Energy and Fuels, AGH University of Science and Technology, Mickiewicza 30, 30-059 Krakow, Poland; sztekler@agh.edu.pl (K.S.); kalawa@agh.edu.pl (W.K.); lmika@agh.edu.pl (Ł.M.); amlonka@agh.edu.pl (A.M.-M.); marcinsowa25@gmail.com (M.S.)
* Correspondence: wnowak@agh.edu.pl

Abstract: The article presents experimental results of the metal-based and carbon nanotube additives influence on sorption kinetics of a silica-gel-based adsorption bed in an adsorption chiller. The purpose of the doping is to improve the efficiency of sorption processes within the bed by use of metallic and non-metallic additives characterized by higher thermal diffusivity than basic adsorption material. The higher the thermal conductivity of the bed, the faster the sorption processes take place, which directly translates into greater efficiency of the refrigerator. In this study, sorption kinetics of pure silica gel sorbent doped with a given amount of aluminum (Al) and copper (Cu) powders and carbon nanotubes (CNT) were analyzed. The tests were performed on DVS Dynamic Gravimetric Vapor Sorption System apparatus used for dynamic vapor sorption measurements. A decrease in the amount of adsorbed water was observed with an increase in the mass share of the additives in the performed studies. Experimental results show that, CNTs seems to be the most promising additive as the sorption process time was reduced with the smallest decrease in water uptake. Any significant reduction of adsorption time was noted in case of the Al addition. Whereas, in case of Cu doping, delamination of the mixture was observed.

Keywords: silica gel; additives; sorption capacity; sorption process time; kinetics sorption

Citation: Sztekler, K.; Kalawa, W.; Mika, Ł.; Mlonka-Medrala, A.; Sowa, M.; Nowak, W. Effect of Additives on the Sorption Kinetics of a Silica Gel Bed in Adsorption Chiller. *Energies* **2021**, *14*, 1083. https://doi.org/10.3390/en14041083

Academic Editor: Giorgio Vilardi
Received: 27 January 2021
Accepted: 12 February 2021
Published: 19 February 2021

1. Introduction

Electricity demand for cooling increases very quickly, high thermal comfort in buildings is essential for human health and well-being. However, increasing number of refrigeration devices affects the peak demand for electricity and becomes a major challenge for the energy system. Despite of the rapid development of solar energy technologies and easier access to renewable sources of energy, the highest cooling demand does not always agree with the solar energy potential during the day. Therefore, increasing demand for electricity used for cooling purposes has to be covered mostly by the power grid. Due to high energy consumption, such systems indirectly enhance environmental pollution through the consumption of fossil fuels in power plants [1,2]. There is an urgent need to move into more environmentally friendly solutions, like adsorption cooling systems [3,4]. In this technology power can by supplied by low-temperature heat extracted from the outside, like waste heat or solar energy [5–7]. Generation of production of heat, electricity, and useful cold (trigeneration) allows to increase the efficiency of the process and reduce the amount and cost of primary energy [1]. Sorption chillers are even a promising technology in residential applications [8].

The adsorption process is performed through the interactions of the adsorbate on the adsorbent's specific surface area [9]. In adsorption chillers, the adsorbate is liquid and the adsorbent is a solid material with a highly developed specific surface area [10]. The sorbent saturation can be achieved through physical or chemical interactions. Physical interactions are weaker than chemical bounds; therefore, the physisorption is a reversible process of removing adsorbed adsorbate vapors and it can be achieved easily at increased temperatures [2,3].

In general, a good adsorbent should be characterized by: high thermal conductivity and low specific heat [11], it has to be non-toxic and non-corrosive [12]. High durability [13] and similar adsorption properties over time, with relatively small price and high water uptake at low temperature [14] are also crucial. In case of adsorbate, it should be characterized by a relatively low evaporation temperature, high evaporation heat and high latent heat value per unit of volume. Small molecular size, low viscosity, together with low specific heat, low specific volume in liquid state, and high thermal conductivity will additionally enhance the cooling device performance. For safety reason, the adsorbate has to be chemically and thermally stable, non-toxic, and non-flammable [4,15–17]. The most common working pairs in adsorption cooling devices are: activated carbon–methanol, activated carbon–ammonia, silica gel–water, and zeolite–water [16,18,19].

The most important parameter describing refrigeration systems is a coefficient of performance (COP). From the definition COP is the ratio between the heat removed in the evaporator and heat supplied for desorption and bed pre-heating process [16]. Adsorption chillers are characterized by a relatively low COP, which is usually between 0.5 and 0.6 [20,21]. Low COP directly affects dimensions and mass of the equipment, making adsorption chillers big and heavy [22]. To improve their competitiveness on the market of refrigeration systems, a huge effort is put on increasing of the COP of the adsorption chillers. Second important refrigerator performance indicator is specific cooling power (SCP), which determines the cooling capacity in relation to the mass of the adsorbent. Higher SCP increases cooling capacity from a given amount of sorbent and it allows to reduce the size of the equipment, thanks to a smaller sorbent consumption [4]. SCP strongly depends on the cooling bed water temperature [23]. There are several methods found in literature to enhance the adsorption chiller efficiency, most basic is to modify the process parameters like: bed regeneration temperature [24,25], one cycle time, number of beds [21,26,27], or reduction of the heat transfer resistance between the sorbent and metal substrate [28,29].

Low thermal conductivity of the bed is one of the most important factors affecting the low performance of adsorption chillers. The bed materials are selected based on their highly porosity and large specific surface area, both parameters determine high sorption capacity of the bed. However, in most cases highly porous materials has also low thermal conductivity coefficient [16].

The heat exchange in the bed can be enhanced by increase of the heat exchange at the boundary layer of the heat exchanger wall and the adsorbent, or by an increase of adsorbent thermal conductivity [20]. In case of silica gel adsorption bad, many hollow spaces filled with still air are found, thermal conductivity coefficient of air is even lower than in case of silica gel (around 0.025 W/(mK)) [20]. Such behavior results in discontinuities in heat transport within the bed. The idea is to fill the gaps with a material characterized by higher thermal conductivity than adsorbent. The additive should not have negative influence on the adsorbent's sorption properties [20]. The most commonly used higher thermal conductivity materials are metallic powders studied in the literature on the example of aluminum, copper, brass, and stainless steel [30,31] or fine carbon materials like carbon nanotubes [32,33]. In case of adsorbents used in sorption chillers, the goal is not only to reduce the porosity of the bed but also to improve its mechanical properties and preserve sorption capacity. In previous studies described in the literature, the main goal was to define the effect of the additives on thermal conductivity of the material. This study deals with another important issue in the subject of new sorption material development and shows the research on addition of high thermal conductivity materials to the silica gel and its influence on water sorption and desorption kinetics. The presented work is a continuation of the study published in [34] and presenting the comparative analysis of the thermal diffusivity of a silica-gel-base adsorption bed with carbon nanotubes and metals additives (Cu or Al). Thermal diffusivity was measured using the laser flash method. As a part of the article [34], the structural properties of silica gel SG and carbon nanotubes (CNTs) were also investigated. The research results showed that the addition of aluminum had the greatest impact on the improvement of thermal diffusivity. The key property of the additives, apart from their thermal diffusivity, was their density, which affected the homogeneity of the mixture. The highest homogeneity was obtained for samples containing silica gel with the addition of aluminum. It has resulted in achieving the highest thermal diffusivity of these mixtures [34]. Such broad comparative investigation including both structural, thermal and sorption properties of silica gel doped with metallic and non-metallic additives was not published yet and gives a new and valuable knowledge for further development of new sorption materials dedicated for adsorption chillers.

2. Materials and Methods

2.1. Materials

In this study silica gel was selected as the adsorbent. It is a highly porous adsorbent, commonly used in adsorption chillers. The main aim of the study was to analyze the effect of adsorbent doping with materials of high thermal conductivity on sorption kinetics of the obtained mixture. The additives were added to the adsorbent to increase the thermal conductivity of the entire bed by reducing the discontinuity of internal heat transport and increase the mechanical strength of the bed. The reference material in this study was pure silica gel. Pure silica gel was doped with three different additives in powder form. The additives selected were: aluminum (Al), carbon nanotubes (CNT), and copper (Cu), which were used in various mass proportions: 5%, 15%, and 25%. The use of additives will result in reduction of sorption capacity of the mixture, as the additives do not show significant sorption properties. Such doping may contribute to a decrease in mass transport within the bed. Therefore, proper sample homogenization is essential and will influence the experimental results. All analyzed samples are summarized in Table 1.

Table 1. Use of additives Aluminum (Al), Carbon nanotubes (CNT), Copper (Cu) for Silica gel (SG).

Sample	Basic Ingredient	Additional Ingredient	Mass share of Additional Component (%)	The Total Mass of the Sample (mg)
SG + 5% Al	Silica gel	Aluminum	5	50
SG + 15% Al	Silica gel	Aluminum	15	50
SG + 25% Al	Silica gel	Aluminum	25	50
SG + 5% CNT	Silica gel	Carbon nanotubes	5	50
SG + 15% CNT	Silica gel	Carbon nanotubes	15	50
SG + 25% CNT	Silica gel	Carbon nanotubes	25	50
SG + 5% Cu	Silica gel	Copper	5	50
SG + 15% Cu	Silica gel	Copper	15	50

2.2. Methods

Sorption kinetics of the tested samples were determined using Dynamic Gravimetric Vapor Sorption System DVS Vacuum. The apparatus measures the mass change of the sample during adsorption and desorption of a certain amount of adsorbent with high sensitivity, equal to 0.1 µg. Temperature stability at 25 °C is equal to ± 0.02 °C and the humidity conditions generated are typically in the range of ± 0.1% with respect to the target value [35]. Temperature range is the most fundamental experimental parameter, the device has the ability to determine the sorption kinetics from 20 °C up to 70 °C.

The DVS Vacuum offers static and dynamic sorption experiments and measures adsorption-desorption isotherms and adsorption-desorption isobars over a broad range of temperatures and various adsorbates. DVS Vacuum device automatically measures changes in the mass of the sample in a closed chamber.In the static method the volumetric sorption method is used to perform conventional vacuum sorption measurements at static mode. In this method vapor goes into a sealed chamber and the analysis is performed under static conditions (without any further gas flow). This method is recommended for higher partial pressures, measurements performed at low partial pressures result in high measurement errors. In the dynamic vacuum technique, flow rate to the chamber and from the chamber and downstream exit rate of vapor are controlled by a sensitive valve. Therefore, this method is recommended for measurements at low partial pressures. The volumetric method may not be able to give accurate results at low pressures, because system leaks may occur. In this method a pressure change is measured, and volumetric change is derived using a standard relationship:

$$PV = NkT \tag{1}$$

where P = Pressure of the gas; V = Volume of the gas; T= Temperature of the gas; N = Number of particles in the gas; and k = Boltzmann constant (1.38066×10^{-23} J/K).

In the volumetric method, measured vapor or gas pressure is not as precise and sensitive as weight change. Accuracy of the volumetric method is worse than in the gravimetric method. In case of volumetric method, to obtain the same accuracy like in gravimetric one, it is necessary to use large sample volume and time required to attain sorption equilibrium increases. DVS Vacuum system can work in both static and dynamic modes.

In this study, silica gel mixtures were adsorbents and distilled water was the adsorbate. The experimental method was composed of 22 stages. First stage was dedicated to sample drying, the sample was heated up to 100 °C for 60 min. The same period of time was set for the entire system temperature stabilization. Next, 20 stages of 20 min, each were set in the experimental method. Each stage differed in the relative pressure P/P_0, which was gradually increased from 10% up to 100% and then decreased again to 10%. Based on experimental results, adsorption and desorption isotherms and sorption kinetics curves were calculated. The water uptake for all samples depending on its saturation pressure was measured. The steam flow rate was constant and equal to 15 sccm (standard cubic centimeters per minute). Each sample was tested at three process temperatures: 25 °C, 40 °C, and 60 °C.

3. Results and discussion

The results of the performed experiments are presented in Figures 1–10. Presented sorption kinetics graphs describe the relation of the sample reference mass change (in percent) in a function of time. The reference mass is the mass of the sample recorded at the beginning of the measurement when the relative pressure increased from 0% to 10%. Based on the results adsorption and desorption processes trends and intensity of both processes were determined. Based on the measurements, it might be concluded which stage of the process its further continuation can negatively affect the overall performance of the refrigerator. On each graph the curves for three process temperatures: 25 °C, 40 °C, and 60 °C are presented. Additional approximation calculations were made using MATLAB software and showed in Figures 1–10, together with experimental results.

Figure 1 shows the sorption kinetics of pure silica gel sample denoted as SG. This sample with no additives was considered as a reference sample.

In Figure 2 a comparison of the adsorption and desorption process for the SG + 5% Al sample at different temperatures is shown. At 60 °C, the sorption cycle is much shorter than in case of other process temperatures. Additionally, the sample tested at the highest process temperature - 60 °C, is characterized by the highest sorption capacity.

Experimental results of sorption kinetics for the sample SG + 15% Al are shown in Figure 3. In this case, at the highest process temperature, the lowest amount of water vapor was adsorbed by the tested sample. Additionally, the desorption process proceeds more intensively in this case. Samples tested at 25 °C and 40 °C are characterized by a similar desorption tendency and a similar water uptake. Higher intensity of the adsorption process was noted at 40 °C.

Figure 4 presents the adsorption and desorption cycles tendencies for the SG + 25% Al sample. At 40 °C, the sample was characterized by the highest adsorption intensity, while the most intense desorption process was observed at 60 °C. The maximal water uptake is similar in all analyzed cases.

Figure 5 shows the sorption kinetics for the sample with the addition of 5% carbon nanotubes to silica gel. The highest sorption capacity was observed for the sample tested at 25 °C. The temperature of 40 °C contributes to faster adsorption and the temperature of 60 °C to faster desorption.

Figure 6 shows the course of sorption kinetics for a sample designated as SG + 15% CNT. Adsorption and desorption were performed most intensively for the sample tested, respectively, at 40 °C and 60 °C. The maximum amount of adsorbed water vapor is similar in both temperatures 40 °C and 60 °C and the lowest for the sample tested at 25 °C.

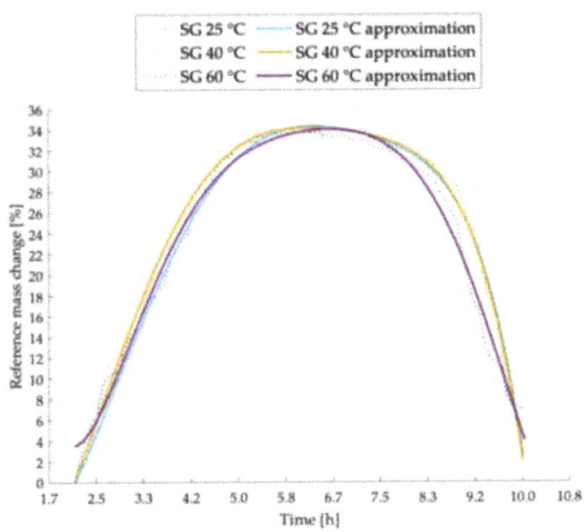

Figure 1. Sorption kinetics graph for SG sample tested at temperatures of 25 °C, 40 °C, 60 °C.

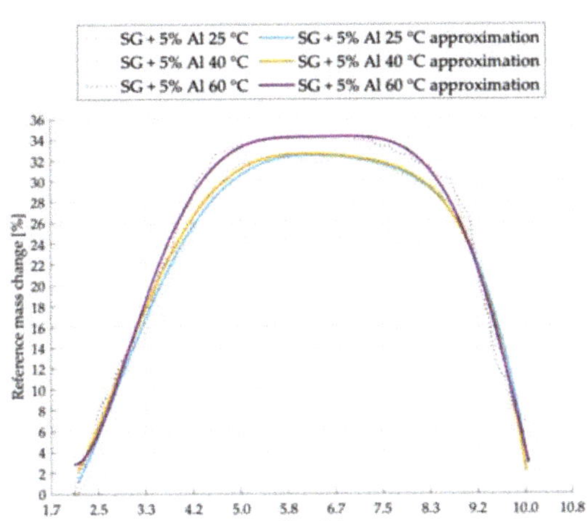

Figure 2. Sorption kinetics graph for SG + 5% Al sample tested at temperatures of 25 °C, 40 °C, 60 °C.

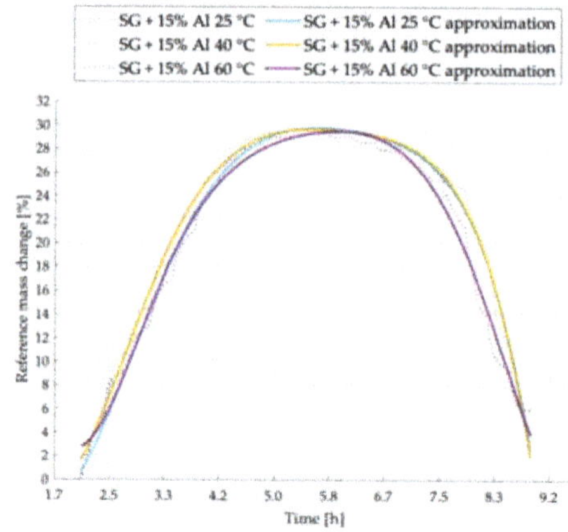

Figure 3. Sorption kinetics graph for SG + 15% Al sample tested at temperatures of 25 °C, 40 °C, 60 °C.

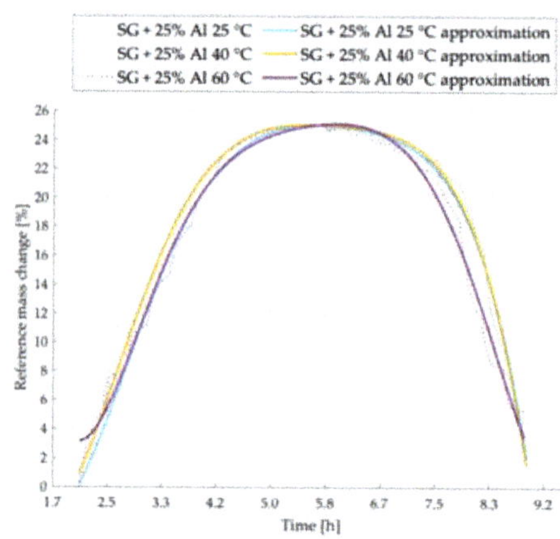

Figure 4. Sorption kinetics graph for SG + 25% Al sample tested at temperatures of 25 °C, 40 °C, 60 °C.

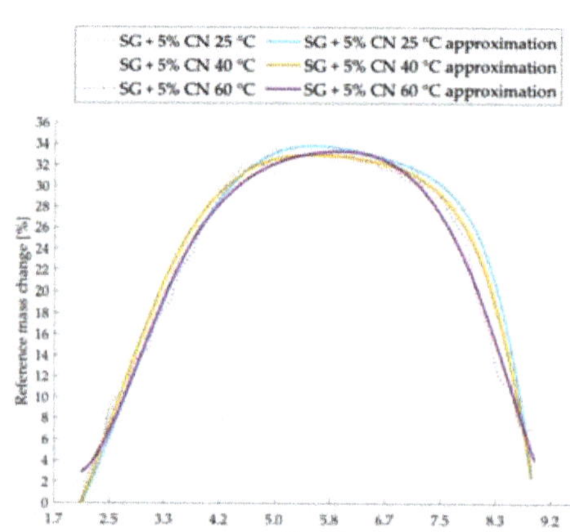

Figure 5. Sorption kinetics graph for SG + 5% CNT sample tested at temperatures of 25 °C, 40 °C, 60 °C.

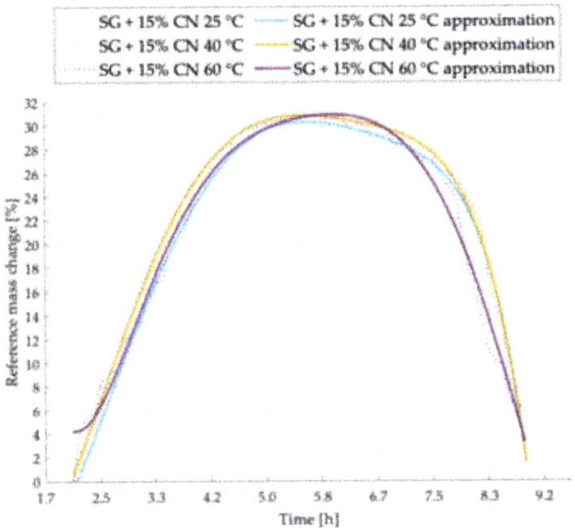

Figure 6. Sorption kinetics graph for SG + 15% CNT sample tested at temperatures of 25 °C, 40 °C, 60 °C.

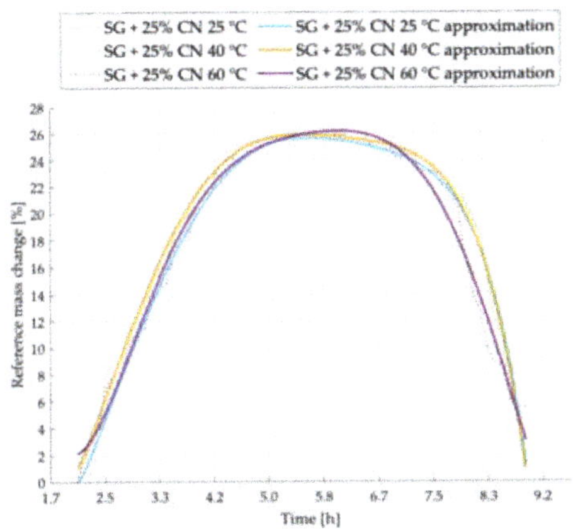

Figure 7. Sorption kinetics graph for SG + 25% CNT sample tested at temperatures of 25 °C, 40 °C, 60 °C.

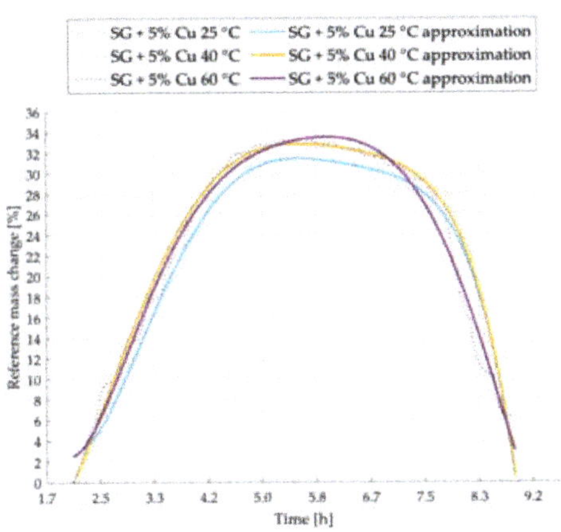

Figure 8. Sorption kinetics graph for SG + 5% Cu sample tested at temperatures of 25 °C, 40 °C, 60 °C.

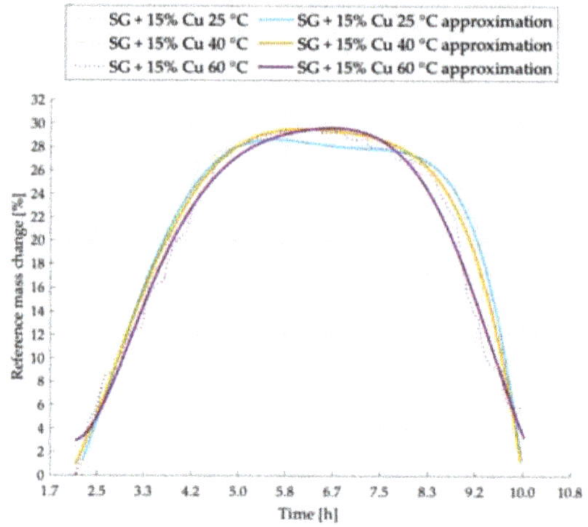

Figure 9. Sorption kinetics graph for SG + 15% Cu sample tested at temperatures of 25 °C, 40 °C, 60 °C.

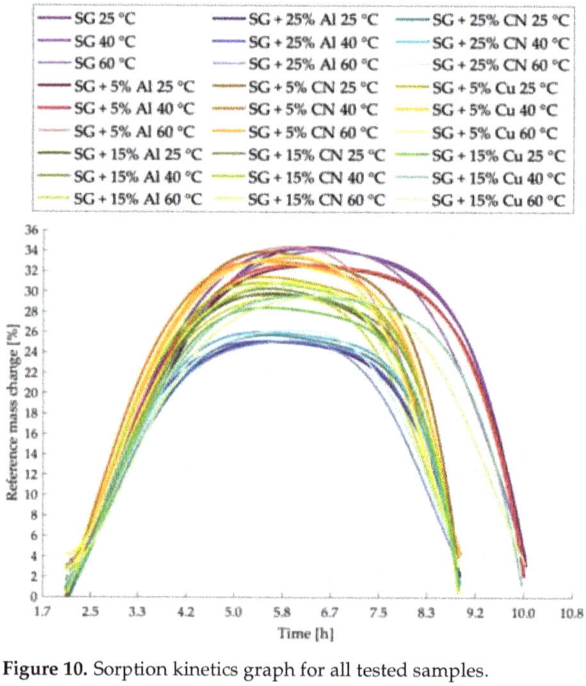

Figure 10. Sorption kinetics graph for all tested samples.

Figure 7 presents the sorption kinetics for the mixture with 25% addition of carbon nanotubes. The increase in desorption intensity for the sample tested at 60 °C and the increase in adsorption intensity at 40 °C are clearly visible. The mass of adsorbed vapor is similar for each of the samples. Its highest value is achieved for the temperature of 60 °C and the lowest for 25 °C.

The sorption kinetics for the sample with 5% copper addition is shown in Figure 8. A clear decrease in the sorption capacity for the sample tested at 25 °C was noticed. Like in the graphs presented earlier, also in this case, the adsorption intensity was the highest at 40 °C and the desorption intensity increased significantly at 60 °C.

For the sample SG + 15% Cu, the sorption kinetics is shown in Figure 9. At the temperature of 25 °C, the sorption capacity of the mixture is the smallest, and the time needed to complete the entire cycle is shorter in comparison to other temperatures. At 60 °C the intensity of desorption process with a similar amount of adsorbed steam was greater than in the case of the sample tested at 40 °C. The duration of the entire cycle for this mixture was the longest in comparison to all other tested samples.

Figure 10 shows sorption kinetics for all samples analyzed in this study to show a relationship not only between process temperatures but also to make a comparison between the materials added. Doping of SG with materials characterized by high thermal conductivity lads to cycle shortening. As expected, doping influenced also water uptake by adsorbent, as the amount of porous material decreases with increased amount of doped additive.

In this study, clear differences in sorption capacity, intensity of adsorption and desorption, and both processes times were observed in all tested mixtures. The differences result from different properties of the doped materials and different mass fractions of additives in the samples.

Table 2 shows the reference mass changes of each sample in % and the differences between the change in the reference sample mass and the mass of the sample with additive examined at the same temperature.

Table 2. Percentage change in reference mass of all tested samples.

Sample Designation	Temperature (°C)	Reference Mass Change (Sorption Capacity) (%)	Difference in Sample Mass Change Towards to the Reference Sample Tested at the Same Temperature (pp)	Time from Obtaining Reference Mass to Achieving a 20% Change in Reference Mass of the Sample for Adsorption Process (min)	Time from the Start of the Desorption Process Until Obtaining a 20% Change in Reference Mass of the Sample for Desorption Process (min)
SG	25	34.35	-	98.55	194.19
SG	40	34.21	-	86.07	193.80
SG	60	33.79	-	97.18	176.20
SG + 5% Al	25	32.50	1.79	96.04	192.65
SG + 5% Al	40	32.63	1.57	86.65	191.16
SG + 5% Al	60	34.31	0.47	93.32	168.35
SG + 15% Al	25	29.77	4.53	93.02	160.16
SG + 15% Al	40	29.67	4.51	85.34	160.15
SG + 15% Al	60	29.29	4.42	93.50	145.11
SG + 25% Al	25	25.01	9.29	110.40	146.39
SG + 25% Al	40	25.11	9.15	104.25	146.54
SG + 25% Al	60	24.98	8.78	109.61	130.88
SG + 5% CNT	25	33.74	0.56	83.10	170.06
SG + 5% CNT	40	32.92	1.27	75.67	166.52
SG + 5% CNT	60	33.21	0.60	84.64	152.03
SG + 15% CNT	25	30.17	4.13	95.78	165.98
SG + 15% CNT	40	30.76	3.43	84.84	166.03
SG + 15% CNT	60	30.85	2.96	91.78	149.03
SG + 25% CNT	25	25.63	8.67	110.63	152.91
SG + 25% CNT	40	25.91	8.28	102.86	154.84
SG + 25% CNT	60	26.19	7.62	107.75	144.33
SG + 5% Cu	25	31.43	2.87	83.44	166.69
SG + 5% Cu	40	32.76	1.43	81.14	166.40
SG + 5% Cu	60	33.26	0.55	86.39	150.31
SG + 15% Cu	25	28.34	5.96	92.43	164.00
SG + 15% Cu	40	29.52	4.67	100.14	185.87
SG + 15% Cu	60	29.51	4.30	108.68	169.18

During samples preparation, some phenomena were observed that may significantly affect the use of porous sorbents with additives as the adsorption chiller bed material. One of the serious issues was achieving a homogeneous mixture of two different loose materials. The purpose of mixing is to obtain a relatively even dispersion of the substances in the mixture. The complexity of the process is observed after a closer analysis, especially when mixing substances with different properties such as porosity and density. Mixing can be defined as the mutual movement of two or more portions of different materials, resulting in the required level of uniformity in the final product. In the case of the silica gel sample with the addition of 5% and 15% Cu, a heterogeneous structure of the sample was observed after mixing, with a clear division between silica gel and copper fractions. The reason for not achieving homogeneity of the sample is too large difference in densities between the copper powder and silica gel, which resulted in copper sinking to the bottom of the measuring vessel. The lack of homogeneity of the material leads to decrease of effectives of the heat exchange process in the bed between the exchanger tubes and the silica gel, the copper powder layer will be deposited at the bottom of the bed and will have a much smaller impact on the heat exchange process between the bulk material and the heat exchanger tubes. As obtaining a homogeneous sample was a challenge in case of both SG + 5% Cu and SG + 15% Cu, it was decided to exclude 25% addition of copper to silica gel in the presented study.

According to the experimental results it might be concluded that the process temperature influences the sorption capacity in minor extent. In major number of cases, the sorption capacity decreases with process temperature increase. Some slight differences in the reference mass change might be explained by measurement uncertainty and sample heterogeneity, described above. As expected, samples with the addition of 5% of the mass fraction showed the closest sorption properties to the reference sample. On average, the decrease in sorption capacity with respect to a silica gel sample is about 1.0% for aluminum addition, 0.8% for carbon nanotubes and 1.6% for copper. Whereas the highest loss of sorption properties of all mixtures in relation to the reference sample was obtained for samples with a 25% mass fraction of the additives. For samples with addition of aluminum the average loss of adsorbed water for all temperatures was 9.1%, while for the addition of carbon nanotubes it was 8.2%. The maximum loss was observed for the sample with aluminum 9.3%. In case of silica gel samples with a 15% of additives, the greatest losses of sorption capacity were observed for mixtures containing copper. The average decrease in the sorption capacity of the mixtures is 4.5% for aluminum additive, 3.5% for carbon nanotubes additive and about 5.0% for copper additive.

The highest reduction of the adsorption process time was observed for the 5% addition of CNTs and Cu and 15% addition of Al to the silica gel. The 25% addition of CNTs and Al increased the adsorption process time noticeably. The shortest adsorption process time was denoted for 40 °C process temperature in all analyzed cases.

The increase of process temperature resulted in reduction of desorption process time; in most cases the shortest desorption time was noticed at 60 °C. Only samples with Cu addition, especially with 15% of Cu in the sample, did not present such a behavior, most probably due to high heterogeneity of the sample. The highest desorption time decrease was observed for samples of SG with 25% addition of Al and CNTs (26% and 20% respectively). However, the smallest addition of the CNTs resulted in very high reduction of the desorption time (12–14%) and this value did not decrease that noticeably in case of Al. When aluminum powder was added to the samples, each increase of the additive amount decreased noticeably the desorption time and the sorption capacity.

4. Conclusions

Low performance, big dimensions, and weight are main obstacles when considering adsorption cooling devices development and commercialization. Most recent research was dedicated to performance improvement by increase of thermal diffusivity of the adsorbents. The greater thermal conductivity of the bed will result in desorption process enhancement, which directly influence chiller performance factors. It also provides stabilization of temperature and chilled water production. Additionally, the reduced desorption time leads to the device energy consumption reduction. In this study the influence of additional component in the silica gel on the amount of adsorbed water vapor was analyzed.

In this study sorption properties of silica gel doped with three additives in powder form: aluminum, carbon nanotubes, and copper were analyzed. Each additive is expected to contribute in the bed sorption properties decrease but also in increase in thermal conductivity of the material (what was presented in previous works of the Authors [34]). As expected, the experimental results showed that the 5% mass fraction in the least way contributes to the loss of sorption properties in comparison to the reference sample, and the loss increases with the increasing amount of the additive in the sample. However, even 25% addition of metal powder or carbon nanotubes did not decrease sorption capacity of the mixture by more than 10%. It was also noticed that carbon nanotubes have the smallest influence on the loss of sorption capacity of the silica gel mixtures doped with a high thermal diffusivity additive. It was also observed that there is a tendency to decrease the loss of sorption properties of the samples with increasing temperature, but in a minor extent.

According to the Authors, the CNTs addition to silica gel seems to be the most promising mixture due to the reduction of the sorption process time with the smallest decrease in water uptake. The most advantageous amount of carbon nanotubes added to the SG is 5%, which reduces the time of adsorption and desorption at 40 °C by about 15% and 14% respectively (given for 20% difference in the reference mass of the sample in). Under these conditions, the amount of water absorbed decreased by about 1.5%. However, to obtain better desorption times a higher temperature and Al addition is required.

A similar reduction in the times of adsorption and desorption process are obtained for SG with 5% Cu addition, but due to the tendency to delamination of the mixture and deposition of copper powder at the bottom of the sample, the authors do not indicate such a mixture as applicable in real conditions. The delamination of the mixture may adversely affect the sorption processes during the operation of such an adsorption bed (lower adsorbent mass, copper accumulated at the bottom of the sample).

The addition of Al to SG did not cause a significant reduction in adsorption time in all tested temperatures. Only 5% addition of Al to SG leads to about 10% reduction of adsorption time (given for 20% difference in the reference mass of the sample) without decrease of the mixture sorption properties. It should be mentioned, however, that in the case of Al to SG, the homogeneity of the mixture can be obtained most easily due to the similar density of both materials. Furthermore, as presented in [34], the mixture containing aluminum gave the best results in terms of sample thermal diffusivity enhancement. The addition of 5% aluminum leaded to a 90% increase in thermal diffusivity in comparison to the raw silica gel.

Author Contributions: Conceptualization, K.S. and Ł.M.; Methodology, K.S. and M.S.; Formal analysis, K.S. and Ł.M., Investigation, W.K.,K.S; Resources, W.K.; Data curation, K.S. and M.S., W.K.; Writing—original draft preparation, K.S., Ł.M., and A.M.-M.; writing—review and editing, Ł.M. and K.S., A.M; Supervision, W.N.; Project administration, K.S.; Funding acquisition, Ł.M. and W.N. All authors have read and agreed to the published version of the manuscript.

Funding: The paper was funded by subsidy for research Faculty of Energy and Fuels number 16.16.210.476.

Conflicts of Interest: The authors declare no conflict of interest.

Abbreviations

COP	Coefficient of performance
SCP	Specific cooling power
Cu	Copper
Al	Aluminum
SG	Silica gel
CNT	Carbon nanotubes

References

1. Sztekler, K.; Kalawa, W.; Stefanski, S.; Krzywanski, J.; Grabowska, K.; Sosnowski, M.; Nowak, W.; Makowski, M. Using adsorption chillers for utilising waste heat from power plants. *Therm. Sci.* **2019**, *23*, S1143–S1151. [CrossRef]
2. Zarei, M.; Davarpanah, A.; Mokhtarian, N.; Farahbod, F. Integrated feasibility experimental investigation of hydrodynamic, geometrical and, operational characterization of methanol conversion to formaldehyde. *Energy Sources Part A Recover. Util. Environ. Eff.* **2020**, *42*, 89–103. [CrossRef]
3. Pan, Q.; Peng, J.; Wang, R. Experimental study of an adsorption chiller for extra low temperature waste heat utilization. *Appl. Therm. Eng.* **2019**, *163*, 114341. [CrossRef]
4. Younes, M.M.; El-Sharkawy, I.I.; Kabeel, A.E.; Saha, B.B. A review on adsorbent-adsorbate pairs for cooling applications. *Appl. Therm. Eng.* **2017**, *114*, 394–414. [CrossRef]
5. Choudhury, B.; Saha, B.B.; Chatterjee, P.K.; Sarkar, J.P. An overview of developments in adsorption refrigeration systems towards a sustainable way of cooling. *Appl. Energy* **2013**, *104*, 554–567. [CrossRef]
6. Jeremias, F.; Fröhlich, D.; Janiak, C.; Henninger, S.K. Water and methanol adsorption on MOFs for cycling heat transformation processes. *New J. Chem.* **2014**, *38*, 1846–1852. [CrossRef]
7. Alizadeh, S.M.; Ghazanfari, A.; Ehyaei, M.A.; Ahmadi, A.; Jamali, D.H.; Nedaei, N.; Davarpanah, A. Investigation the integration of heliostat solar receiver to gas and combined cycles by energy, exergy, and economic point of views. *Appl. Sci.* **2020**, *10*, 307. [CrossRef]
8. Esfandi, S.; Baloochzadeh, S.; Asayesh, M.; Ehyaei, M.A.; Ahmadi, A.; Rabanian, A.A.; Das, B.; Costa, V.A.F.; Davarpanah, A. Energy, Exergy, Economic, and Exergoenvironmental Analyses of a Novel Hybrid System to Produce Electricity, Cooling, and Syngas. *Energies* **2020**, *13*, 6453. [CrossRef]
9. Li, T.; Wang, R.; Wang, L. High-efficient thermochemical sorption refrigeration driven by low-grade thermal energy. *Chinese Sci. Bull.* **2009**, *54*, 885–905. [CrossRef]
10. Shabir, F.; Sultan, M.; Miyazaki, T.; Saha, B.B.; Askalany, A.; Ali, I.; Zhou, Y.; Ahmad, R.; Shamshiri, R.R. Recent updates on the adsorption capacities of adsorbent-adsorbate pairs for heat transformation applications. Renew. *Sustain. Energy Rev.* **2020**, *119*, 109630. [CrossRef]
11. Bu, X.; Wang, L.; Huang, Y. Effect of pore size on the performance of composite adsorbent. *Adsorption* **2013**, *19*, 929–935. [CrossRef]
12. Zeng, T.; Huang, H.; Kobayashi, N.; Li, J. Performance of an Activated Carbon-Ammonia Adsorption Refrigeration System. *Nat. Resour.* **2017**, *08*, 611–631. [CrossRef]
13. Li, A.; Ismail, A.B.; Thu, K.; Ng, K.C.; Loh, W.S. Performance evaluation of a zeolite—Water adsorption chiller with entropy analysis of thermodynamic insight. *Appl. Energy* **2014**, *130*, 702–711. [CrossRef]
14. Entezari, A.; Ge, T.S.; Wang, R.Z. Water adsorption on the coated aluminum sheets by composite materials (LiCl + LiBr)/silica gel. *Energy* **2018**, *160*, 64–71. [CrossRef]
15. Goyal, P.; Baredar, P.; Mittal, A.; Siddiqui, A.R. Adsorption refrigeration technology—An overview of theory and its solar energy applications. Renew. Sustain. *Energy Rev.* **2016**, *53*, 1389–1410. [CrossRef]
16. Fernandes, M.S.; Brites, G.J.V.N.; Costa, J.J.; Gaspar, A.R.; Costa, V.A.F. Review and future trends of solar adsorption refrigeration systems. Renew. Sustain. *Energy Rev.* **2014**, *39*, 102–123. [CrossRef]
17. Shmroukh, A.N.; Hamza, A.; Ali, H.; Ookawara, S. Adsorption working pairs for adsorption cooling chillers: A review based on adsorption capacity and environmental impact. Renew. *Sustain. Energy Rev.* **2015**, *50*, 445–456. [CrossRef]
18. Valizadeh, K.; Davarpanah, A. Environmental Effects Design and construction of a micro-photo bioreactor in order to dairy wastewater treatment by micro-algae: Parametric study. Energy Sources, Part A Recover. *Util. Environ. Eff.* **2020**, *42*, 611–624. [CrossRef]
19. Davarpanah, A.; Zarei, M.; Valizadeh, K.; Mirshekari, B. CFD design and simulation of ethylene dichloride (EDC) thermal cracking reactor. Energy Sources Part A Recover. *Util. Environ. Eff.* **2019**, *41*, 1573–1587. [CrossRef]
20. Grabowska, K.; Krzywanski, J.; Nowak, W.; Wesolowska, M. Construction of an innovative adsorbent bed configuration in the adsorption chiller—Selection criteria for effective sorbent-glue pair. *Energy* **2018**, *151*, 317–323. [CrossRef]
21. Sztekler, K.; Kalawa, W.; Nowak, W.; Mika, L.; Gradziel, S.; Krzywanski, J.; Radomska, E. Experimental Study of Three-Bed Adsorption Chiller with Desalination Function. *Energies* **2020**, *13*, 5827. [CrossRef]
22. Askalany, A.A.; Henninger, S.K.; Ghazy, M.; Saha, B.B. Effect of improving thermal conductivity of the adsorbent on performance of adsorption cooling system. *Appl. Therm. Eng.* **2017**, *110*, 695–702. [CrossRef]

23. Bahrehmand, H.; Khajehpour, M.; Bahrami, M. Finding optimal conductive additive content to enhance the performance of coated sorption beds: An experimental study. *Appl. Therm. Eng.* **2018**, *143*, 308–315. [CrossRef]
24. Vodianitskaia, P.J.; Soares, J.J.; Melo, H.; Gurgel, J.M. Experimental chiller with silica gel: Adsorption kinetics analysis and performance evaluation. *Energy Convers. Manag.* **2017**, *132*, 172–179. [CrossRef]
25. Tso, C.Y.; Chan, K.C.; Chao, C.Y.H.; Wu, C.L. Experimental performance analysis on an adsorption cooling system using zeolite 13X/CaCl2 adsorbent with various operation sequences. *Int. J. Heat Mass Transf.* **2015**, *85*, 343–355. [CrossRef]
26. Krzywanski, J.; Grabowska, K.; Herman, F.; Pyrka, P.; Sosnowski, M.; Prauzner, T.; Nowak, W. Optimization of a three-bed adsorption chiller by genetic algorithms and neural networks. *Energy Convers. Manag.* **2017**, *153*, 313–322. [CrossRef]
27. Sztekler, K.; Kalawa, W.; Nowak, W.; Mika, L.; Grabowska, K.; Krzywanski, J.; Sosnowski, M.; Al-Harbi, A.A. Performance evaluation of a single-stage two-bed adsorption chiller with desalination function. *J. Energy Resour. Technol.* **2020**. [CrossRef]
28. Chang, K.S.; Chen, M.T.; Chung, T.W. Effects of the thickness and particle size of silica gel on the heat and mass transfer performance of a silica gel-coated bed for air-conditioning adsorption systems. *Appl. Therm. Eng.* **2005**, *25*, 2330–2340. [CrossRef]
29. Sztekler, K.; Kalawa, W.; Medrala, A.M.; Nowak, W.; Mika, Ł.; Krzywanski, J.; Grabowska, K.; Sosnowski, M.; Debniak, M. The Effect of Adhesive Additives on Silica Gel Water Sorption Properties. *Entropy* **2020**, *22*, 327. [CrossRef]
30. Demir, H.; Mobedi, M.; Ülkü, S. The use of metal piece additives to enhance heat transfer rate through an unconsolidated adsorbent bed. *Int. J. Refrig.* **2010**. [CrossRef]
31. Rezk, A.; Al-Dadah, R.K.; Mahmoud, S.; Elsayed, A. Effects of contact resistance and metal additives in finned-tube adsorbent beds on the performance of silica gel/water adsorption chiller. *Appl. Therm. Eng.* **2013**, *53*, 278–284. [CrossRef]
32. Wu, H.; Al-Rashed, A.A.A.A.; Barzinjy, A.A.; Shahsavar, A.; Karimi, A.; Talebizadehsardari, P. Curve-fitting on experimental thermal conductivity of motor oil under influence of hybrid nano additives containing multi-walled carbon nanotubes and zinc oxide. *Phys. A Stat. Mech. Appl.* **2019**. [CrossRef]
33. Hemmat Esfe, M.; Motahari, K.; Sanatizadeh, E.; Afrand, M.; Rostamian, H.; Reza Hassani Ahangar, M. Estimation of thermal conductivity of CNTs-water in low temperature by artificial neural network and correlation. *Int. Commun. Heat Mass Transf.* **2016**. [CrossRef]
34. Kulakowska, A.; Pajdak, A.; Krzywanski, J.; Grabowska, K.; Zylka, A.; Sosnowski, M.; Wesolowska, M.; Sztekler, K.; Nowak, W. Effect of metal and carbon nanotube additives on the thermal diffusivity of a silica-gel-based adsorption bed. *Energies* **2020**, *16*, 1391. [CrossRef]
35. Surface Measurement Systems Company Materials. Operating Manual of DVS Vacuum, Book 2019.

energies

MDPI

Article

Energy and Exergy Analyses of Adsorption Chiller at Various Recooling-Water and Dead-State Temperatures

Ahmad A. Alsarayreh [1,*], Ayman Al-Maaitah [2], Menwer Attarakih [3] and Hans-Jörg Bart [4]

1 Precision Industries, P.O. Box 37448, Dubai, United Arab Emirates
2 Wahaj Investment L.L.C., 24B St, Comm-365, Ind 2-Dubai, United Arab Emirates; ayman@wahajsolar.com
3 Department of Chemical Engineering, University of Jordan, Queen Rania St, Amman 11942, Jordan; attarakih67@gmail.com
4 Thermische Verfahrenstechnik, TU Kaiserslautern, 67653 Kaiserslautern, Germany; bart@mv.uni-kl.de
* Correspondence: ahmadsarayrah@ymail.com; Tel.: +971-543061810

Abstract: We conducted energy and exergy analyses of an adsorption chiller to investigate the effect of recooling-water temperatures on the cooling capacity and Coefficient of Performance (COP) with variable cycle modes. We investigated both the effect of the recooling-water temperature and the dead state temperature on the exergy destruction in the chiller components. Our results show that there is an optimum reheat cycle mode for each recooling-water temperature range. For the basic single stage cycle, the exergy destruction is mainly accrued in the desorber (49%), followed by the adsorber (27%), evaporator (13%), condenser (9%), and expansion valve (2%). The exergy destruction for the preheating process is approximately 35% of the total exergy destruction in the desorber. By contrast, the precooling process is almost 58% of the total exergy destruction in the adsorber. The exergy destruction decreases when increasing the recooling-water and the dead state temperatures, while the exergy efficiency increases. Nonetheless, the exergy efficiency decreases with an increase in the recooling-water temperature at fixed dead state temperatures. The effect of the mass recovery time in the reheat cycle on exergy destruction was also investigated, and the results show that the exergy destruction increases when the mass recovery time increases. The exergy destruction in the adsorbent beds was the most sensitive to the increase in mass recovery time.

Keywords: adsorption; exergy; dead state; adsorption cooling; reheat cycle, mass recovery

Citation: Alsarayreh, A.A.; Al-Maaitah, A.; Attarakih, M.; Bart, H.-J. Energy and Exergy Analyses of Adsorption Chiller at Various Recooling-Water and Dead-State Temperatures. *Energies* **2021**, *14*, 2172. https://doi.org/10.3390/en14082172

Academic Editor: Jaroslaw Krzywanski

Received: 18 March 2021
Accepted: 9 April 2021
Published: 13 April 2021

1. Introduction

The worldwide demand for cooling equipment is growing rapidly due to population growth, increased living standards, and rising temperatures due to climate change. Space cooling accounted for about 8.5% of world electricity consumption in 2019, and the electricity demand for space cooling is expected to increase by 50% by 2030 if there is no major efficiency improvement [1]. The vapor–compression refrigeration cycles driven by electricity have the main global market share for air conditioning [2].

Thermally driven cooling technologies, which can also be driven by solar thermal energy and/or waste heat as absorption or adsorption cooling, are among the most common energy-saving solutions for cooling applications. In addition, the thermally driven cooling technology reduces greenhouse gas (GHG) emissions and ensures a higher indoor air quality [3].

The term "thermally driven chiller cycle" describes the thermodynamic cycle in which heat is absorbed at one pressure level and raised to a higher level where it is rejected. The basic procedure of the thermally driven chiller cycle is as follows:

- The cold air is obtained as the heat is rejected from the chilled water in the evaporator of the chiller.
- Heat input is the heat required in the generator to drive the process, which is delivered either by the solar system or by backup heat sources.

- Heat rejection is ideally the summation of the useful load and driving heat; it can be dissipated normally by a cooling tower or dry cooler.

Figure 1 illustrates the thermodynamic principle of the thermally driven chiller.

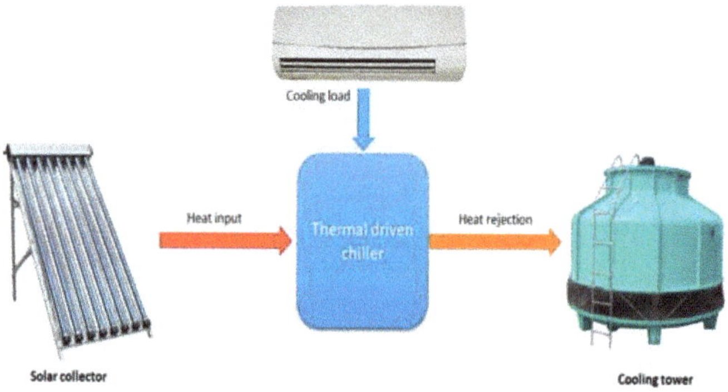

Figure 1. Thermally driven system and its energy flow.

Adsorption cooling is considered a promising thermal cooling technology, as adsorption chillers are normally driven by low-grade heat below 100 °C with minimum electrical energy consumption [4]. The temperature can be easily achieved by the solar collector or recovered easily from the heat rejected from a diesel genset.

The principle of adsorption is based on the interaction of gases and solids. It employs the physical uptake of refrigerant (adsorbate) on the surface of the adsorbents such as silica gel, zeolite, and activated carbon due to van der Waal's or polar bonding forces [5]. Figure 2a shows the adsorption–cooling cycle consisting of four main processes: isosteric heating, desorption, isosteric cooling, and adsorption [6]. Figure 2b illustrates a simple adsorption–cooling system, which includes four main components: an adsorbent bed, condenser, evaporator, and expansion valve. The adsorbent bed is typically cooled or heated using circulated media during the adsorption and desorption processes, respectively, to maintain a continuous adsorption–desorption cycle.

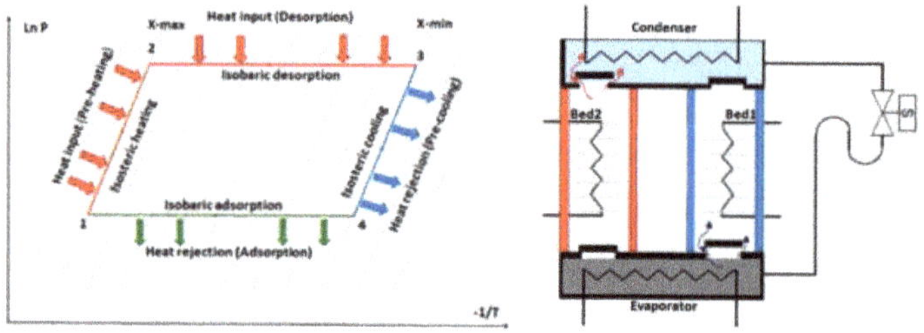

Figure 2. (**a**) Clapeyron diagram of the simple adsorption–cooling cycle. (**b**) Schematic diagram of the adsorption–cooling system.

The energy performance analyses of adsorption cooling based on the first law of thermodynamics were previously investigated [7–9]. Many ideas to improve the system performance using adsorbent material, heat and mass recovery, multi-bed, and multi-stage technologies were introduced [10]. In general, three main working temperatures affect the adsorption cooling system performance: hot water, recooling water, and chilled water temperatures. The Coefficient of Performance (COP) and cooling capacity increase when the hot water and chilled water temperatures increase and decrease with the increase in the recooling-water temperature [11].

The first law of thermodynamics fails to account for energy quality and does not measure the irreversibility loss [12]. Moreover, energy analysis cannot help identify the irrational use of energy sources, as is the case in exergy analysis [13]; therefore, the exergy analysis of adsorption cooling systems is important in assessing the system components' losses and efficiency. The exergy efficiency is measured as the ratio of exergy input to output in the system as useful output [14]. The remaining exergy is lost due to the irreversibility of the system [15]. The total exergy includes physical, chemical, kinetic, and potential exergies. As the major part of the exergy is physical exergy represented by the exergies due to heat, work, and mass flow, the other exergies can be neglected [16].

Not many researchers have investigated the exergy efficiency of the adsorption cooling system. Baiju and Chandrasekharan [16] performed an exergy destruction and efficiency study for solar adsorption cooling system using artificial neural network models; the predicted results of the Artificial Neural Network (ANN) model were compared with the calculated values at R close to 1 and low Root-mean-square (RMS) values. Their results showed that the overall exergy destruction increased with the increase in heat source temperature. In contrast, the overall exergy efficiency increased until it reached the optimum operating temperature and its maximum and then decreased.

Baiju and Muraleedharan [17] carried out an experimental study of a solar hybrid adsorption refrigeration system. The exergy analysis results showed that the main exergy was destroyed in the adsorbent bed due to material constraints. The expansion valve exhibited the highest exergetic efficiency, followed by the evaporator, condenser, and adsorbent bed with 79.8%, 54.7%, 42.3%, and 11%, respectively.

Ogueke and Ndeke [18] analyzed the exergy performance of an adsorption solar refrigerator by applying an exergy balance for each refrigerator component. The maximum exergy destruction occurred in the preheating and desorption process (collector/generator) followed by the evaporator and condenser.

Cao et al. [19] conducted a transient two-dimensional numerical exergy analysis of adsorption cooling cycles for three scenarios: a mass recovery cycle, heat recovery cycle, and a combination of these two cycles. Results showed that as the heat input temperature increased, the total exergy loss increased, while the exergy efficiency decreased. The mass recovery cycle improved the exergy efficiency. It reduced the total loss by 2.76% compared to that of the basic cycle, whereas the heat recovery significantly improved the exergy efficiency by 23.20%. The combined heat and mass recovery cycle showed an improvement in the exergy efficiency of 11.30% and 37.12% compared to those of the heat recovery and basic cycle, respectively.

All of the above-mentioned work on the investigation and analysis of the exergy of the adsorption chiller was mainly based on theoretical modeling results while ignoring the recooling-water temperature and assuming the dead state temperature to be constant. As the recooling temperature cannot be maintained at a low level, especially with a dry-cooled adsorption chiller, we aimed to investigate the energy and exergy of adsorption chillers at different recooling-water temperatures with different modes based on experimental results. We investigated four modes: single-stage mode, 25% of reheat mode (25% of the cooling duration), 50% reheat mode (50% of the cooling duration), and 75% reheat mode (75% of the cooling duration).

2. Methodology and Mathematical Model

The energy and exergy performance analyzed in this study is based on the experimental results for a proposed prototype adsorption chiller with a reheating process. Figure 3a illustrates the proposed chiller components, while Figure 3b shows a Clapeyron diagram of the thermodynamic processes of the adsorption chiller with the reheating process.

The adsorption chiller with the reheating process consists of two adsorbent beds, a condenser, an expansion valve, an evaporator, four non-return valves, and a motorized ball valve. The adsorbent beds work as adsorbers and desorbers. The motorized valve opens once the adsorption/desorption processes end, which allows the high-pressure hot refrigerant vapor to flow from the desorber to the adsorber (mass recovery process). The non-return valve controls the refrigerant flow and allows it to flow from the evaporator to the adsorber, and from the desorber to the condenser (Figure 3a).

The chiller works via six thermodynamics processes (Figure 3b): mass recovery with pressurization when the two beds are connected (A,B), preheating (B,C), the desorption process (C,D), mass recovery with depressurization (D,E), precooling (E,F), and the adsorption process (F,A).

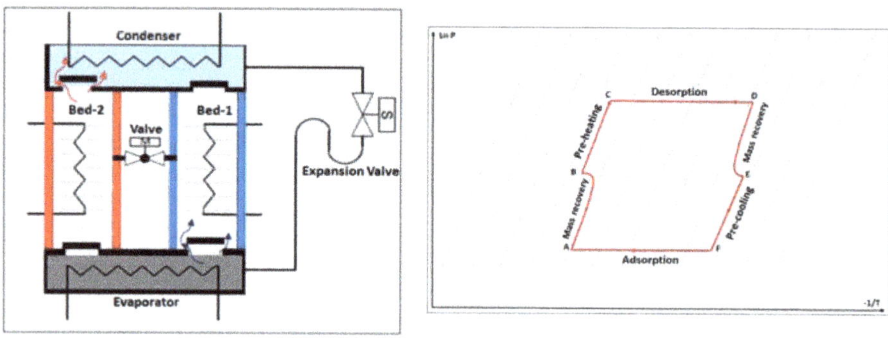

(a) (b)

Figure 3. (**a**) Schematic diagram of the adsorption–cooling system. (**b**) Clapeyron diagram of the adsorption–cooling cycle with reheat mode.

2.1. Energy Performance

The cooling capacity Q_{ch}, heat input capacity Q_{hot}, and the COP were calculated using Equations (1)–(3), respectively. The COP represents the ratio of the cooling output to the required amount of heat input; the COP of the chiller is calculated based on the total cycle period (tcy), which includes the periods of precooling, adsorption, preheating, and desorption.

$$Q_{ch} = \frac{\int_0^{tcy} \dot{m}_{chw} \times Cp_{chw} \times (T_{chw\_in} - T_{chw\_out})dt}{tcy} \tag{1}$$

$$Q_{hot} = \frac{\int_0^{tcy} \dot{m}_{hw} \times Cp_{hw} \times (T_{hw\_in} - T_{hw\_out})dt}{tcy} \tag{2}$$

$$COP = \frac{Q_{ch}}{Q_{hot}} \tag{3}$$

where:

$\dot{m}_{chw}$—mass flow rate of chilled water (kg·s^{-1});
Cp_{chw}—specific heat capacity of chilled water (kJ·kg^{-1}·K^{-1});
T_{chw_in}—inlet temperature of chilled water (K);
T_{chw_out}—outlet temperature of chilled water (K);
$\dot{m}_{hw}$—mass flow rate of hot water (kg·s^{-1});
Cp_{hw}—specific heat capacity of hot water (kJ·kg^{-1}·K^{-1});
T_{hw_in}—inlet temperature of hot water (K);
T_{hw_out}—outlet temperature of hot water (K).

2.2. Exergy Analysis

The exergy balance equation in rate form can be written as Equation (4) [20]

$$\dot{E}x_Q - \dot{E}x_w = \sum_{ext} \dot{m}_{ext} \times ex_{ext} - \sum_{in} \dot{m}_{in} \times ex_{in} + \dot{E}x_D \tag{4}$$

where:
$\dot{E}x_Q$—heat exergy (KW);
$\dot{E}x_w$—work exergy (KW);
$\dot{m}_{ext}$—refrigerant outlet mass flow rate ($Kg{\cdot}s^{-1}$);
m_{in}—refrigerant inlet mass flow rate ($Kg{\cdot}s^{-1}$);
ex—mass flow exergy ($kJ{\cdot}kg^{-1}$);
$\dot{E}x_D$—exergy destruction (kJ).
$\dot{E}x_Q$ is the heat exergy and can be written as Equation (5)

$$\dot{E}x_Q = \sum\left(1 - \frac{T_0}{T_h}\right) \times \dot{Q} \quad (5)$$

where:
T_0—dead state temperature (K);
T_h—heat input temperature (K);
$\dot{Q}$—heat transfer rate (kW).
ex is the mass flow exergy given by [21]

$$ex = (h - h_0) - T_0 \times (s - s_0) \quad (6)$$

where:
h—specific enthalpy ($kJ{\cdot}Kg^{-1}$);
h_0—specific enthalpy at dead state temperature ($kJ{\cdot}Kg^{-1}$);
s—specific entropy ($kJ{\cdot}kg^{-1}{\cdot}K^{-1}$);
s_0—specific entropy at dead state temperature ($kJ{\cdot}kg^{-1}{\cdot}K^{-1}$).
In this study, the subscript '0' represents the ambient condition.

The exergy destruction ($\dot{E}x_D$) in the adsorption chiller components can be calculated as per Equations (7)–(12):

- Exergy destruction in the adsorber ($\dot{E}x_{D.ads}$):

$$\dot{E}x_{D.ads} = \sum\left(1 - \frac{T_0}{T_{ads}}\right) \times \dot{Q}_{ads} + \sum \dot{m}_{ref,ads} \times [(h_{v.eva} - h_{l.ads}) - T_0 \times (s_{v.eva} - s_{l.ads})] \quad (7)$$

where:
$\dot{E}x_{D.ads}$—exergy destruction in the adsorber (kJ);
T_0—dead state temperature (K);
T_{ads}—adsorbent bed temperature (K);
$\dot{Q}_{ads}$—adsorption heat capacity (kW);
$\dot{m}_{ref,ads}$—refrigerant mass flow rate in the adsorber bed ($Kg{\cdot}s^{-1}$);
$h_{v.eva}$—specific enthalpy of the refrigerant vapor in the evaporator ($kJ{\cdot}Kg^{-1}$);
$h_{l.ads}$—specific enthalpy of the refrigerant liquid in the adsorber bed ($kJ{\cdot}Kg^{-1}$);
$s_{v.eva}$—specific entropy of the refrigerant vapor in the evaporator ($kJ{\cdot}kg^{-1}{\cdot}K^{-1}$).
$s_{l.ads}$—specific entropy of the refrigerant liqued in the adsorbent bed ($kJ{\cdot}kg^{-1}{\cdot}K^{-1}$).

- Exergy destruction in the desorber ($\dot{E}x_{D.des}$):

$$\dot{E}x_{D.des} = \sum\left(1 - \frac{T_0}{T_{des}}\right) \times \dot{Q}_{des} + \sum \dot{m}_{ref,des} \times [(h_{l.des} - h_{v.des}) - T_0 \times (s_{l.des} - s_{v.des})] \quad (8)$$

where:
$\dot{E}x_{D.des}$—exergy destruction in the desorber (kJ).
T_0—dead state temperature (K);
T_{des}—desorber bed temperature (K);

$\dot{Q}_{des}$—desorption heat capacity (kW).
$\dot{m}_{ref,des}$—refrigerant mass flow rate in the desorber bed (Kg·s^{-1});
$h_{l.des}$—specific enthalpy of the refrigerant liquid in the desorber bed (kJ·Kg^{-1});
$h_{v.des}$—specific enthalpy of the refrigerant vapor in the desorber bed (kJ·Kg^{-1});
$s_{l.des}$—specific entropy of the refrigerant liquied in the desorber bed (kJ·kg^{-1}·K^{-1}).
$s_{v.des}$—specific entropy of the refrigerant vapor in the desorber bed (kJ·kg^{-1}·K^{-1}).

- Exergy destruction in the condenser ($\dot{E}x_{D.cond}$):

$$\dot{E}x_{D.cond} = \sum\left(1 - \frac{T_0}{T_{cond}}\right) \times \dot{Q}_{cond} + \sum \dot{m}_{ref,cond} \times [(h_{v.des} - h_{l.cond}) - T_0 \times (s_{v.des} - s_{l.cond})] \quad (9)$$

where:
$\dot{E}x_{D.cond}$—exergy destruction in the condenser (kJ);
T_0—dead state temperature (K);
T_{cond}—condenser temperature (K);
$\dot{Q}_{cond}$—condensation heat capacity (kW);
$\dot{m}_{ref,cond}$—refrigerant mass flow rate in the condenser (Kg·s^{-1});
$h_{v.des}$—specific enthalpy of the refrigerant vapor in the desorber bed (kJ·Kg^{-1});
$h_{l.cond}$—specific enthalpy of the refrigerant liquid in the condenser (kJ·Kg^{-1});
$s_{v.des}$—specific entropy of the refrigerant vapor in the desorber bed (kJ·kg^{-1}·K^{-1});
$s_{l.cond}$—specific entropy of the refrigerant liquid in the condenser (kJ·kg^{-1}·K^{-1}).

- Exergy destruction in the evaporator ($\dot{E}x_{D.eva}$):

$$\dot{E}x_{D.eva} = \sum\left(1 - \frac{T_0}{T_{eva}}\right) \times \dot{Q}_{eva} + \sum \dot{m}_{ref,eva} \times [(h_{l.eva} - h_{v.eva}) - T_0 \times (s_{l.eva} - s_{v.eva})] \quad (10)$$

where:
$\dot{E}x_{D.eva}$—exergy destruction in the evaporator (kJ);
T_0—dead state temperature (K);
T_{eva}—evaporator temperature (K);
$\dot{Q}_{eva}$—evaporation heat capacity (kW);
$\dot{m}_{ref,eva}$—refrigerant mass flow rate in the evaporator (Kg·s^{-1});
$h_{l.eva}$—specific enthalpy of the refrigerant liquid in the evaporator (kJ·Kg^{-1});
$h_{v.eva}$—specific enthalpy of the refrigerant vapor in the evaporator (kJ·Kg^{-1});
$s_{l.eva}$—specific entropy of the refrigerant liquid in the evaporator (kJ·kg^{-1}·K^{-1});
$s_{v.eva}$—specific entropy of the refrigerant vapor in the evaporator (kJ·kg^{-1}·K^{-1}).

- Exergy destruction in the expansion ($\dot{E}x_{D.expa}$):

$$\dot{E}x_{D.expa} = \sum \dot{m}_{ref,cond} \times [(h_{l.cond} - h_{l.eva}) - T_0 \times (s_{l.cond} - s_{l.eva})] \quad (11)$$

where:
$\dot{E}x_{D.expa}$—exergy destruction in the expansion (kJ);
T_0—dead state temperature (K);
$\dot{m}_{ref,cond}$—refrigerant mass flow rate in the condenser (Kg·s^{-1});
$h_{l.cond}$—specific enthalpy of the refrigerant liquid in the condenser (kJ·Kg^{-1});
$h_{l.eva}$—specific enthalpy of the refrigerant liquid in the evaporator (kJ·Kg^{-1});
$s_{l.cond}$—specific entropy of the refrigerant liquid in the condenser (kJ·kg^{-1}·K^{-1});
$s_{l.eva}$—specific entropy of the refrigerant liquid in the evaporator (kJ·kg^{-1}·K^{-1}).

- Total exergy destruction ($\dot{E}x_{D.tot}$):

$$\dot{E}x_{D.tot} = \dot{E}x_{D.ads} + \dot{E}x_{D.des} + \dot{E}x_{D.cond} + \dot{E}x_{D.eva} + \dot{E}x_{D.expa} \quad (12)$$

$\dot{m}$ is assumed to be zero during the preheating and precooling of the adsorbent bed, evaporator, condenser, and expansion valve and also during the reheat phase for the condenser, evaporator, and expansion value.

The cyclic exergy efficiency, which is the ratio of exergy output of the system to the exergy input, can be expressed as in Equation (13)

$$\eta_{ex} = \frac{Ex_{(Q_{eva})}}{Ex_{(Q_{des})}} \tag{13}$$

We have the following assumptions:

- The water vapor (refrigerant) behaves as an ideal gas.
- The pressure and temperature inside the adsorbent bed are uniforms.
- Potential, kinetic, and chemical effects are neglected.
- The expansion process is isenthalpic.
- The pressure drop in the non-return values is neglected.
- The chiller is well insulated, and there are no heat losses to the surroundings.

3. Experimental Setup

The proposed prototype adsorption chiller tested in this study consisted of two beds; each one has 40 kg of silicoaluminophosphates zeolite (SAPO-34) coated on copper heat exchangers. The experiment was set under the following conditions:

- Hot water inlet temperature (Th_in) of 90 ± 0.5 °C maintained by an electric water heater.
- Recooling-water inlet temperature (Tre_in) of 30–45 °C maintained by a dry-cooler with a 100 L buffer tank.
- Chilled water inlet temperature (Tch_in) of 18 ± 0.5 °C maintained by an electrical heater.
- Flow rates for hot, recooling, and chilled water are 1.2, 1.2, and 0.71 L/s, respectively.
- Cooling cycle time duration of 550 s.
- Four modes were investigated: single-stage mode, 25% reheat cycle mode (25% mass recovery time of the cooling duration), 50% reheat cycle mode (50% mass recovery time of the cooling duration), and 75% reheat cycle mode (75% mass recovery time of the cooling duration)

The following instrumentation setup was used to measure the experimental performance of the chiller during the reheat cycle, as shown in Figure 4. The instrument locations are described as follows:

- Three electromagnetic flowmeters (FM_h, FM_ch, and FM_re) (manufactured by ALIA GROUP) with accuracies of ± 0.4% were used to measure the water flow rate at the hot water loop, chilled water loop, and recooling-water loop, respectively.
- Eight platinum resistance thermometers (PT100 Class A, Pico Technology, St Neots, UK) with two temperature measuring data loggers (PT-104 is a four-channel logger) having a resolution of 0.001 °C and an accuracy of 0.015 °C were used to measure the inlet and outlet temperatures for the hot water, chilled water, and recooling water, in addition to the temperature of the storage tanks.

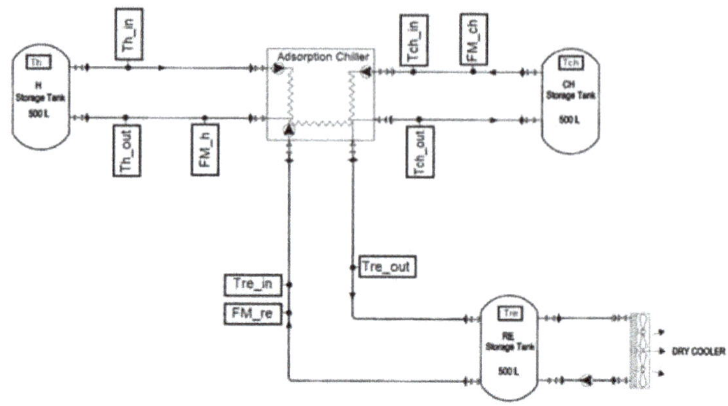

Figure 4. Schematic of prototype adsorption chiller testing equipment.

4. Results and Discussion

The cooling capacity and COP are calculated based on the experimental results as per Equations (1)–(3). The experimental temperature profile for the adsorption chiller components was exported to an excel sheet where the exergy was analyzed using Equations (4)–(13). The refrigerant properties were imported from CoolPack software (https://www.ipu.dk/products/coolpack/, accessed on 20 January 2021).

Figure 5 shows the cyclic cooling capacity at different recooling-water temperatures and modes with a fixed cyclic cooling duration of 530 s based on the experimental results for single stage and the variable reheat cycle time. The single-stage mode exhibited a better capacity at recooling temperatures below 32 °C compared to other modes. The 25% mode exhibited a higher cooling capacity within a recooling-water temperature range of 32 to 43 °C, while the 50% mode exhibited a higher capacity at temperatures higher than 43 °C.

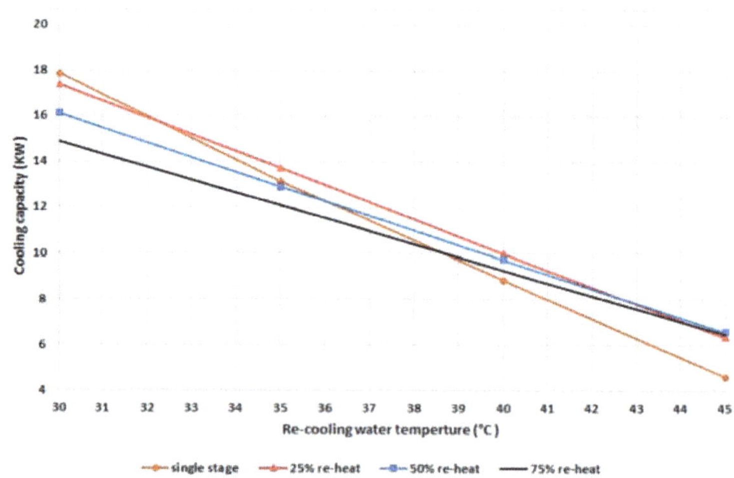

Figure 5. Cooling capacity at various recooling-water temperatures and modes.

With the increase in the recooling-water temperature, the adsorber's uptake is decreased, which directly influences the cooling capacity. To achieve a higher capacity with the increased recooling-water temperature, the system requires a longer mass recovery time; therefore, the uptake of the adsorber is increased and that of the adsorbate in the desorber is decreased, which can be realized by increasing the pressure in the condenser as well as the recooling-water temperature and decreasing the adsorption pressure. As evident from Figure 4, there is an optimum mode for each recooling-water temperature related to the optimum water-vapor uptake and the condensation pressure. To maintain the maximum cooling capacity at different recooling-water temperatures, the chiller should be operated in the variable mode instead of conventional single-mode systems.

The experimental results of COP are shown in Figure 6 for a fixed cycle time of 530 s. The results show that the single-stage mode exhibited the highest COP at various recooling-water temperatures up to 42 °C, followed by the 25% mode. In general, the COP decreases when the recooling-water temperature increases due to the decrease in water-vapor uptake in the adsorbent bed, which leads to a decrease in the cooling capacity that is more than the heat input.

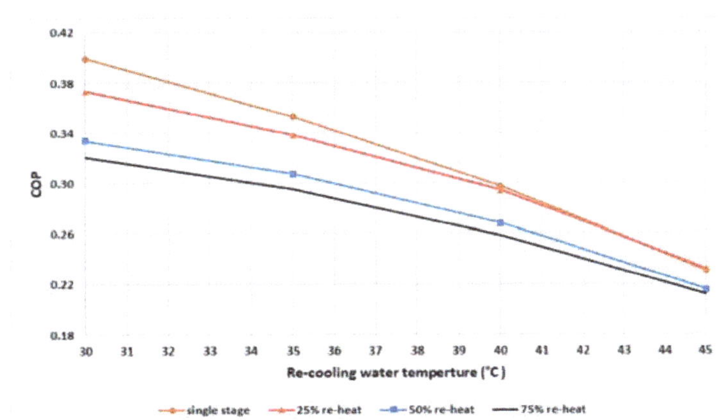

Figure 6. COP at various recooling-water temperatures and modes.

The COP of the single-stage mode is higher than that of other modes up to 42 °C, becomes equal to the COP of the 25% mode from 42 to 44 °C, and becomes slightly lower from 44 °C.

Figure 7 shows the breakdown of the exergy destruction in the proposed chiller components at 30 °C recooling-water temperature and single-stage cycle as per Equations (6)–(10). It can be noted that the highest exergy destruction is accrued in the desorber, followed by the adsorber, evaporator, condenser, and expansion valve, with proportions of 49%, 27%, 13%, 9%, and 2%, respectively. The desorber has the highest exergy destruction because of the high difference between the desorber and dead state temperatures, which are proportionally related. Another reason for higher consumption in the adsorbent bed (desorber and adsorber) is the exergy destruction in the preheating and precooling processes; as shown in Figure 8, the preheating share is around 35% of the exergy destruction in the desorber, whereas the precooling share is about 58% of that in the adsorber. The results in Figure 6; Figure 7 indicate the importance of focusing on the adsorbent bed design by minimizing the heat exchange material and improving the heat efficiency to reduce the exergy destruction in the preheating and precooling processes.

Figure 9 shows the total exergy destruction and exergy efficiency at various recooling-water temperatures (assuming the dead state temperature increases linearly with the recooling-water temperature). The total exergy destruction decreases as the input exergy of the refrigerant decreases when the dead state temperatures increase because of the decreased precooling and preheating exergy destruction, where the difference in temperatures between the desorber and adsorber is decreased. The exergy efficiency increases with increasing recooling-water temperatures because the increase in recooling-water temperature is associated with increasing dead state temperatures, which decreases the irreversibility as the desorber bed works at a temperature close to the dead state temperature.

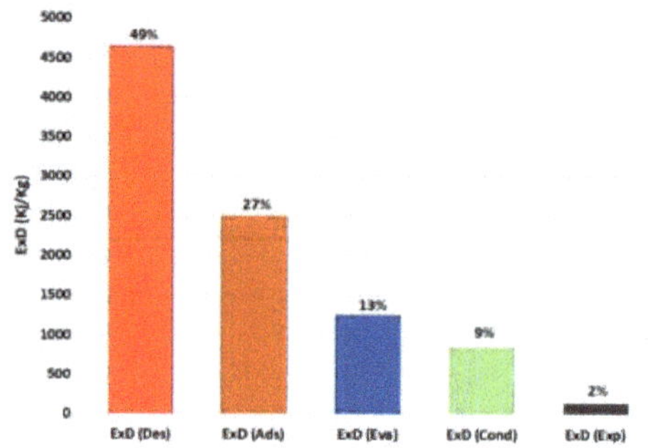

Figure 7. Exergy destruction in the chiller component at 30 °C recooling-water temperature. (Ex: exergy destruction, Des: desorber, Ads: adsorber, Eva: evaporator, Cond: condenser, Exp: expansion valve).

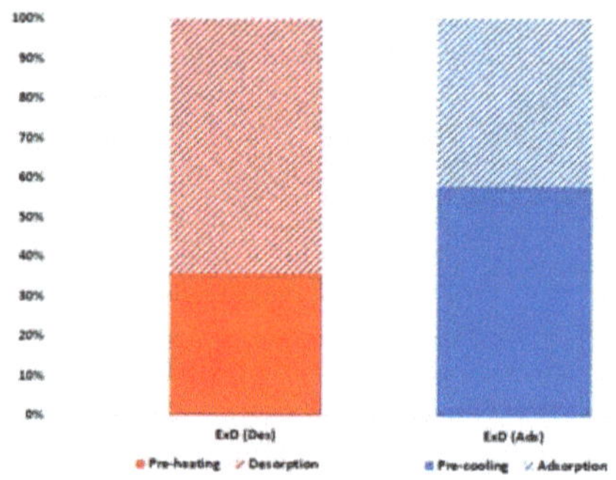

Figure 8. Exergy destruction in the adsorbent beds at 30 °C recooling-water temperature.

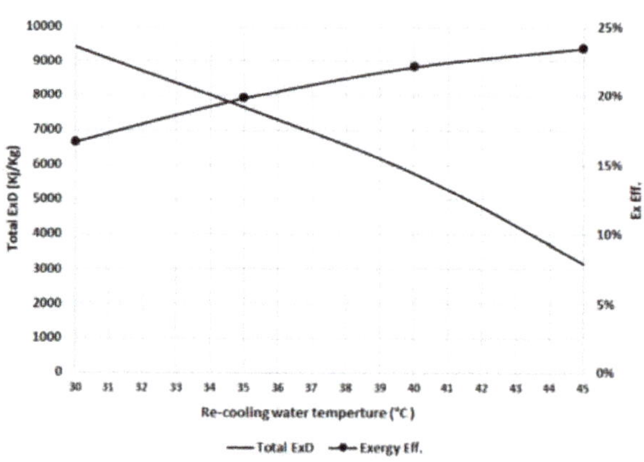

Figure 9. Total exergy destruction and exergetic efficiency in the chiller at various recooling-water temperatures.

Figure 10 shows the exergy destruction in the chiller component at various recooling-water temperatures. It can also be noted that the desorber is most sensitive to the recooling-water temperature, followed by the adsorber, condenser, evaporator, and expansion because the preheating exergy destruction decreases proportionally with the recooling-water temperature. The adsorber and condenser exhibit the same trend as they both have a constant temperature difference from the dead state temperature. The evaporator has less sensitivity to recooling-water temperature as the difference with the dead state temperature increases. The results from curves 8 and 9 clearly indicate that inefficient energy use occurs at lower temperatures and that there is a high potential for improvement at low recooling water temperatures, mainly in the adsorbent bed, via the use of a heat exchanger and time cycle optimization.

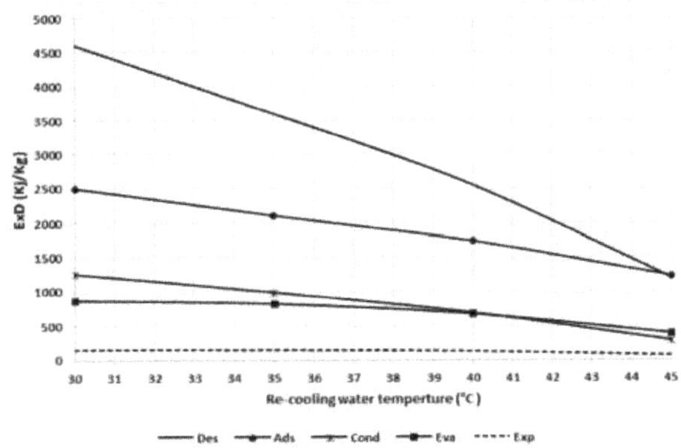

Figure 10. Exergy destruction in the chiller components at various recooling-water temperatures.

To separately investigate the effect of the recooling-water temperature on the exergy destruction, the dead state temperature is assumed constant at 25 °C regardless of the recooling-water temperature, and the result is compared to the variable dead state temperature (dead state temperature is assumed to be 5 °C less than the recooling-water temperature). Figure 11a,b show the total exergy destruction and exergetic efficiency, respectively. It can be noted that the exergy destruction decreases with increasing recooling-water temperature for fixed and variable dead state temperature. As shown in Figure 11a, it is clear that the total exergy destruction decreases with an increasing dead state temperature. The exergetic efficiency curves of Figure 11b show that the fixed dead state temperature exhibits the opposite trend compared to the variable dead state temperature.

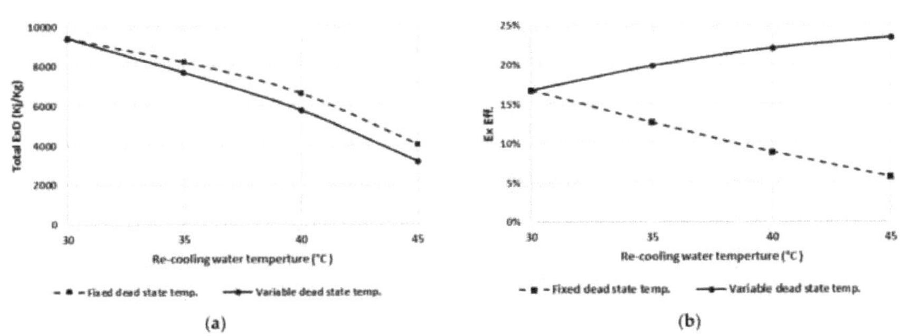

Figure 11. (**a**) Total exergy destruction, and (**b**) efficiency at fixed and variable dead state temperatures.

The exergetic efficiency decreases when increasing the recooling-water temperature at a fixed dead state temperature of 25 °C, mainly because the input exergy destruction increases as the difference between dead state and recooling temperatures increases; however, the exergetic efficiency increases as the dead state temperature increases even at higher recooling temperatures.

Figure 12 shows that the desorber, adsorber, and condenser have more exergy destruction at fixed dead state temperatures than those at the variable dead state temperature; however, the evaporator has lower exergy destruction at the fixed dead state temperature than the variable dead state temperature, as the difference is constant between the evaporator and dead state temperatures at fixed temperatures. The results from Figures 10 and 11 indicate that it is better to work at a recooling water temperature near to the ambient or dead state temperature if achievable. Table 1 summarizes the recooling water increase effect on the exergy destruction and efficiency for Figures 8–12.

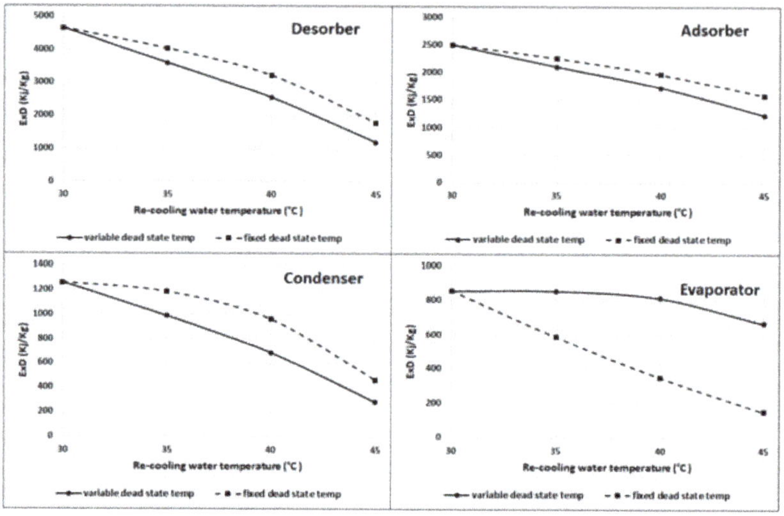

Figure 12. Exergy destruction in the desorber, adsorber, condenser, and evaporator at fixed and variable dead state temperatures.

Table 1. Effect of recooling water increase on the exergy destruction and efficiency.

<table>
<tr><th rowspan="2">Components</th><th colspan="2">Variable Dead State Temperature (5 °C Less Than the Recooling Water Temperature)</th><th colspan="2">Fixed Dead State Temperature (25 °C)</th></tr>
<tr><th>Exergy Destruction</th><th>Total Exergy Efficiency</th><th>Exergy Destruction</th><th>Total Exergy Efficiency</th></tr>
<tr><td>Adsorber</td><td>Decreasing</td><td rowspan="4">Increasing</td><td>Decreasing</td><td rowspan="4">Decreasing</td></tr>
<tr><td>Desorber</td><td>Decreasing</td><td>Decreasing</td></tr>
<tr><td>Condenser</td><td>Decreasing</td><td>Decreasing</td></tr>
<tr><td>Evaporator</td><td>Decreasing</td><td>Decreasing</td></tr>
</table>

Next, Figure 13 shows the effect of the reheating cycle time for the exergy destruction of the chiller component. The exergy destruction increases with an increase in the reheat cycle time as more exergy will be added during the mass recovery. Adsorbent beds (adsorber and desorber) are more sensitive to this effect as they are directly affected by more exergy input in the bed during the mass recovery process. The evaporator and condenser are less sensitive than adsorbent beds because the effect is due to the cooling capacity and exergy output.

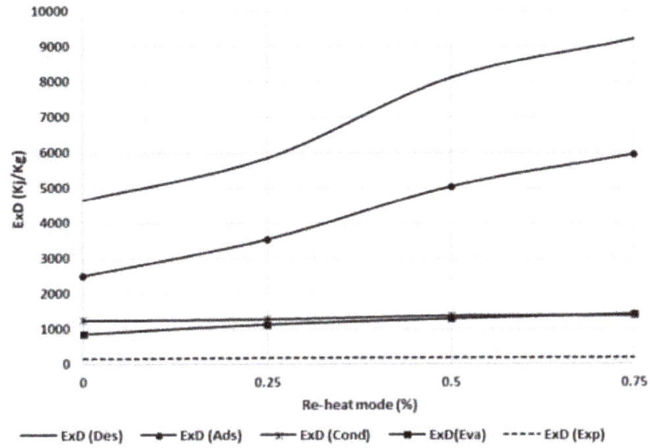

Figure 13. Exergy destruction in the chiller components at different reheating modes.

Figure 14 demonstrates that the exergy efficiency increases with recooling-water temperature, as explained previously when discussing Figure 11. The exergy efficiency increases with an increase in the recooling-water temperature, mainly due to increased dead state temperature. The exergy efficiency increases sharply with recooling-water temperature up to 40 °C. Higher recooling-water temperatures have a lower effect as the refrigerant uptake decreases sharply for this type of zeolite. As such, this will increase the exergy destruction ratio. The results from Figures 13 and 14 indicate that while increasing the mass recovery time improves the cooling capacity, this, on the other hand, also increases the exergy destruction and reduces the exergy efficiency. As such, an optimization analysis of the mass recovery time and cooling capacity should be conducted in order to select the optimum mode.

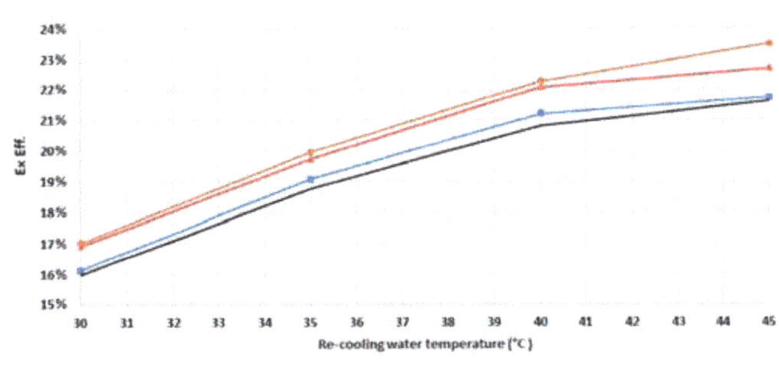

Figure 14. Exergetic efficiency for the chiller components at different modes.

5. Conclusions

The energy and exergy analyses of the adsorption cooling cycle at various recooling-water temperatures were investigated based on the experimental results with different cycle modes. The cooling capacity and COP were found to decrease with increasing recooling-water temperatures. An optimum mode for each recooling-water temperature range was found depending on the cooling capacity and COP. In addition, exergy destruction for each component of the chiller was analyzed based on the experimental data. The highest exergy destruction occurred in the desorber, followed by the adsorber, evaporator, condenser, and expansion valve with 49%, 27%, 13%, 9%, and 2%, respectively, at nominal working conditions.

The exergy destruction for the preheating process was 35% of the total exergy destruction in the desorber; however, the exergy destruction for the precooling process was 58% of the total destruction in the adsorber.

The exergy destruction and efficiency were both found to decrease with increasing recooling-water temperatures at fixed dead state temperatures. The exergy destruction was found to decrease with an increasing dead state and recooling-water temperatures; however, the exergy efficiency increases with an increasing dead state temperature and recooling-water temperature. The results show that the exergy destruction increases with increasing the mass recovery time in the reheat cycle. The exergy destruction in the adsorbent beds was the most sensitive for the increase in mass recovery time.

Author Contributions: Conceptualization, A.A.A.; methodology, A.A.A. and A.A.-M.; software, A.A.-M.; validation, A.A.A. and A.A.-M.; formal analysis, A.A.A.; investigation, A.A.A.; resources, A.A.-M.; data curation, A.A.A.; writing—original draft preparation, A.A.A.; writing—review and editing, A.A.-M.; visualization, A.A.A.; supervision, A.A.-M.; M.A. and H.-J.B. All authors have read and agreed to the published version of the manuscript.

Funding: This research received no external funding.

Institutional Review Board Statement: Not applicable.

Informed Consent Statement: Not applicable.

Data Availability Statement: Not applicable.

Acknowledgments: Not applicable.

Conflicts of Interest: The authors declare no conflict of interest.

References

1. IEA. *Cooling*; IEA: Paris, France, 2020.
2. Ayou, D.S.; Coronas, A. New Developments and Progress in Absorption Chillers for Solar Cooling Applications. *Appl. Sci.* **2020**, *10*, 4073. [CrossRef]
3. Narayanan, R. Chapter Seven—Heat-Driven Cooling Technologies. In *Clean Energy for Sustainable Development*; Rasul, M.G., Azad, A.K., Sharma, S.C., Eds.; Academic Press: Cambridge, MA, USA, 2017; pp. 191–212.
4. Rucha, P.; Salibaa, S.; Onga, C.L.; Al-Shehrib, Y.; Al-Rihailib, A.; Al-Mogbelb, A.; Michela, B. Heat-Driven Adsorption Chiller Systems for Sustainable Cooling Applications. In Proceedings of the 11th IEA Heat Pump Conference 2014 Conference Proceedings, Montreal, QC, Canada, 12–16 May 2014.
5. Thu, K. Adsorption Desalination: Theory & Experiments. Ph.D. Thesis, National University of Singapore, Singapore, 2010.
6. Ambarita, H.; Kawai, H. Experimental study on solar-powered adsorption refrigeration cycle with activated alumina and activated carbon as adsorbent. *Case Stud. Therm. Eng.* **2016**, *7*, 36–46. [CrossRef]
7. Núñez, T.; Mittelbach, W.; Henning, H.-M. Development of an adsorption chiller and heat pump for domestic heating and air-conditioning applications. *Appl. Therm. Eng.* **2007**, *27*, 2205–2212. [CrossRef]
8. Wang, R. Performance improvement of adsorption cooling by heat and mass recovery operation. *Int. J. Refrig.* **2001**, *24*, 602–611. [CrossRef]
9. Restuccia, G.; Freni, A.; Vasta, S.; Aristov, Y. Selective water sorbent for solid sorption chiller: Experimental results and modelling. *Int. J. Refrig.* **2004**, *27*, 284–293. [CrossRef]
10. Alahmer, A.; Ajib, S.; Wang, X. Comprehensive strategies for performance improvement of adsorption air conditioning systems: A review. *Renew. Sustain. Energy Rev.* **2019**, *99*, 138–158. [CrossRef]
11. Ghilen, N.; Gabsi, S.; Benelmir, R.; El Ganaoui, M. Performance Simulation of Two-Bed Adsorption Refrigeration Chiller with Mass Recovery. *J. Fundam. Renew. Energy Appl.* **2017**, *7*. [CrossRef]
12. Bhatia, S.C. (Ed.) 14—Geothermal power generation. In *Advanced Renewable Energy Systems*; Wood-Head Publishing India: Delhi, India, 2014; pp. 334–388.
13. Evola, G.; Costanzo, V.; Marletta, L. Exergy Analysis of Energy Systems in Buildings. *Buildings* **2018**, *8*, 180. [CrossRef]
14. Szargut, J.; Morris, D.; Steward, F. *Exergy Analysis of Thermal, Chemical, and Metallurgical Processes*; Hemisphere Publishing Corporation: New York, NY, USA, 1988.
15. Vijayaraghavan, S.; Goswami, D.Y. On Evaluating Efficiency of a Combined Power and Cooling Cycle. *J. Energy Resour. Technol.* **2003**, *125*, 221–227. [CrossRef]
16. Baiju, V.; Muraleedharan, C. Exergy Assessment of Single Stage Solar Adsorption Refrigeration System Using ANN. *ISRN Mech. Eng.* **2012**, *2012*, 1–10. [CrossRef]

17. Baiju, V.; Muraleedharan, C. Energy and exergy analysis of solar hybrid adsorption refrigeration system. *Int. J. Sustain. Eng.* **2013**, *6*, 289–300. [CrossRef]
18. Ogueke, N.; Ndeke, C. Exergy based performance analysis of a solid adsorption solar refrigerator. *Int. J. Renew. Energy Res.* **2014**, *4*, 363–370.
19. Cao, N.V.; Duong, X.Q.; Lee, W.S.; Park, M.Y.; Lee, S.S.; Chung, J.D. Exergy Analysis of Advanced Adsorption Cooling Cycles. *Entropy* **2020**, *22*, 1082. [CrossRef] [PubMed]
20. Rezaie, B.; Javan, S.; Mohammadi, V.; Ahmadi, P. Performance Assessment and Optimization of a Combined Cooling, Heating and Power (CCHP) System for Residential Application using Low Grade Heat of an Internal Combustion Engine. *Curr. Altern. Energy* **2017**, *2*, 1. [CrossRef]
21. Cihan, A.; Hacıhafızog˘lu, O.; Kahveci, K. Energy-exergy analysis and modernization suggestions for a combined-cycle power plant. *Int. J. Energy Res.* **2006**, *30*, 115–126. [CrossRef]

energies

MDPI

Article

Optimisation of Operation of Adsorption Chiller with Desalination Function

Karol Sztekler

Faculty of Energy and Fuels, AGH University of Science and Technology, Mickiewicza 30, 30-059 Krakow, Poland; sztekler@agh.edu.pl

Abstract: The demand for electricity is growing rapidly along with economic development and increasing population. At present, its production is mainly based on non-renewable sources, which has negative impacts on the environment and contributes to global warming. A large proportion of the produced electricity is consumed by refrigeration equipment. Climate change and the progress of civilisation are additionally increasing the demand for cooling, with increasing electricity consumption as a consequence. One of the options for obtaining eco-friendly cooling is the use of adsorption chillers. These devices are powered by low-temperature heat and their operation only requires a small amount of electrical energy. The source of low-temperature heat can be, e.g., waste heat generated in many industrial processes. Its use allows one to increase energy efficiency and achieve additional financial benefits. However, adsorption chillers are characterised by low coefficients of performance. This paper presents possibilities to improve their performance. It also presents the results of tests carried out on a three-bed adsorption chiller with desalination function. The aim of the investigation was to determine the effect of the cycle time on the coefficient of performance (COP) and specific cooling power (SCP). The working pair was silica gel and water. The results confirmed the effect of the duration of adsorption and desorption on the COP and SCP of the adsorption chiller. Increasing the duration of the cycle led to an increase in the COP.

Keywords: adsorption; chiller; desalination

Citation: Sztekler, K. Optimisation of Operation of Adsorption Chiller with Desalination Function. *Energies* **2021**, *14*, 2668. https://doi.org/10.3390/en14092668

Academic Editor: David Borge-Diez

Received: 29 March 2021
Accepted: 4 May 2021
Published: 6 May 2021

1. Introduction

Dynamic economic development, population growth and observed climatic changes imply that the demand for cooling will be growing every year. Cooling processes are used in many industries, such as in the food industry for food production and storage, the pharmaceutical industry, heavy industry and power generation [1]. Refrigeration units using thermal compression and using adsorption and desorption processes are modern alternatives to the most commonly used conventional electrically driven compressor refrigeration systems. The need to seek innovative solutions for the refrigeration industry has been dictated, among other things, by the necessity to save electricity and reduce the environmental impact.

The market for refrigeration equipment is growing steadily. Data from the International Institute of Refrigeration (IIR) show that there are currently over 5 billion refrigeration units and heat pumps in operation globally [2]. In 2016, more than 1.6 billion air conditioning units were in use, of which more than half were located in the United States and China alone [3]. Middle Eastern countries are a particularly important market. The demand for cooling is extremely high there due to the very warm climate. It is estimated that almost 100% of households in the Middle East are equipped with air conditioning [3]. The data quoted above clearly show the scale of the use of cooling devices and, consequently, the energy consumption related to their operation. In this context, it is particularly important that the vast majority of the units in use are compressor systems driven by electricity [1]. According to the report of the International Institute of Refrigeration in Paris, the refrigeration sector consumes approximately 20% of the world electricity production [2]. On the other hand, estimates of the International Energy Agency (IEA) indicate that the share of air conditioning equipment in the global electricity demand was approximately 10% in 2016 [3]. During peak cooling demand in summer, this value can increase significantly [3]. Taking into account that global electricity production is mainly based on non-renewable sources, it can be concluded that the refrigeration sector indirectly contributes to accelerating the rate of adverse climate change [4].

Adsorption chillers can be powered by the low-temperature waste heat with temperatures in the range of 60 °C to 90 °C or by the heat from solar energy, and therefore they are characterised by very low electricity consumption [1,5,6]. Furthermore, adsorption chillers can operate at low temperatures, which additionally contributes to the reduction in thermal energy losses. Many processes in various industries emit energy which is lost to the atmosphere unless it is recovered. In thermal power plants, approximately 50–70% of the energy value of the fuel is rejected as waste heat [6]. In such cases, one of the possibilities of using waste heat is the production of cold energy in adsorption chillers [1]. The use of this type of equipment offers the possibility of creating polygeneration systems capable of producing electricity, heat and cold simultaneously [7]. In addition to reducing the negative impact of greenhouse gas emissions, energy recovery can also bring significant financial benefits [1]. A very important feature of adsorption chillers in the context of environmental care is the use of non-ozone-depleting refrigerants such as water [5]. It is a cheap, widely available, safe and the most thermally stable adsorbate which extends the applicability of adsorption chillers to water desalination [8,9].

Moreover, in addition to several previously mentioned advantages, adsorption chillers are characterised by easy and quiet operation and a small number of moving mechanical parts, which makes them especially valuable in applications where reliability of the devices is very important [8,10]. Replacing an electric compressor with a thermal compressor has the advantage of no vibration, thus opening up many new application possibilities [11]. Despite there being so many advantages of using adsorption chillers, there are also some limitations. These include the low coefficient of performance (COP) compared to conventional refrigeration systems, which is usually between 0.5 and 0.6, the low specific cooling power (SCP), and the small mass and heat flow inside the bed [8,12]. All of these are related to the large mass and dimensions of adsorption chillers [6,13]. Continuous research is being conducted to improve the efficiency of adsorption chillers so as to make them more attractive and competitive in the market. In a later part of the paper, a number of methods to improve the performance parameters of adsorption chillers are presented.

One of the disadvantages of the materials used is that the highly porous adsorbents have relatively lower thermal conductivity [8]. At the same time, this is a field of searching for new solutions that take into account the balance between good thermal conductivity and sufficiently high sorption capacity [8,14]. Sorption materials that can work with water are, e.g., silica gel, zeolite and activated alumina [15]. Some very important characteristics of the sorbents are the ability to adsorb a maximum amount of adsorbate per unit mass and a low specific heat [8,15]. Particularly important from the point of view of the heat source and chiller operation is the bed regeneration temperature. This is also one of the characteristics of the sorption materials and can be different for different materials [12,15]. Furthermore, adsorbents must be non-toxic, durable, readily available and cheap [8,14,15]. The high porosity of such materials is an important characteristic from the point of view of sorption properties, but it negatively affects heat conduction [15].

An adsorption chiller with desalination function can be characterised by several key parameters. These include the coefficient of performance (COP), the specific cooling power (SCP) and the specific daily water production (SDWP) [16,17]. These indicators depend on many factors. Some of them, such as bed regeneration temperature, chilled water temperature or cycle time, can be easily modified in order to obtain the optimal values. These will vary depending on the design of the chiller, the resources available and the desired effect. This paper analyses the impacts of these factors on the COP. In addition to the above-mentioned possibilities of enhancing performance, there are also some other factors which are strictly related to the design of the chiller and the properties of the adsorbent bed. These include improving the heat exchange efficiency inside the bed and the utilisation of heat and mass recovery in the system [6,7,14]. The former method can be divided according to the three keyways of its implementation:

- Improving heat transfer between the heat exchanger and the sorbent [12,18];
- Improving heat transfer in the sorbent [7];
- Using modern porous materials as sorbents [12].

A typical material that heat exchangers are made of is copper, characterised by a high thermal conductivity coefficient of 380 W/(mK) [12]. Additionally, the most commonly used adsorbent is silica gel, which has a thermal conductivity coefficient of merely 0.17 W/(mK) [12]. The significant difference in the value between these two materials means that heat transport in the boundary zone is hindered. One possible way to improve heat transfer in this area is, e.g., by coating the silica gel layer in contact with the heat exchanger surface with a binder [12]. In addition, increasing the heat transfer area as a result of using different exchanger designs also results in greater heat transfer between the exchanger and the adsorbent. Various design options for heat exchangers in an adsorbent bed have been presented in the paper [18].

An effective way to enhance heat transfer inside the bed is by doping the sorbents with materials with a higher coefficient of thermal conductivity [7,12]. Increasing the thermal conductivity of the bed can shorten the cycle time [19]. These can be, e.g., metal powders with a smaller grain size than the sorbent grains, which additionally allows one to reduce the amount of voids filled with still air in the bed [12]. The results of research on increasing the coefficient of thermal conductivity of the bed by doping silica gel with aluminium, copper and carbon nanotube powders are presented, among others, in the article [7]. The aim of that study was to check the influence of additives characterised by a high coefficient of thermal conductivity on the sorption kinetics of the mixture obtained [7]. Improved heat exchange in the bed can also be achieved by using modern sorption materials which combine good thermal conductivity with high sorption capacity. An example of this type of materials can be hybrid MOFs (metal–organic frameworks) [12].

Improved chiller performance can be achieved by using a number of methods related to heat recovery [6,20]. The idea of heat regeneration is to use the heat from the warmer bed being cooled in order to preheat the next bed. Likewise, the fluid circulating in the cooling circuit can be used to preheat another bed after the cooling of one bed [20]. Many advanced heat recovery cycles have been proposed in various publications. These include, e.g., the thermal wave cycle, the forced convection thermal wave cycle and the cascade cycle [6].

The right cycle time is one of the most important parameters of an adsorption chiller. Its length determines the amount of adsorbate flowing from the evaporator to the bed or from the bed to the condenser. It is absolutely vital to prevent a situation where this time will be too short because too little adsorbate will then evaporate into the bed or into the condenser. Too low a time of adsorption and desorption results in the inadequate temperature of chilled water and reduced production of desalinated water. On the other hand, too long desorption time may lead to unnecessary energy consumption for heating the bed and its components without resulting in more desorbed adsorbate. As a result, the number of cycles that can be performed during one day is also reduced [21]. COP increases with longer cycle time, because the increasing time is associated with the reduction in heat consumption, which powers the adsorption chiller [22,23]. Moreover, longer cycle times contribute to achieving a lower cooling capacity (CC) [24]. The effect of the cycle time on the value of the cooling capacity was investigated by Saha et al. [24]. In this study, the highest value of CC was obtained for the duration of adsorption/desorption equal to 180–300 s [24]. One of the ways of influencing the efficiency of the adsorption chiller by manipulating the cycle time is to shorten the bed regeneration time in relation to the time allocated for adsorption, due to the fact that it can run 2–3 times faster than adsorption (for monolayers) [25]. The influence of the ratio of desorption time to adsorption time on the efficiency of the refrigerator was investigated by El-Sharkawy et al. [26]. It was indicated that the optimal value of the coefficient *f* will be variable for various operating conditions and the design of adsorption chillers. However, it has been noticed that the efficiency will increase if this coefficient is lower than one [26].

Adsorption chillers are the solution that can not only become a source of eco-friendly cooling but also contribute to reducing the problem of freshwater scarcity. The shortage of potable water is becoming an increasingly common problem globally. It is estimated that freshwater available to humans only accounts for approximately 0.75% of the Earth's total water supply [27]. The UN predicts that nearly 5.7 billion people could be affected by water scarcity by 2050 [28]. Continued population growth, the expanding economy and ongoing climate change are all contributing to an increase in water demand [28]. Water treatment is therefore required to meet this demand. The oldest desalination technologies have been used for decades [29]. Nowadays, however, conventional desalination processes are considered the most energy-intensive water treatment options [30]. For this reason, it is necessary to develop modern technologies that do not require such high energy inputs. One such possibility is the use of adsorption chillers for water desalination. Table 1 shows the specific energy consumption (SEC) for different desalination methods. The adsorption desalination technology is characterised by the lowest electricity consumption compared to conventional methods such as multi-stage flash (MSF), multi-effect distillation (MED) and seawater reverse osmosis (SWRO) [16,31].

Table 1. Comparison of energy consumption for selected desalination technologies. AD: adsorption; MSF: multi-stage flash; MED: multi-effect distillation; SWRO: seawater reverse osmosis.

Desalination Method	Electrical Energy Consumption (kWh/m^3)	Thermal Energy Consumption (kWh/m^3)	Ref.
AD	<1.5	39.8 (waste or solar heat)	[16,31]
MSF	3.5–5	69.44–83.33	[16]
MED	1.5–5	41.67–61.11	[16]
SWRO	4–6	-	[31]

Using saline water as adsorbate, after condensation in a condenser, purified and desalinated water is obtained, which is almost free of dissolved solids [32]. One example of an adsorption desalination plant in operation is the world's largest adsorption chiller with desalination function located in Solar Village near the capital of Saudi Arabia, Riyadh [33]. It has a desalinated water production capacity of 100 m^3/day. The outstanding achievement in comparison to other desalination technologies in this case is the very low unit price of water, which is below USD 0.40/m^3 of water. This is due to the very low power consumption of the plant, which amounts to about 1.2 kWh/m^3 [34]. The cooling capacity of the unit is 1070 kW [33].

This paper presents the results of an experimental tests carried out for a laboratory three-bed adsorption chiller with desalination function in which the first and the second beds were synchronised carrying out the same adsorption or desorption processes at the same time while the third bed worked in the opposite phase. The study was designed to test the effect of cycle time on the value of the coefficient of performance (COP) and specific cooling power (SCP).

2. Materials and Methods

2.1. Adsorption Chiller with Desalination Function

The investigation was carried out on a three-bed adsorption chiller with desalination function located at the Energy Centre of the AGH University of Science and Technology in Krakow, Poland. The system investigated can operate in either a cooling or a desalination mode. The adsorption chiller can operate with, both, two-bed or three-bed mode. Such a design of the device allows one to carry out tests in various configurations of beds with different beds synchronised with each other in each variant. The use of multi-bed systems increases the stability of the chiller operation and enables the adsorption and desorption to be carried out in a cyclic manner. It has been proven that the use of a 4-bed chiller increases their recovery efficiency by 70% as compared with a 2-bed chiller operating under the same conditions [20]. At the same time, a six-bed design has a 40% higher recovery efficiency as compared with the four-bed design [20].

Table 2 summarises the basic parameters of the adsorption chiller investigated. The chiller can produce chilled water with a temperature of up to 7 °C and up to 1.5 kW of cooling. The pressure in the evaporator, condenser and beds reaches values in the range of 0.5–3 kPa, 3–7 kPa and 0.5–7 kPa, respectively.

Table 2. Basic nominal parameters of the investigated three-bed adsorption chiller with desalination function.

Chiller Components	Parameter	Value	Unit
Evaporator	Cooling capacity	1.10	kW
	Chilled water inlet temperature	12	°C
	Chilled water outlet temperature	7	°C
	Chilled water mass flow	0.052	kg/s
Condenser	Capacity	2.00	kW
	Cooling water inlet temperature	20	°C
	Cooling water outlet temperature	22	°C
	Cooling water mass flow	0.250	kg/s
	Daily distillate production	40	kg

The basic adsorbent used in the unit concerned is silica gel, which in this case is provided by KD Corporation from the Republic of Korea while the adsorbate is water regardless of the mode of operation of the chiller. Table 3 shows the basic properties of silica gel, while Figure 1 shows the water vapor adsorption–desorption isotherms of the tested silica gel, determined from the DVS Vacuum apparatus data.

Table 3. Properties of silica gel.

Sorbent	Granulation (μm)	Bulk Density (g/mL)	Specific Surface Area (m^2/g)	Pore Volume (cm^3/g)	Pore Diameter (Avg. Value) (nm)	Thermal Conductivity (W/(mK))
KD Corporation—silica gel	700–800	0.73	750–800	0.44	1.0–3.0	0.177

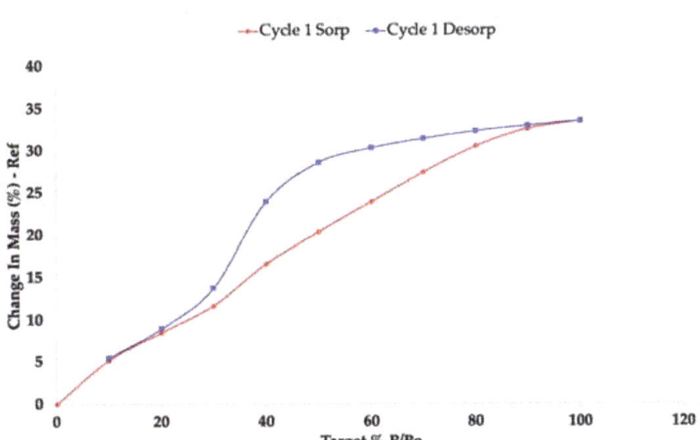

Figure 1. Water vapor adsorption–desorption isotherms on the SG sorbent for the temperature of 35 °C.

Figure 2 shows a simplified diagram of the investigated adsorption chiller with desalination function. The diagram shows its most important components, circuits and points representing measuring devices responsible for the measurements of temperature, pressure and flow rate.

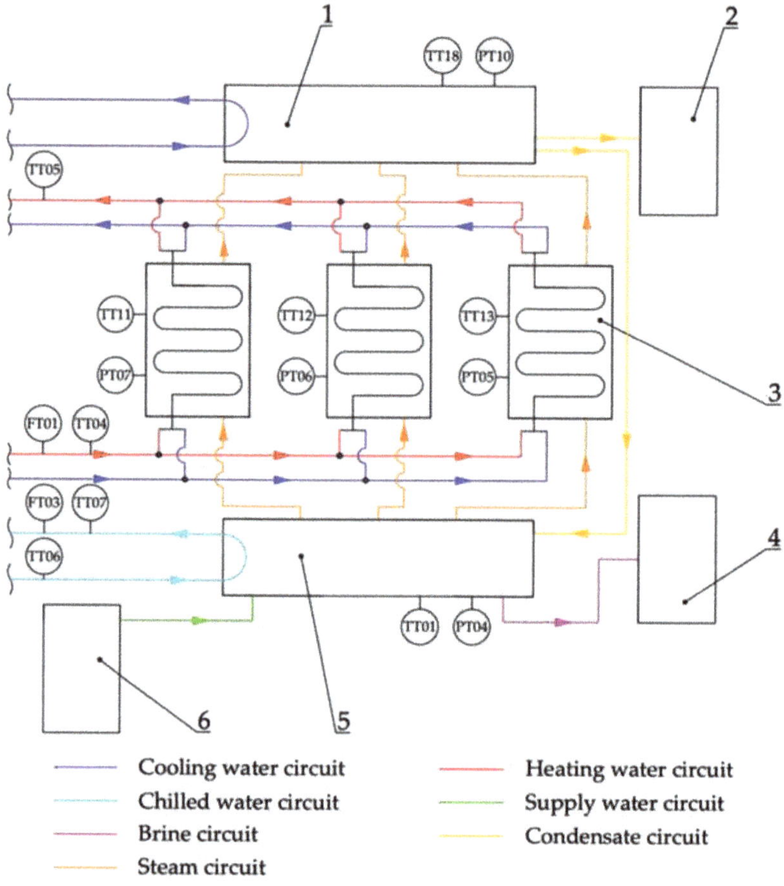

Figure 2. Simplified diagram of the three-bed adsorption chiller with desalination function used in the investigation. 1—condenser; 2—distillate tank; 3—adsorbent bed; 4—brine tank; 5—evaporator; 6—deaerator; TT01—temperature in the evaporator; TT04—hot water inlet temperature; TT05—hot water outlet temperature; TT06—chilled water inlet temperature; TT07—chilled water outlet temperature; TT11—temperature in bed 1; TT12—temperature in bed 2; TT13—temperature in bed 3; TT18—temperature in the condenser; PT04—pressure in the evaporator; PT07—pressure in bed 1; PT06—pressure in bed 2; PT05—pressure in bed 3; PT10—pressure in the condenser; FT01—hot water flow; FT03—chilled water flow.

The main components of the chiller are three adsorption beds (3), an evaporator (5) and a condenser (1). Each bed is equipped with a finned Tichelmann lamella exchanger with a total length of 880 mm and a total width of 450 mm. The construction of the evaporator and condenser is cylinder-shaped with a length and diameter of 1500 and 350 mm and 1450 and 220 mm, respectively. In addition, the system is equipped, among other things, with a distillate tank (2), a brine tank (4), a deaerator (6) and an expansion valve not shown in the figure. Fluids in various forms and under different conditions flow through the system in seven main circuits.

In the first stage, the salted water is deaerated in the deaerator and flows to the evaporator. In the evaporator, there is a system of pipes responsible for absorbing the heat from the outlet chilled water and transferring it to the supplied salt water. As a result, water in the evaporator changes its state of matter and evaporates. This process takes place at a low temperature, which is made possible by the appropriately low absolute pressure inside the evaporator. The evaporated water is transported from the evaporator to the adsorbent bed, thus starting the adsorption phase. Adsorption is an exothermic process. Therefore, the heat generated in the bed must be dissipated into the cooling water. This ensures that the correct adsorption temperature is maintained so that vapour adsorption occurs most efficiently [32]. Once adsorption is complete, the valve between the evaporator and the bed is closed and the adsorber pre-heating phase begins. Three-way valves are set so as to allow heating water to flow through the bed. The heating water reaches the required temperature in an 18-kW electric boiler simulating a source of low-temperature heat. The heated water flows into the buffer tank, from which it is then led to the beds in the regeneration phase. The boiler works periodically, keeping the water temperature at the set value. When the temperature in the bed reaches the required value, the desorption process takes place. Water in the heating circuit is still supplied to the heat exchanger in the bed. Water previously adsorbed on the surface of the silica gel desorbs and increases its pressure due to the thermal energy supplied. The valve between the bed and the condenser is opened and the resulting gas phase flows into the condenser, where it condenses by giving up its heat on the surface of the exchanger, through which cooling water flows. Depending on whether the adsorption chiller is in the cooling or the desalination mode, the condensate formed in the lower part of the condenser can flow through the expansion valve back to the evaporator to form a closed circuit, or it can be collected in the distillate tank. The expansion valve is responsible for maintaining the pressure difference between the condenser and the evaporator during the flow of condensate. The bed is then cooled before the next cycle. The cooler allows the heat to be recovered from the water that heats the adsorber during desorption. When the three-way valve is set correctly, cooling of the bed with cooling water begins. Then, water is supplied for some time to the heating water circuit so that as little as possible of the heat still gathered inside the bed is lost. Figure 3 shows photos of the investigated adsorption chiller with desalination function.

2.2. Measuring Instruments

The system in question is equipped with a system of measuring apparatus which allow the necessary parameters to be measured. During the testing, the temperature, pressure and flow rate were measured. Figure 2 schematically shows the location of the measuring points and they are marked with a description. Table 4 presents the parameters of the measuring instruments used. The measurements of temperature in the evaporator TT01, hot water inlet temperature at the beds TT04, hot water outlet temperature at the beds TT05, chilled water inlet temperature at evaporator TT06, chilled water outlet temperature the evaporator TT07, temperature in the beds TT11–TT13 and in the condenser TT18 were made using PT-1000 temperature sensors. Pressure values in the evaporator PT04, in the condenser PT10 and in the beds PT05–PT07 were measured using a pressure transducer. Flow rates of chilled water FT03 and hot water FT01 were measured using an electromagnetic flow meter.

Table 4. Characteristics of measuring apparatus.

Parameter Measured	Instrument	Operating Range	Accuracy
Pressure	Pressure transducer	0–99 kPa	0.50%
Flow rate	Electromagnetic flow meter	1–100 L/min	0.50%
Temperature	Temperature sensor PT-1000	−80 °C to 150 °C	0.1 °C

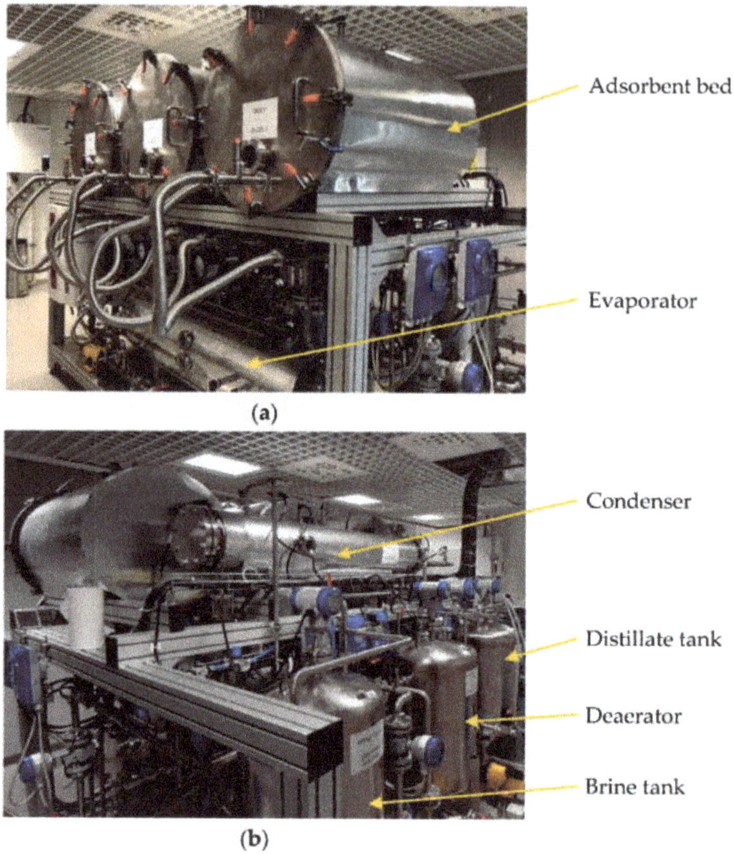

Figure 3. Three-bed adsorption chiller with desalination function used in the investigation: (**a**) front view, (**b**) back view.

All parameters were recorded during the steady-state operation of the adsorption chiller. The time from switching on the device to achieving steady-state parameters was approximately 1 h. The values displayed by the measuring instruments were saved every 5 s using the SCADA control system on a personal computer. Analysis of the results obtained was possible by exporting them to csv format in tabular form and then running them in Microsoft Excel.

The measuring devices and transducers are powered from a special switchgear. The system is equipped with control system software which enables the users to control all the pumps and valves and records the measurements. In addition, it is possible to set the duration of each phase, including half-cycle time and heat recovery time. The system includes manual ball valves (1/2-inch, 3/4-inch) and pneumatic valves between the evaporator and the beds, as well as the beds and the condenser. In addition, circulation pumps with a capacity of 0–500 L/h for heating, cooling and chilled water, are present in the plant.

2.3. Conditions for Conducting the Experiment

The study was designed to analyse the effect of different cycle times on the performance of the adsorption chiller. The experiment was conducted on a chiller using all three of the available beds. Beds 1 and 2 were operated in synchronisation with each other, while bed 3 was operated in the opposite phase. This paper presents the results of tests for heating water temperature of 80 ± 3 °C and for various cycle times conducted under the same conditions. This study is focused on the determination of COP and SCP for different cycle times. In article [5], the effect of different heating water temperatures in the range of 55 °C to about 80 °C on the COP and SCP of a three-bed adsorption chiller was investigated. The COP and SCP increased with increasing heat source temperature and have the high value for the temperature of heat water of about 80 °C [5]. Therefore, for this temperature, tests were performed. Thus, adsorption occurred at the same time in the first and second beds while desorption was carried out in bed 3, or the other way around. Six different cycle times were analysed: 100, 200, 300, 500, 600 and 900 s. The dependence of the COP, SCP and other parameters of the adsorption chiller on the cycle time was tested.

The COP and SCP were calculated, based on [35], according to Equations (1) and (5) which are shown below:

$$\mathrm{COP} = \frac{\mathrm{CC}}{\mathrm{HP}} \tag{1}$$

$$\mathrm{CC} = \dot{m}_c \cdot c_{p,c} \cdot \Delta T_c = \dot{m}_c \cdot c_{p,c} \cdot (T_{i,c} - T_{o,c}) = \mathrm{FT03} \cdot c_{p,c} \cdot (\mathrm{TT06} - \mathrm{TT07}) \tag{2}$$

$$\mathrm{HP} = \dot{m}_h \cdot c_{p,h} \cdot \Delta T_c = \dot{m}_h \cdot c_{p,h} \cdot (T_{i,h} - T_{o,h}) = \mathrm{FT01} \cdot c_{p,h} \cdot (\mathrm{TT04} - \mathrm{TT05}) \tag{3}$$

$$c_{p,h} = c_{p,c} = 4200 \ \frac{\mathrm{J}}{\mathrm{kg \cdot K}} \tag{4}$$

$$\mathrm{SCP} = \frac{\mathrm{CC}}{m_a} \tag{5}$$

Equations (1)–(5) use symbols with the following meanings: CC—cooling capacity, (W); HP—heating power supplied to the system, (W); m—flow rate, (kg/s); c_p—specific heat of water; T—temperature, (°C); m_a— the mass of sorbent, which was 12 kg. The indices used in the equations are: c—chilled water; h—heating water; i—inlet; o—outlet.

The cooling capacity is determined from the parameters of chilled water flowing through the evaporator. A 2 °C difference in evaporator inlet and outlet temperatures allows one to achieve a cooling capacity of 1.5 kW.

3. Results

Figures 4–9 show the distribution of pressures in all of the three beds, the condenser and the evaporator for each cycle time tested.

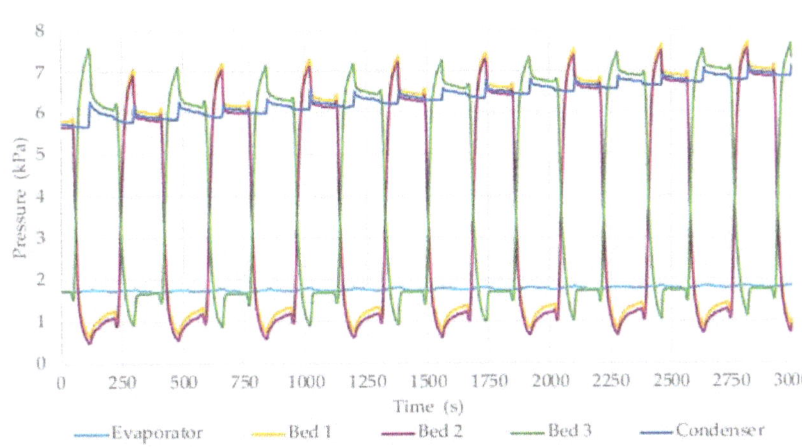

Figure 4. Pressure for cycle time = 100 s.

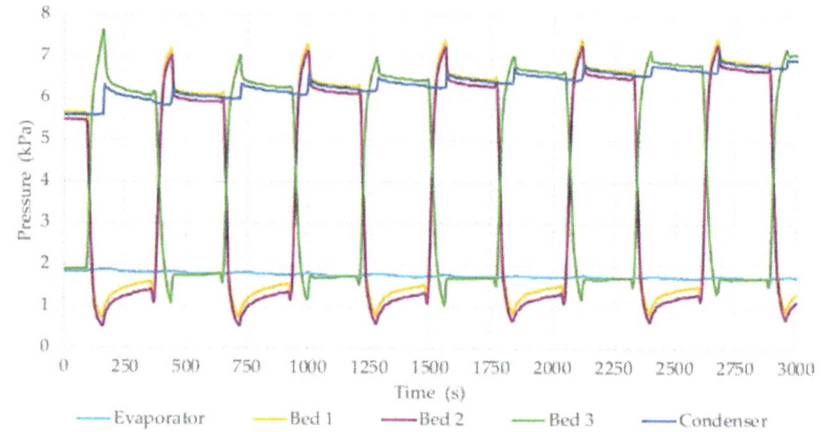

Figure 5. Pressure for cycle time = 200 s.

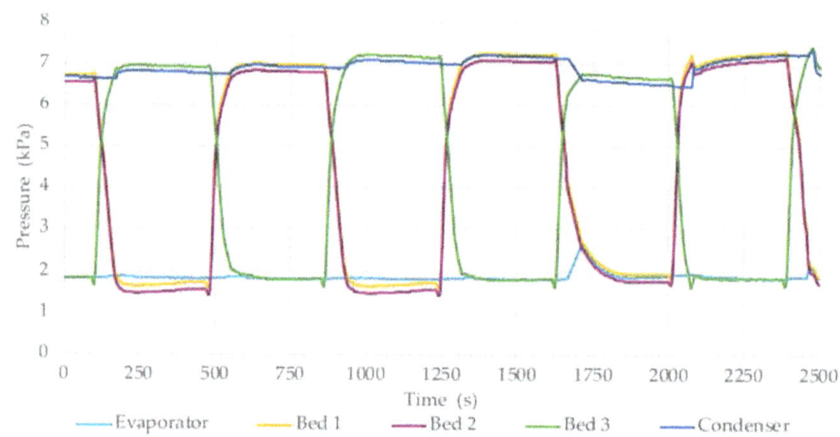

Figure 6. Pressure for cycle time = 300 s.

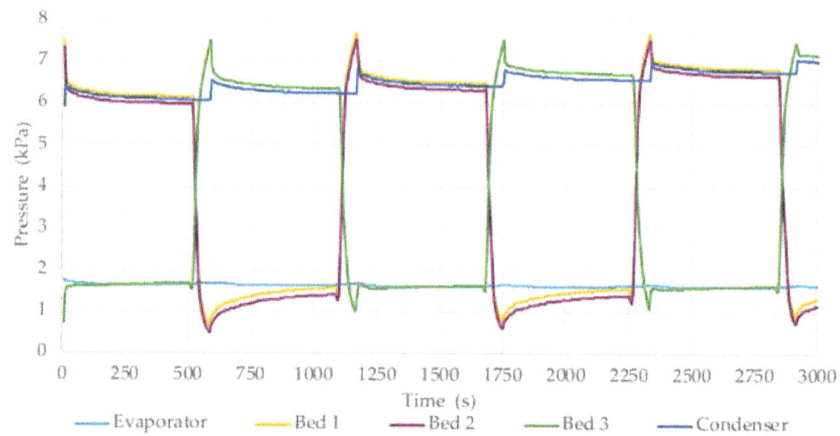

Figure 7. Pressure for cycle time = 500 s.

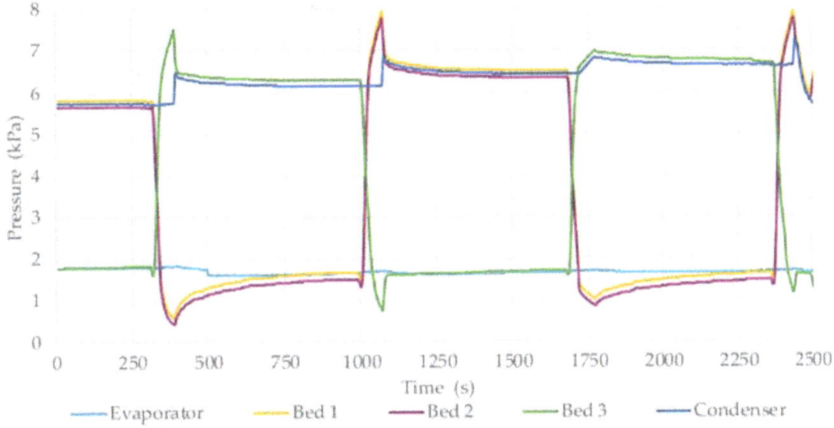

Figure 8. Pressure for cycle time = 600 s.

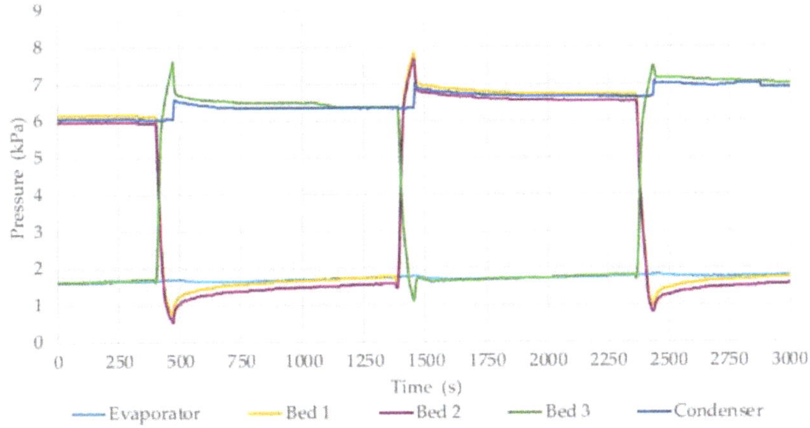

Figure 9. Pressure for cycle time = 900 s.

The beds marked with numbers 1 and 2 were synchronised with each other, which means that adsorption and desorption occurred at the same time. From the graphs above, it can be seen that the pressures in these two beds were similar. Bed 3 operated in the opposite way to the other beds. While adsorption was taking place in beds 1 and 2, desorption was taking place in bed 3. From the above curves, it can be deduced that the moment when adsorption started, the pressure in the respective bed increased and its value approached that of the pressure in the evaporator. In the case of the curves in Figures 4 and 5, for beds 1 and 2, the cycle time was too short for the bed pressure to approach the value of the evaporator pressure. When the bed was switched to the state proper for the desorption process, the pressure in the bed was equal to the condenser pressure.

Figures 10–15 show temperature changes in the beds, the evaporator and the condenser, and the changes in hot water inlet temperature at the beds for each cycle time.

The temperature curves clearly indicate the moments at which the adsorption and desorption phases in the beds start. These processes occur rapidly, which is confirmed by the dynamic temperature changes. In the case of adsorption, cooling water flows through the adsorbent bed, thus lowering its temperature. In this way, the course of the process becomes easier. During the desorption phase, hot water flows through the heat exchanger in the bed, causing an increase in temperature inside the bed and evaporation of water molecules. The temperature of the water that heated the bed was 80 ± 3 °C.

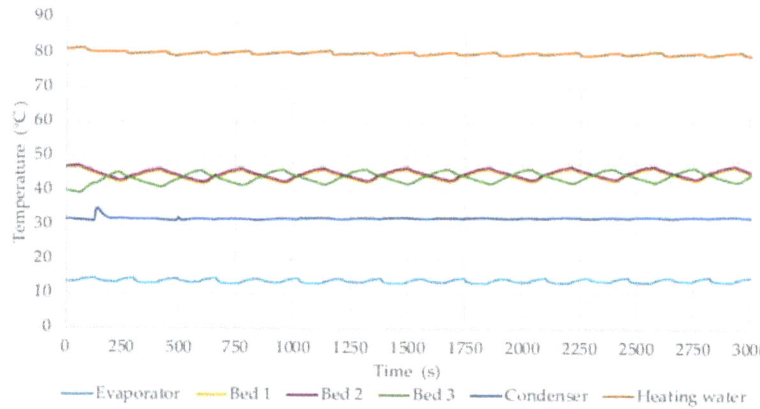

Figure 10. Temperature for cycle time = 100 s.

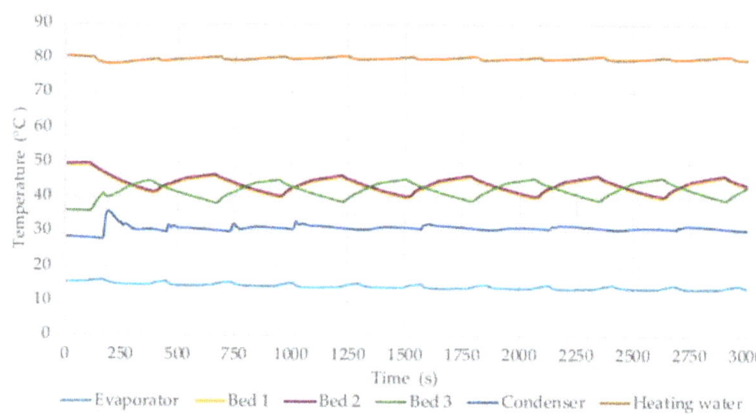

Figure 11. Temperature for cycle time = 200 s.

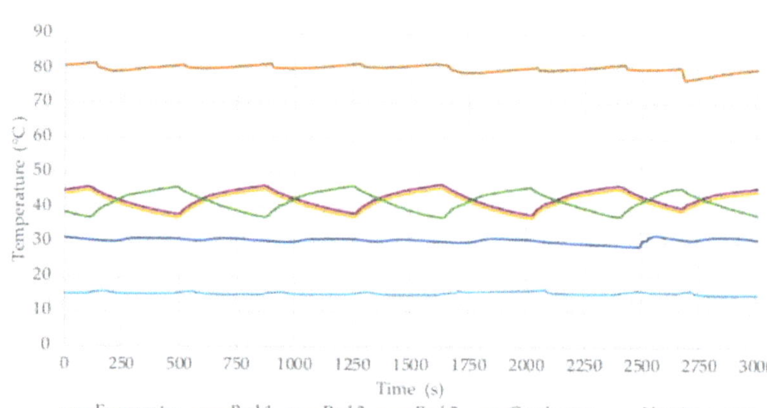

Figure 12. Temperature for cycle time = 300 s.

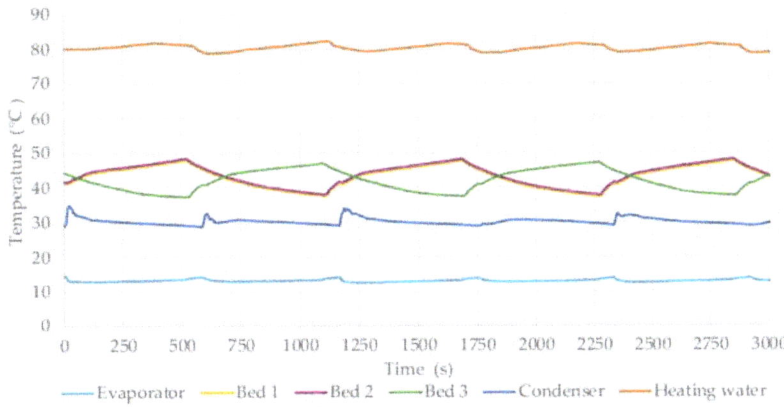

Figure 13. Temperature for cycle time = 500 s.

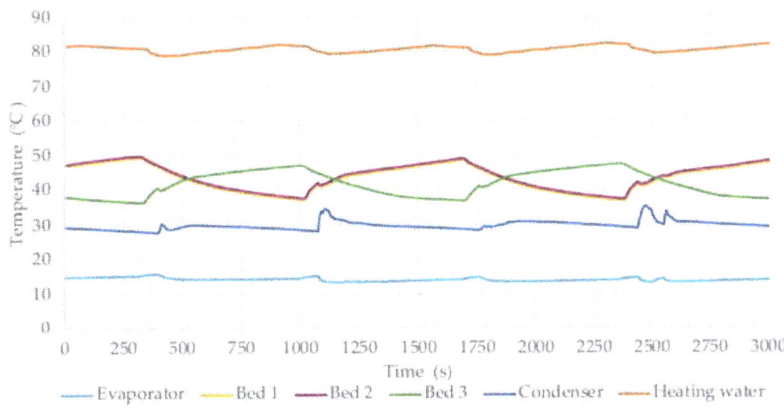

Figure 14. Temperature for cycle time = 600 s.

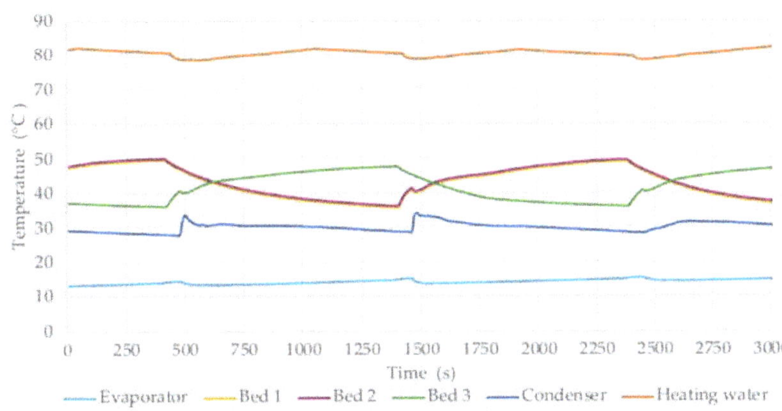

Figure 15. Temperature for cycle time = 900 s.

Figures 16–21 show the chilled water temperature at the evaporator inlet and outlet and the chilled water flow rate for each cycle time.

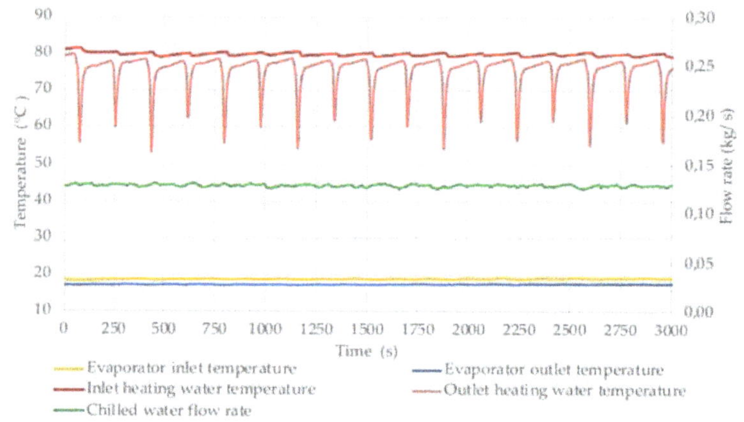

Figure 16. Evaporator inlet and outlet temperature and chilled water flow rate for cycle time = 100 s.

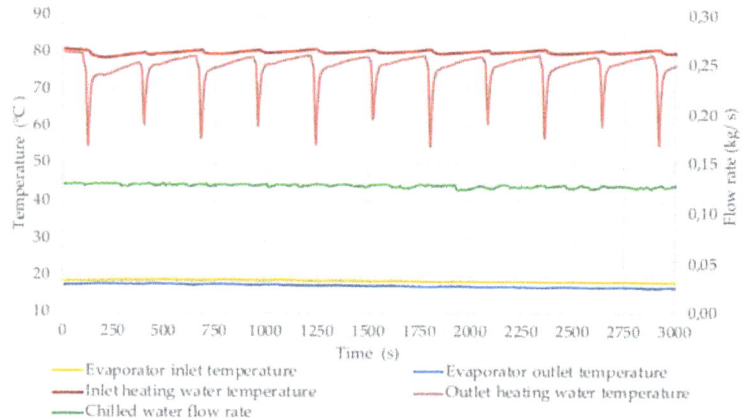

Figure 17. Evaporator inlet and outlet temperature and chilled water flow rate for cycle time = 200 s.

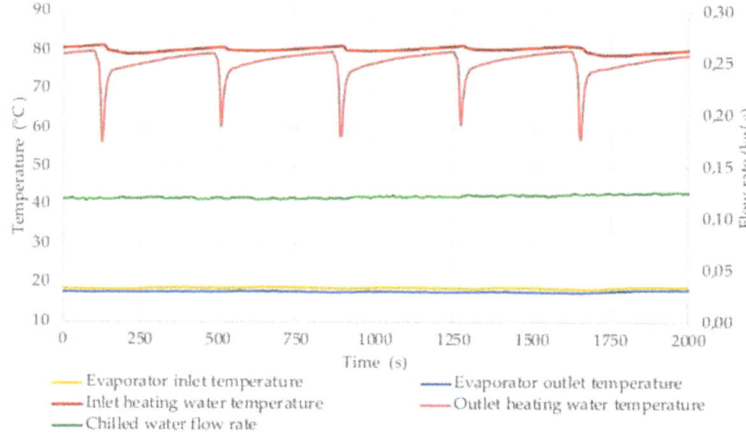

Figure 18. Evaporator inlet and outlet temperature and chilled water flow rate for cycle time = 300 s.

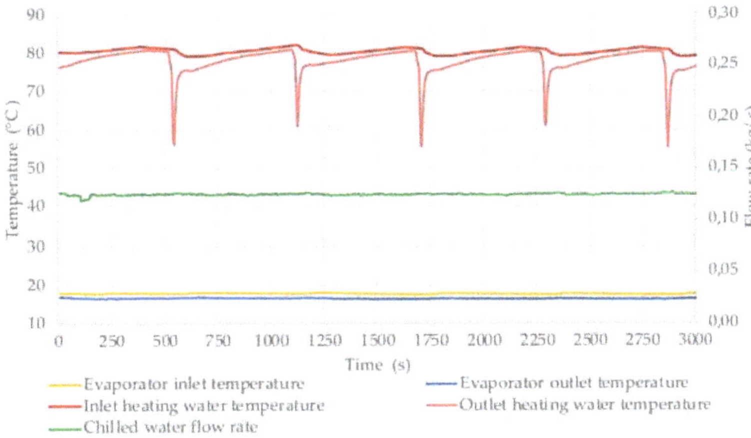

Figure 19. Evaporator inlet and outlet temperature and chilled water flow rate for cycle time = 500 s.

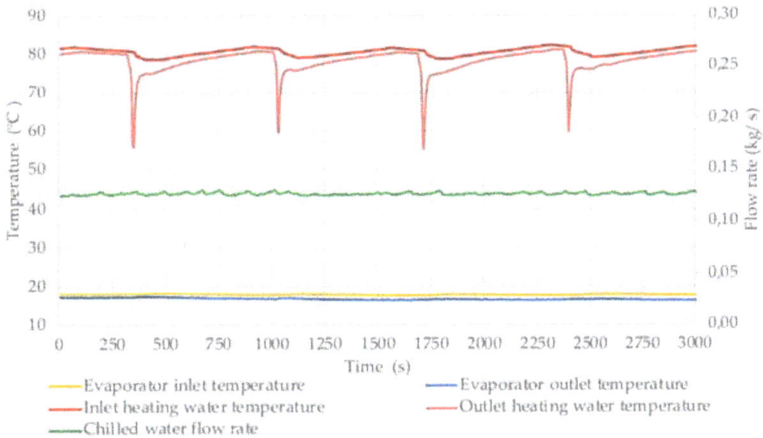

Figure 20. Evaporator inlet and outlet temperature and chilled water flow rate for cycle time = 600 s.

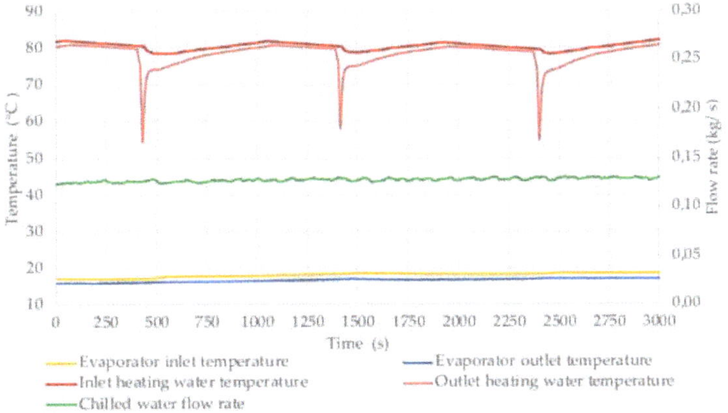

Figure 21. Evaporator inlet and outlet temperature and chilled water flow rate for cycle time = 900 s.

During the testing process, the evaporator was supplied with chilled water at a rate of approximately 0.13 kg/s. The temperature of chilled water at the evaporator inlet is maintained depending on the pressure value in the evaporator. When flowing through the evaporator, chilled water lowers its temperature as a result of giving up thermal energy to the adsorbate which then changes its state of matter and evaporates. Then, the temperature in the evaporator equals the value corresponding to the evaporation temperature of the adsorbate.

4. Discussion

From the results obtained, the effect of the cycle time duration on the COP and SCP can be assessed. An analysis of Figures 10–15 leads to the conclusion that longer cycle times meant higher temperatures in the bed during desorption and lower temperatures during adsorption. With longer cycle times, the heating and cooling times of the bed also increase, which is confirmed by the graphs. For cycle time of 100 s, the temperature during desorption phase is approximately 45 °C and approximately 40 °C during adsorption while for cycle time of 900 s, the corresponding temperatures are approximately 50 °C and 36 °C, respectively.

The lowest value of chilled water temperature obtained during this investigation was 16 °C. The difference between the evaporator inlet and outlet temperatures is approximately 1.5 °C and is stable throughout the cycle.

A graphical representation of the COP values depending on the cycle time is shown in Figure 22. The COP values were calculated according to Equations (1)–(4) as the quotient of the cooling capacity (CC) and the heating power (HP).

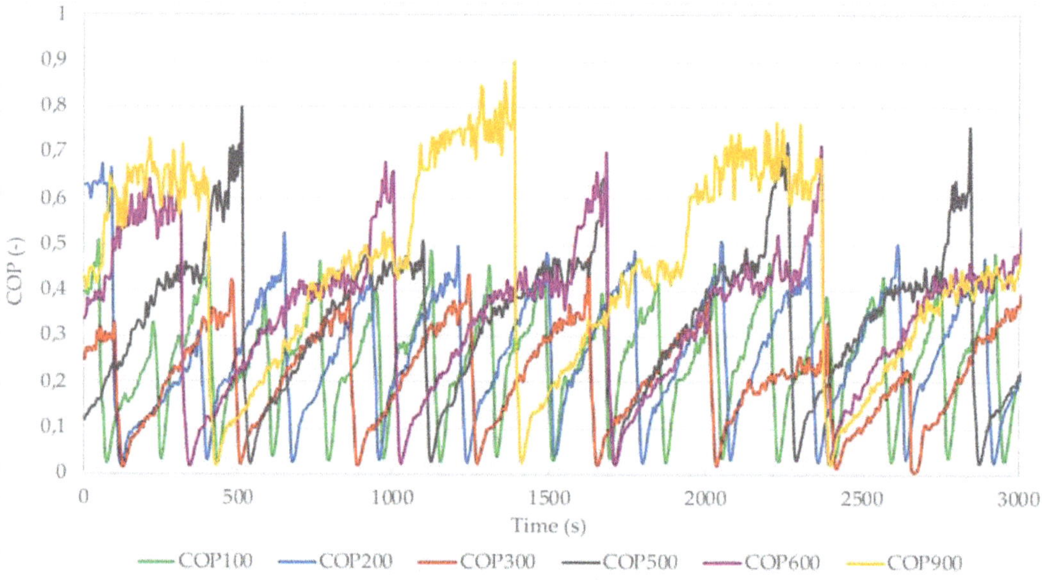

Figure 22. Comparison of the changes in the COP values for all cycle times tested.

The characteristic points in the above graph are the extremes of the curves corresponding to the particular cycle times. The COP reaches its minimum values with the start of the desorption process. At this point, the most intense heat exchange takes place between the hot water and the bed. This is the so-called bed pre-heating phase during which the temperature difference between the bed inlet and outlet is the greatest and reaches up to 35 °C. This is due to the fact that during adsorption, the temperature inside the adsorbent bed is much lower than the temperature required for the desorption phase. This difference decreases as the bed is heated, so the cooling of the water becomes less and less intense. At the end of desorption, when the temperature difference between the bed inlet and outlet is the smallest, the heating power (HP) also reaches minimum values. Then, according to Equation (1), the COP is highest. In addition, it has been noted that for the longer cycles when the bed heating time is longer, the system reaches higher maximum values of the COP.

Figure 23 shows a comparison of COPs for cycles of 100 and 900 s, i.e., the extreme values.

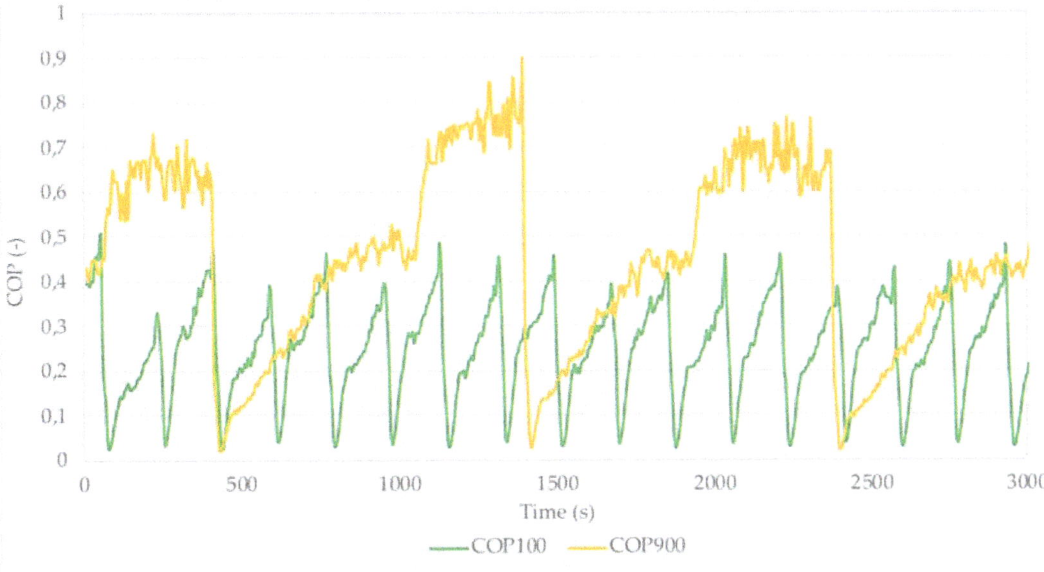

Figure 23. Comparison of changes in COP values for cycle times of 100 s and 900 s.

For a cycle of 100 s, this time is too short for the adsorption and desorption processes to occur in the most efficient manner. The short cycle time also means that it is not possible for the chiller operation to become steady-state. The adsorbent bed is not sufficiently heated and cooled then.

The longest cycle time contributes noticeably to the stable operation of the chiller. Increasing the adsorption and desorption times results in a greater amount of adsorbate being adsorbed and also a greater amount being desorbed from the surface of the sorption material.

Figure 24 shows influence of cycle time on SCP and COP. In accordance with the studies of other authors, COP increases with increasing time, while SCP increases until reaching a certain maximum value, and then decreases with increasing cycle time duration. The lowest values of SCP and COP were obtained for the cycle that lasted for 100 s and they are, respectively, 115 W/kg and 0.37. The highest SCP value of 154 W/kg was obtained for 500 s. The COP for this cycle duration is 0.55. The maximum obtained COP value during these studies is 0.64. It was achieved for the longest cycle duration of 900 s and SCP for this time is about 130 W/kg. However, such a long cycle time is disadvantageous since it is important for this type of device to achieve the highest possible efficiency with the shortest possible cycle time. Based on Figure 24, it can be concluded that the optimal cycle duration for the conditions in which the experiment was conducted is 500 s.

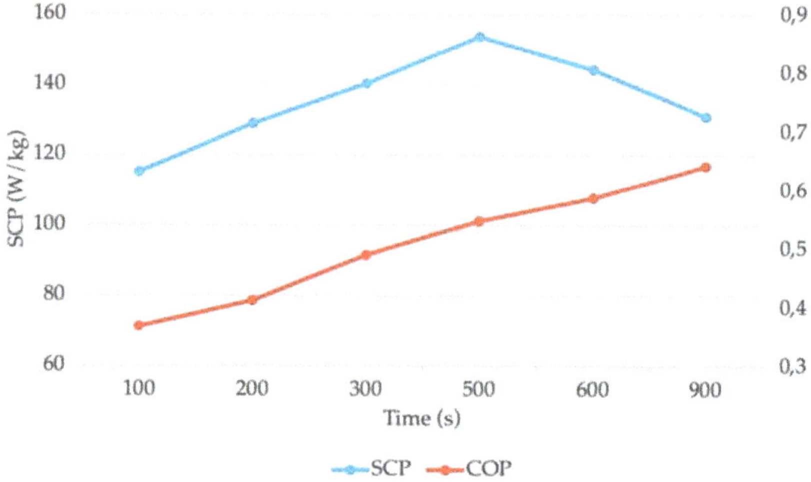

Figure 24. Influence of cycle time on the SCP and COP.

5. Conclusions

This paper presents various possibilities for improving the performance of an adsorption chiller. Experimental results relating to one of the methods are presented. The aim of the study was to investigate the effect of the duration of the adsorption and desorption processes on the COP and SCP of an adsorption chiller with desalination function. An analysis of the results obtained for the test device located at the Energy Centre of the AGH University of Science and Technology in Krakow confirmed that the duration of the entire cycle affects the performance of the adsorption chiller. The results obtained show that an increase in the cycle time can lead significant increase in the COP value. The SCP grows up to the maximum value for a cycle duration of 500 s. Then, the SCP value drops. Furthermore, the course of the cycle for longer adsorption and desorption times is characterised by greater stability of operation. It has also been noted that adsorption and desorption are most effective at the beginning of the cycle.

In the operation of the adsorption chiller, the key is to reduce the mass of the device and one of the ways is to optimally select the time of adsorption/desorption cycles where an appropriate balance between the SCP and COP value is achieved. Based on the data from the experiment, it can be summarised that the time of 100 s is too short because the heat exchanger was not heated evenly and not all of the water during the desorption process was desorbed. On the other hand, during the adsorption process, not all of the water vapor produced in the evaporator was adsorbed, which limits the evaporator's power and thus the device's power COP. When the adsorption/desorption time is increased above 600 s to 900 s, an increase in the COP of the device is observed; however, SCP decreases. This means that too long adsorption/desorption time of cycle causes that because after saturation of the sorbent water, no adsorption occurs and extending the time does not increase the cooling power of the device. However, after desorption of all the water from the bed, an increase in the cycle time does not cause an increase in the chilled power but the use of energy for desorption increases and the process becomes very energy consuming. Therefore, based on the experimental data for a three-bed chiller, it can be concluded that the cycle time of 500 s is the most appropriate because a good balance between COP and SCP is maintained, and the desorption/adsorption process is the most effective. It is important to shorten the cycle time for adsorption–desorption to achieve the highest cooling capacity and COP. The process of adsorption/desorption can be carried out in a shorter time than 900 s. For example, when the cycle time is 500 s, similar COPs and a higher SCP compared to 900 s are achieved.

The search for the optimal parameters of the adsorption chiller is crucial to improve its performance. The application of the right duration of the adsorption and desorption phases greatly affects the performance coefficients. Improving the efficiency of the operation of adsorption chillers will ultimately lead to an increase in their competitiveness in the market.

Funding: The paper was funded by subsidy for research Faculty of Energy and Fuels number 16.16.210.476.

Data Availability Statement: The data that support the results reported in this study are available from the corresponding author upon request.

Conflicts of Interest: The authors declare no conflict of interest.

References

1. Sztekler, K.; Kalawa, W.; Mika, L.; Krzywanski, J.; Grabowska, K.; Sosnowski, M.; Nowak, W.; Siwek, T.; Bieniek, A. Modeling of a Combined Cycle Gas Turbine Integrated with an Adsorption Chiller. *Energies* **2020**, *13*, 515. [CrossRef]
2. Dupont, J.L.; Domanski, P.; Lebrun, P.; Ziegler, F. *The Role of Refrigeration in the Global Economy*; Informatory Note of the International Institute of Refrigeration; The International Institute of Refrigeration: Paris, France, 2019. [CrossRef]
3. *The Future of Cooling: Opportunities for Energy-Efficient Air Conditioning*; Report of the International Energy Agency; International Energy Agency: Paris, France, 2018. Available online: https://www.iea.org/reports/the-future-of-cooling (accessed on 13 April 2021).
4. Key World Energy Statistics. *Report of the International Energy Agency*; International Energy Agency: Paris, France, 2020. Available online: https://www.iea.org/reports/key-world-energy-statistics-2020 (accessed on 13 April 2021).
5. Sztekler, K.; Kalawa, W.; Nowak, W.; Mika, L.; Gradziel, S.; Krzywanski, J.; Radomska, E. Experimental Study of Three-Bed Adsorption Chiller with Desalination Function. *Energies* **2020**, *13*, 5827. [CrossRef]
6. Choudhury, B.; Saha, B.B.; Chatterjee, P.K.; Sarkar, J.P. An overview of developments in adsorption refrigeration systems towards a sustainable way of cooling. *Appl. Energy* **2013**, *104*, 554–567. [CrossRef]
7. Sztekler, K.; Kalawa, W.; Mika, Ł.; Mlonka-Medrala, A.; Sowa, M.; Nowak, W. Effect of Additives on the Sorption Kinetics of a Silica Gel Bed in Adsorption Chiller. *Energies* **2021**, *14*, 1083. [CrossRef]
8. Fernandes, M.S.; Brites, G.J.V.N.; Costa, J.J.; Gaspar, A.R.; Costa, V.A.F. Review and future trends of solar adsorption refrigeration systems. *Renew. Sustain. Energy Rev.* **2014**, *39*, 102–1123. [CrossRef]
9. Ng, K.C.; Thu, K.; Kim, Y.; Chakraborty, A.; Amy, G. Adsorption desalination: An emerging low-cost thermal desalination method. *Desalination* **2013**, *308*, 161–179. [CrossRef]
10. Tzabar, N.; Grossman, G. Analysis of an activated-carbon sorption compressor operating with gas mixtures. *Cryogenics* **2012**, *52*, 491–499. [CrossRef]
11. Luberti, M.; Di Santis, C.; Santori, G. Ammonia/ethanol mixture for adsorption refrigeration. *Energies* **2020**, *13*, 983. [CrossRef]
12. Grabowska, K.; Krzywanski, J.; Nowak, W.; Wesolowska, M. Construction of an innovative adsorbent bed configuration in the adsorption chiller-Selection criteria for effective sorbent-glue pair. *Energy* **2018**, *151*, 317–323. [CrossRef]
13. Younes, M.M.; El-Sharkawy, I.I.; Kabeel, A.E.; Saha, B.B. A review on adsorbent-adsorbate pairs for cooling applications. *Appl. Therm. Eng.* **2017**, *114*, 394–414. [CrossRef]
14. Goyal, P.; Baredar, P.; Mittal, A.; Siddiqui, A.R. Adsorption refrigeration technology-An overview of theory and its solar energy applications. *Renew. Sustain. Energy Rev.* **2016**, *53*, 1389–1410. [CrossRef]
15. Stefański, S.; Nowak, W.; Sztekler, K.; Kalawa, W.; Siwek, T. Applicability of adsorption cooling/desalination systems driven by low- temperature waste heat. *IOP Conf. Ser.* **2019**, *214*, 012126. [CrossRef]
16. Elsayed, E.; AL-Dadah, R.; Mahmoud, S.; Anderson, P.; Elsayed, A. Experimental testing of aluminium fumarate MOF for adsorption desalination. *Desalination* **2020**, *475*, 114170. [CrossRef]
17. Thu, K.; Saha, B.B.; Chua, K.J.; Ng, K.C. Performance investigation of a waste heat-driven 3-bed 2-evaporator adsorption cycle for cooling and desalination. *Int. J. Heat Mass Transf.* **2016**, *101*, 1111–1122. [CrossRef]
18. Stefański, S.; Mika, Ł.; Sztekler, K.; Kalawa, W.; Lis, Ł.; Nowak, W. Adsorption bed configurations for adsorption cooling application. *E3S Web Conf.* **2019**, *108*, 01010. [CrossRef]
19. Askalany, A.A.; Henninger, S.K.; Ghazy, M.; Saha, B.B. Effect of improving thermal conductivity of the adsorbent on performance of adsorption cooling system. *Appl. Therm. Eng.* **2017**, *110*, 695–702. [CrossRef]
20. Chua, H.T.; Ng, K.C.; Malek, A.; Kashiwagi, T.; Akisawa, A.; Saha, B.B. Multi-bed regenerative adsorption chiller—Improving the utilization of waste heat and reducing the chilled water outlet temperature fluctuation. *Int. J. Refrig.* **2001**, *24*, 124–136. [CrossRef]
21. Thu, K.; Ng, K.C.; Saha, B.B.; Chakraborty, A.; Koyama, S. Operational strategy of adsorption desalination systems. *Int. J. Heat Mass Transf.* **2009**, *52*, 1811–1816. [CrossRef]
22. Pan, Q.W.; Wang, R.Z. Study on operation strategy of a silica gel-water adsorption chiller in solar cooling application. *Sol. Energy* **2018**, *172*, 24–31. [CrossRef]
23. Zajaczkowski, B. Optimizing performance of a three-bed adsorption chiller using new cycle time allocation and mass recovery. *Appl. Therm. Eng.* **2016**, *100*, 744–752. [CrossRef]

24. Saha, B.B.; Koyama, S.; Lee, J.B.; Kuwahara, K.; Alam, K.C.A.; Hamamoto, Y.; Akisawa, A.; Kashiwagi, T. Performance evaluation of a low-temperature waste heat driven multi-bed adsorption chiller. *Int. J. Multiph. Flow* **2003**, *29*, 1249–1263. [CrossRef]
25. Glaznev, I.S.; Aristov, Y.I. The effect of cycle boundary conditions and adsorbent grain size on the water sorption dynamics in adsorption chillers. *Int. J. Heat Mass Transf.* **2010**, *53*, 1893–1898. [CrossRef]
26. El-Sharkawy, I.I.; AbdelMeguid, H.; Saha, B.B. Towards an optimal performance of adsorption chillers: Reallocation of adsorption/desorption cycle times. *Int. J. Heat Mass Transf.* **2013**, *63*, 171–182. [CrossRef]
27. Elsaid, K.; Taha Sayed, E.; Yousef, B.A.A.; Kamal Hussien Rabaia, M.; Ali Abdelkareem, M.; Olabi, A.G. Recent progress on the utilization of waste heat for desalination: A review. *Energy Convers. Manag.* **2020**, *221*, 113105. [CrossRef]
28. Ahmed, F.E.; Hashaikeh, R.; Hilal, N. Hybrid technologies: The future of energy efficient desalination—A review. *Desalination* **2020**, *495*, 114659. [CrossRef]
29. Eke, J.; Yusuf, A.; Giwa, A.; Sodiq, A. The global status of desalination: An assessment of current desalination technologies, plants and capacity. *Desalination* **2020**, *495*, 114633. [CrossRef]
30. Nassrullah, H.; Anis, S.F.; Hashaikeh, R.; Hilal, N. Energy for desalination: A state-of-the-art review. *Desalination* **2020**, *491*, 114569. [CrossRef]
31. Sarai Atab, M.; Smallbone, A.J.; Roskilly, A.P. A hybrid reverse osmosis/adsorption desalination plant for irrigation and drinking water. *Desalination* **2018**, *444*, 44–52. [CrossRef]
32. Thu, K.; Yanagi, H.; Saha, B.B.; Ng, K.C. Performance analysis of a low-temperature waste heat-driven adsorption desalination prototype. *Int. J. Heat Mass Transf.* **2013**, *65*, 662–669. [CrossRef]
33. New Energy Transfer Supplies Polish Sorption Technology to Saudi Arabia. Available online: http://newenergytransfer.com/ (accessed on 13 April 2021).
34. This Desalination Process Uses Much Less Energy than RO. Available online: https://www.chemengonline.com/desalination-process-uses-much-less-energy-ro/ (accessed on 13 April 2021).
35. Sztekler, K.; Kalawa, W.; Nowak, W.; Mika, L.; Grabowska, K.; Krzywanski, J.; Sosnowski, M.; Al-Harbi, A.A. Performance evaluation of a single-stage two-bed adsorption chiller with desalination function. *J. Energy Resour. Technol.* **2020**, 1–22. [CrossRef]

energies

MDPI

Article

MIL-160 as an Adsorbent for Atmospheric Water Harvesting

Marina Solovyeva [1], Irina Krivosheeva [1,2], Larisa Gordeeva [1,*] and Yuri Aristov [1]

1 Boreskov Institute of Catalysis, Ac. Lavrentiev av. 5, 630055 Novosibirsk, Russia; solovyeva@catalysis.ru (M.S.); i.krivosheeva@g.nsu.ru (I.K.); aristov@catalysis.ru (Y.A.)

2 Department of Natural Sciences, Novosibirsk State University, 630090 Novosibirsk, Russia

* Correspondence: gordeeva@catalysis.ru; Tel.: +7-383-330-9573

Abstract: Nowadays, the rapidly growing population, climate change, and environment pollution put heavy pressure on fresh water resources. The atmosphere is the immense worldwide and available water source. The Adsorptive Water Harvesting from the Atmosphere (AWHA) method is considered a promising alternative to desalination technologies for remote arid regions. The development of novel adsorbents with advanced water-adsorption properties is a prerequisite for practical realization of this method. Metal–organic frameworks (MOFs) are a novel class of porous crystalline solids that bring a great potential for AWHA due to their extremely high specific surface area, porosity, and tailored adsorption properties. This work addresses MIL-160 as a water adsorbent for AWHA. The water-adsorption equilibrium of MIL-160 was studied by volumetric method, the isosteric heat of adsorption was calculated, and finally, the potential of MIL-160 for AWHA was evaluated for climatic conditions of the deserts of Saudi Arabia, Mongolia, the Sahara, Atacama, and Mojave as reference arid regions. MIL-160 was shown to ensure a maximum specific water productivity of 0.31–0.33 g_{H2O}/g_{ads} per cycle. High fractions of water extracted (0.90–0.98) and collected (0.48–0.97) could be achieved at a regeneration temperature of 80 °C with natural cooling of the condenser by ambient air. The specific energy consumption for water production varied from 3.5 to 6.8 kJ/g, which is acceptable if solar heat is used to drive the desorption. The AWHA method employing MIL-160 is a promising way to achieve a fresh water supply in remote arid areas.

Keywords: adsorptive water harvesting from the atmosphere; metal–organic frameworks; MIL-160; water vapor adsorption; specific water productivity; specific energy consumption

Citation: Solovyeva, M.; Krivosheeva, I.; Gordeeva, L.; Aristov, Y. MIL-160 as an Adsorbent for Atmospheric Water Harvesting. *Energies* **2021**, *14*, 3586. https://doi.org/10.3390/en14123586

Academic Editor: Jaroslaw Krzywanski

Received: 20 May 2021
Accepted: 12 June 2021
Published: 16 June 2021

1. Introduction

Climate change leading to the desertification of a vast area, environmental pollution, and a rapidly growing world population make the fresh water supply one of the major issues of our time [1]. Although the worldwide resources of fresh water are quite abundant ($1 \cdot 10^5$ km^3), most parts of them are in the form of hard-to-reach glaciers and deep underground water. The rest (rivers, lakes, and shallow underground water) is distributed very unevenly, which puts enormous pressure on water resources in arid regions. Nowadays, 1.5 billion people are facing portable water scarcity, and by 2025 this number is expected to grow to 3 billion. The most water-scarce regions include North Africa, the Near and Middle East, Northern China, India, Eastern Australia, Mexico, northeastern Brazil, and the west coast of South America [2,3].

At the same time, the total amount of moisture in the Earth's atmosphere—about 13,000 km^3—significantly exceeds the world's requirements [4,5]. Due to its worldwide availability, interest in atmospheric water harvesting has been observed since ancient times. Artificial springs and ponds were described in [6] that accumulated moisture from the air during dewfall. Currently, water harvesting from the air is considered a promising method for water supply in arid and inland regions [7]. There are two basic types of atmospheric water extraction: passive, which does not require additional energy input (rainwater, fog, and dew collection), and active, which can operate only with an external input of electric or thermal energy. The active method can be based on: (i) the air cooling below its dew point to condense water [8], and (ii) the sorption of moisture by solid [9,10] or liquid desiccants [11,12].

The Adsorptive Water Harvesting from the Atmosphere (AWHA) method is founded on the absolute humidity in arid regions, which is comparatively stable during the day. The air temperature and, consequently, the relative humidity opposite vary greatly. [13]. The AWHA process involves two main stages: (i) the adsorption of moisture on an adsorbent at night when relative humidity is high, and (ii) the desorption of water from the adsorbent during the day driven by an external heat source (e.g., solar heat) with subsequent vapor condensation on a cold surface [14]. The main performance indexes of the AWHA are the specific water productivity (SP) per cycle, or the ratio of <the mass of water collected>/<the adsorbent mass>, and the specific energy consumption (SEC), equal to <the energy input>/<the mass of water collected>. Considering a quite low moisture content of 4–15 g per 1 m^3 of air in arid regions [15–18], hundreds of cubic meters of air have to be passed through the adsorbent to extract several liters of water. This emphasizes the primary importance of the efficiencies of water extraction during adsorption and collection during the desorption/condensation stages. These efficiencies can be expressed as fractions: δ_{ex} = <amount of water adsorbed>/<amount of vapor passed through the adsorber> of water extraction, and δ_{col} = <amount of water collected>/<amount of water desorbed> of water collection [13].

The performance of AWHA is closely related to the used adsorbent, or, to be more precise, to the harmonization of its properties with the climatic conditions of the region where the process is implemented. A thermodynamic analysis showed that to enhance the performance of AWHA, the adsorbent should have adsorption centers with different affinities for water [13]. Adsorption sites with a strong affinity for water vapor effectively adsorb moisture at low relative pressure during the adsorption stage, and ensure high efficiency of water extraction. Adsorption sites with weak affinity easily release vapor during desorption at high relative pressure, and promote high water collection efficiency. The affinity should be distributed over a wide range, corresponding to the climate conditions of the region where the AWHA is realized. Such an adsorbent exhibits a divariant adsorption equilibrium, with uptake rising between relative humidity (*RH*) values, corresponding to the adsorption and desorption stages.

A number of adsorbents have been proposed for AWHA, including common porous solids like silica gel [19,20], crystalline alumino-phosphate (zeotype) AQSOA Z01 [21], composite sorbents involving hygroscopic salts embedded inside solid matrix [10,14,22,23]. A low water-adsorption capacity of silica gels and a high regeneration temperature of conventional zeolites have restricted the practical realization of AWEA. Several advanced zeotypes possess S-shaped adsorption isotherms, and can be regenerated at a low temperature. However, an analysis showed that their adsorption properties did not match the climatic conditions of the arid regions very well. Their drawbacks were a quite low amount of water exchanged under conditions of the AWHA cycle in arid regions (below 0.23 g/g, see Section 3.3.1) and expansive synthesis [24]. The "salt/matrix" composites possess a large adsorption capacity and variable adsorption properties. However, their use in open AWHA systems may be obstructed by the potential leakage of the salt solution, which is formed during the moisture adsorption, from the pore space [25]. This could lead to the adsorbent degradation and the adsorber corrosion.

Metal–organic frameworks (MOFs) are a novel family of crystalline compounds that consist of metal ions or clusters and bridging polydentate organic linkers. MOFs are characterized by one-, two-, or three-dimensional frameworks with a large fraction of voids [26–28]. Due to their extraordinarily surface area and pore volume, and the presence of both hydrophilic and hydrophobic parts, MOFs possess unique adsorption properties such as a large water-adsorption capacity and a great diversity of water-adsorption isotherms. [29,30]. Furthermore, the possibility to control MOFs' adsorption properties by modifying both organic linkers and inorganic clusters makes it possible to tailor the

MOFs with required sorption properties and hydrolytic stability [31,32], which inspired high hopes for enhancement of AWHA performance [33].

Seo et al. [34] first proposed mesoporous MIL-100 and MIL-101 as desiccants for AWHA. Since then, the study of AWHA employing MOFs has been attracting steadily increasing interest [4,13,35,36]. For a humid climate at the relative humidity of ambient air RH > 40–50%, MIL-100(Cr) and MIL-101(Cr) are promising adsorbents that display a high working capacity of up to 1.2 mL/g [37]. A Co_2Cl_2BTDD designed for AWHA was able to deliver 0.82 mL/g of water under moderately humid conditions at RH = 30% [38]. The performance of MOFs as adsorbents for arid climates is essentially lower. Kim et al. [39] showed that at the ambient RH = 20%, potentially ~0.2 mL/g per cycle can be harvested with MOF-801 by utilizing solar thermal energy to drive desorption. A microporous MOF-303 has recently been developed and tested at AWHA [40], and the water harvester operating for several cycles per day generated 0.7 and 1.3 L/kg per day at 10 and 32% relative humidity, respectively. MIL-160(Al) was suggested as a promising adsorbent for AWHA regions with arid climates [13]. Chang et al. [41] evaluated the potential of application of MIL-160(Al) as a material for water harvesting in a bench-scale fixed-bed unit, and demonstrated its capable of maximum water productivity of 305 L/(day·ton) at regeneration and condensation temperatures of 80 °C and 10 °C, respectively.

This research focuses on the study of MIL-160 as an adsorbent for the AWHA process and evaluation of its potential under climatic conditions of extremely arid regions (the Mojave and Sahara Deserts, Chile, Algeria, the central part of Saudi Arabia, and Mongolia). The performance characteristics of AWHA employing MIL-160 were estimated in terms of the specific consumption of thermal energy for the water production, the specific mass of the water harvested, and the fractions of water extracted from the air and water collected. These fractions are important performance indexes, as they determine the specific electric energy consumption for air blowing through the fixed-bed adsorber, which can be significant due to a small moisture content in the atmosphere in arid regions.

The structure of MIL-160 is formed by inorganic AlO_6 octahedra, which form cis-corner-sharing chains linked via carboxylate groups of the ligand (2,5-furandicarboxylic acid) (Figure 1) [42]. It possesses both strong and weak adsorption sites, which allows AWHS to promote high efficiencies of water extraction and collection.

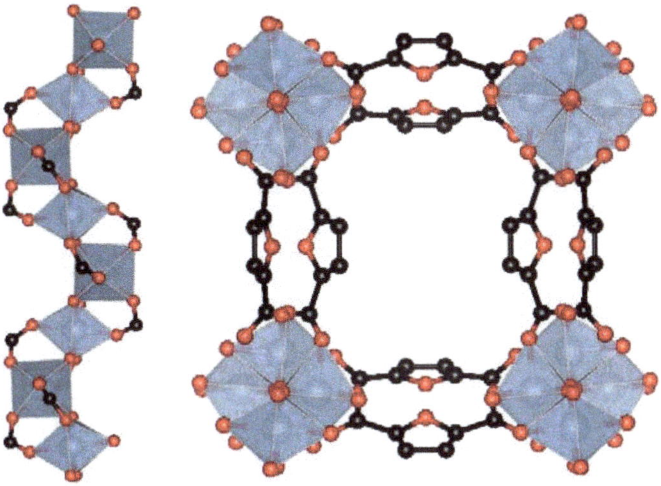

Figure 1. Scheme of the MIL-160 structure: helical inorganic chains formed by AlO6 octahedra (**left**); channels in framework constructed from inorganic chains linked via carboxylate groups of the 2,5-furandicarboxylic acid (**right**). AlO_6 octahedra (light gray), atoms C (black), O (red).

2. Materials and Methods

2.1. Sample Synthesis

MIL-160 was synthesized by a hydrothermal method according to a slightly modified procedure described in [42]. 2,5-furandircaboxilic acid (4.680 g, 30 mmol, >99%), $AlCl_3 \cdot 6H_2O$ (7.240 g, 30 mmol), and NaOH (1.212 g, 30 mmol) were dissolved in distilled H_2O (60 mL) with stirring using a magnetic stirrer (~500 rpm) for 15 min. The obtained mixture of reagents was placed in a Teflon container inside a 100 mL cylindrical stainless-steel autoclave and heated at temperature T = 120 °C for 24 h in a convection oven. After cooling down to the ambient temperature, the mixture was separated from the solution with a small-pore paper filter using a water-jet pump and washed with hot water to remove the unreacted organic ligand. Finally, the obtained white solid precipitate was activated at 150 °C for 15 h under continuous evacuation.

2.2. Characterization

The structure of the obtained MIL-160 sample was confirmed by PXRD analysis using a Bruker D8 ADVANCE diffractometer with an XRK-900reaction chamber. The measurements were carried out using CuK_α-radiation (λ = 1.54 Å) in scanning mode (2θ = 5–60°), with a step width of 2θ = 0.02° and an accumulation time of 10 s at each point.

IR-spectra were recorded on an Agilent Cary 660 FT-IR spectrometer using an attenuated total reflection (ATR) attachment, the 025-2018 MIRacle ZnSe Perf Crystal Plate, at 25 °C in the 500–4000 cm^{-1} range.

The BET surface area, the micropore volume, and total pore volume of the synthesized MIL-160 were determined by low-temperature nitrogen adsorption at 77 K using a Quantachrome Nova 1200e gas sorption analyzer. The samples were degassed in a vacuum at 150 °C for 3 h before measurement of N_2 sorption isotherms. The specific surface area was calculated by the BET analysis of the adsorption branch of the isotherm in a relative pressure range of 0.01–0.025. The total pore volume was obtained from the amount of N_2 adsorbed at a relative pressure close to unit P/P_0 = 0.99. The micropore volume was calculated using the statistical thickness analysis of the isotherm adsorption branch and de Boer's t-method. The pore-size distribution was calculated using the Dubinin–Astakhov method.

2.3. Water-Adsorption Equilibrium

The water-adsorption isotherms of MIL-160 were measured by a volumetric method at T = 50, 35, and 20 °C and in the water-vapor pressure range P = 0.8–40.0 mbar. Before the measurements, a preliminary degassing was carried out at 80 °C for 3 h under continuous evacuation (the residual pressure was 0.1 mbar) for drying the MIL-160 sample. The dry sample weight m_0 was 0.50216 ± 0.00002 g. Afterward, the sample was cooled down to a fixed temperature, which was controlled by a thermostat connected to the adsorber with an accuracy of ±0.1 K. The temperature was measured with a K-type thermocouple placed at the bottom of the adsorber. Then, the measuring cell with the adsorbent was connected to the buffering vessel filled with water vapor, which resulted in a jump in water-vapor pressure over the sample. The vapor pressure was measured using an MKS Baratron® Type 626a pressure sensor with an accuracy of ± 0.01 mbar. The vapor-pressure jump initiated the adsorption, which led to a gradual decrease in the pressure. The sample was maintained at a fixed temperature for 1–4 h to reach equilibrium. The amount of water vapor sorbed (Δw (g/g)) was calculated from the temporal pressure dependence $P(t)$ using the ideal gas equation:

$$\Delta w(t) = \frac{M_{H2O}V}{m_0 RT}(\Delta P(t)) \tag{1}$$

where M_{H2O}—the water molar mass, V—the system volume, —the universal gas constant, and T—the steam temperature. Subsequently, based on the data obtained, a set of water-sorption isotherms $w(P)$ of the MIL-160 sample was calculated.

3. Results and Discussions

3.1. Structure Characterization of as-Prepared MIL-160

The texture characteristics of MOF-801 determined from the N_2 sorption isotherm at 77 K (Figure 2a) belonged to the I-type according to the IUPAC classification, and the hysteresis loop was absent, which is typical for microporous solids. The synthesized MIL-160 possessed the specific surface area S_{BET} = 830 m^2/g and the total pore volume V_p = 0.40 cm^3/g, which agreed with the literature data [42]. The micropore volume V_μ coincided with the V_p. Hence, the synthesized MIL-160 was microporous without meso- or macropores. The pore-size distribution obtained by the Dubinin–Astakhov method (Figure 2b) demonstrated a narrow peak in the micropore-size region. As shown in the curve obtained, the average pore diameter is 0.42 nm, which is consistent with the data presented in [42].

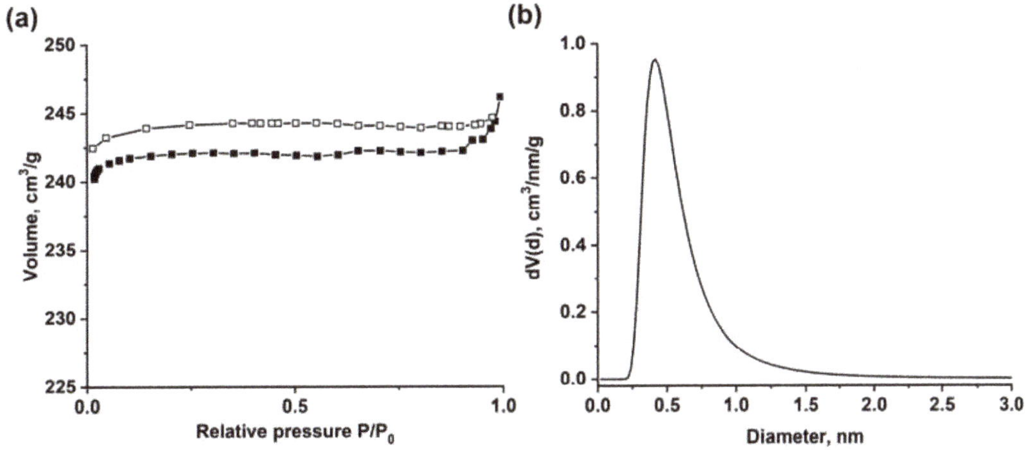

Figure 2. N_2 adsorption (solid symbols) and desorption (open symbols) isotherms (**a**) and pore-size distribution using DA model (**b**) of MIL-160.

The XRD pattern (Figure 3a) of the synthesized sample was consistent with that reported in the literature [42,43]. The FT-IR spectrum of the prepared MIL-160 sample (Figure 3b) agreed well with the literature data [41,43–45]. The characteristic bands with high intensity were attributed to asymmetric (1649 cm^{-1}) and symmetric stretching (1405 cm^{-1}) vibrations of carboxylate groups, along with the C=C bonds' stretching vibrations of the furan ring at 1583 cm^{-1} [41,44,45]. The strong band at 630 cm^{-1} corresponded to the vibrations of the Al–O bond in the MIL-160 structure [41]. The peaks in the range of 1000–1300 cm^{-1} could be attributed to the asymmetric and symmetric stretching vibrations of the C–O–C in the furan rings [41]. A band at 782 cm^{-1} ascribed to the out-of-plane deformation vibrations of C–H bonds in the furan ring also was observed [44]. The broad band around 3500 cm^{-1} was assigned to the vibrations of the water molecules adsorbed from the ambient air, while the weak peak around 3200 cm^{-1} represented the vibration of bridging hydroxyl groups μ-OH in the metal–oxide clusters AlO_6 [41,44]. It is worth noting that the lack of a band at 1700 cm^{-1} indicated the absence of residual free 2,5-furandicarboxylic acid in the pore-synthesized MIL-160 [44,45]. Thus, the characterization of the synthesized MIL-160 by XRD, nitrogen adsorption, and FTIR analysis verified its genuine structure and high purity.

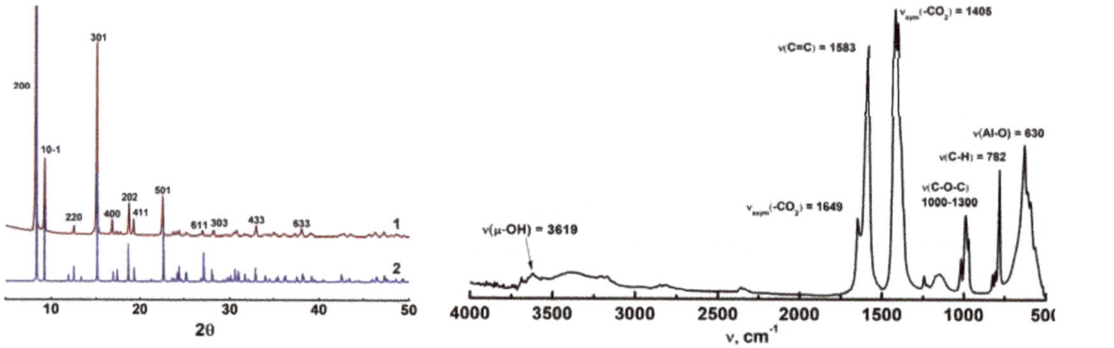

(a) (b)

Figure 3. XRD patterns (**a**) of the synthesized MIL-160 (2) and the curve calculated from CIF-file (1); FT-IR spectra (**b**) of MIL-160.

3.2. Water-Vapor Adsorption on MIL-160

The isotherms of water adsorption on the MIL-160 were S-shaped curves (Figure 4) showing a small water uptake (0.02 g_{H2O}/g_{ads}) at low pressures, followed by a steep increase in uptake up to 0.25–0.30 g_{H2O}/g_{ads} in a narrow pressure range, which depended on the temperature. Upon further increase in vapor pressure, the uptake gradually rose to 0.34 g_{H2O}/g_{ads}, which was somewhat smaller than the specific volume of micropores V_μ = 0.40 cm^3/g. This indicated that a fraction of the pore space in the vicinity of hydrophobic sites (organic linker) remained unoccupied by water molecules.

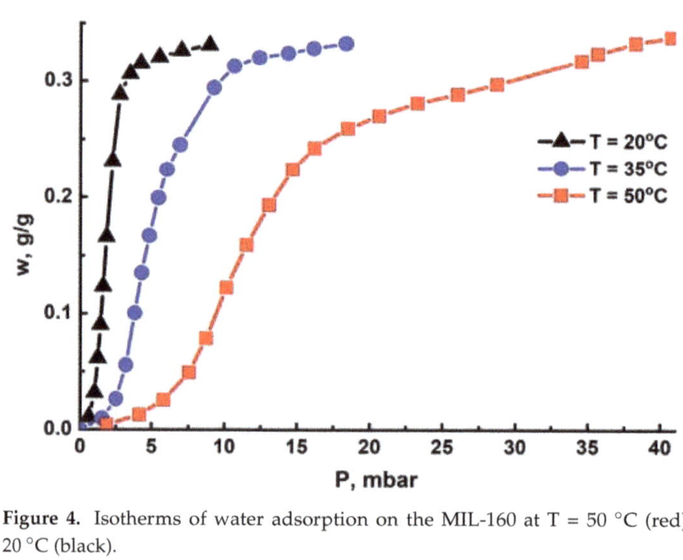

Figure 4. Isotherms of water adsorption on the MIL-160 at T = 50 °C (red), 35 °C (blue), and 20 °C (black).

Based on the obtained data on water adsorption equilibrium on MIL-160, isosteres of water adsorption were plotted in $\ln(P) - 1/T$ coordinates (Figure 5a). The isosteric heat ΔH_{ads} of the water adsorption was calculated according to the Clausius–Clapeyron equation:

$$\ln P = Q_{a\partial c}/(RT) + \text{Const} \quad (2)$$

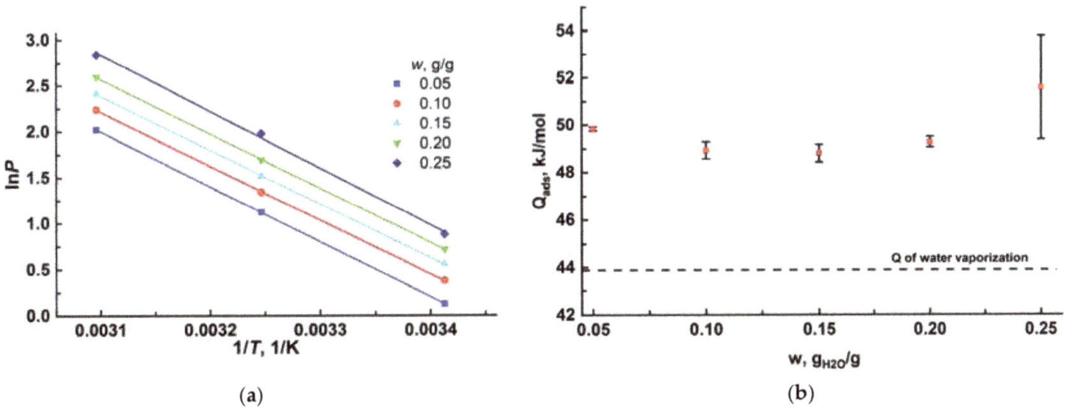

Figure 5. Water adsorption isosteres (**a**) and isosteric enthalpy ΔH_{ads} (**b**) for the MIL-160.

The adsorption heat Q_{adc} varied in the range of 49 to 52 ± 1 kJ/mol at w = 0.05–0.25 g/g (Figure 5b). The obtained values of adsorption heat somewhat exceeded the latent condensation heat of water ΔL = 43.9 kJ mol^{-1} at 303 K, which pointed to a moderately strong interaction of water molecules with the MIL-160's surface.

Experimentally measured water adsorption isobars on the MIL-160 presented as a function of the Polanyi adsorption potential $\Delta F = -RT\ln[P/P_0(T)]$, where P is the partial vapor pressure, P_0 is the saturation vapor pressure at temperature T, merged into one characteristic curve $w(\Delta F)$ (Figure 6). This meant that the water adsorption equilibrium of the MIL-160 obeyed the Polanyi principle of temperature invariance [46]. The characteristic curve $w(\Delta F)$ was approximated by an empiric equation:

$$w = \frac{A\Delta F + B}{1 + e^{(-k_1-(\Delta F - x_{01})}} + \frac{C}{1 + e^{(-k_2-(\Delta F - x_{02})}} \quad (3)$$

with the fitting parameters $A = -1.321{\cdot}10^{-3}$, $B = 1.451{\cdot}10^{-2}$, $C = 0.321$, $k_1 = -10.00$, $k_2 = -1.470$, $x_{01} = 6.875$, and $x_{02} = 6.268$.

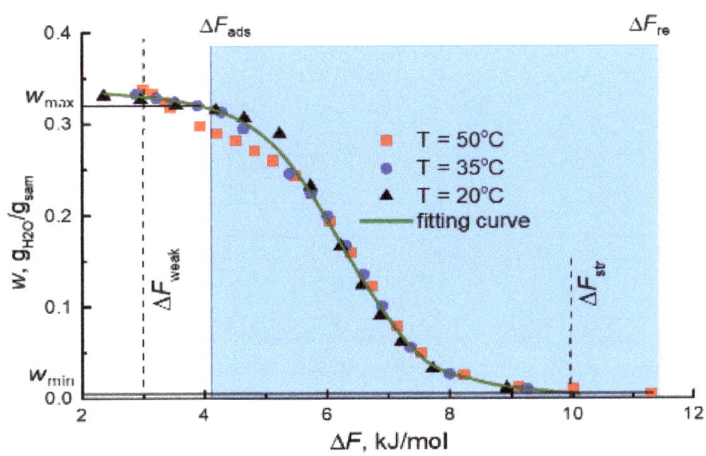

Figure 6. Characteristic curve of water adsorption on the MIL-160 (symbols) and the fitting curve (solid line). Dashed lines indicate the values of adsorption potential, corresponding to the weakest (ΔF_{weak}) and strongest (ΔF_{str}) adsorption sites. The blue area shows the ΔF range for the boundary conditions of the AWHA cycle in Riyadh-Old during the dry season (July).

3.3. Performance of AWHA Employing MIL-160

3.3.1. Specific Water Productivity

The characteristic curve $w(\Delta F)$ was used to evaluate the performance of AWHA employing MIL-160 for climatic conditions of several arid regions located on different continents (Figure 7). The Mojave Desert in North America; the Atacama in Chile, South America; Tamanrasset in Algeria; the Sahara Desert; Riyadh-Old in the central part of Saudi Arabia; and Noyon in Mongolia were selected as reference regions, characterized by their extremely arid climates. The climatic data (the average temperatures T_n and T_d, and the average relative humidity RH_n and RH_d during night and day, respectively) were collected from the Meteonorm Global Climate Database. They were used to calculate the values of adsorption potential ΔF_{ads} and ΔF_{re} (Table 1), corresponding to the boundary conditions of adsorption and desorption stages, respectively:

$$\Delta F_{ad} = -RT_n[\ln P_{am}/P_0(T_n)] \quad (4)$$

$$\Delta F_{re} = -RT_{re}\,[\ln P_{am}/P_0(T_{re})], \quad (5)$$

where T_{re} is the regeneration temperature.

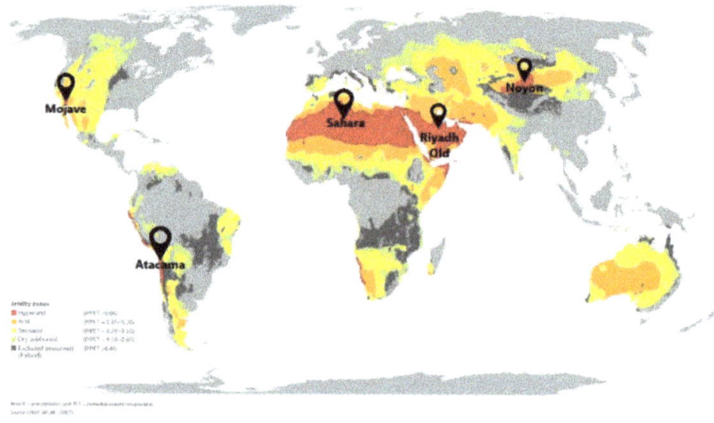

Figure 7. Selected reference regions with extremely arid climates [47].

Table 1. Climatic data for the selected regions.

Region	T_n, °C	RH_n, %	T_d, °C	RH_d, %	P_{am}, mbar	ΔF_{ad}, kJ/mol	ΔF_{re}, kJ/mol (T_{re} = 80 °C)
					January		
Riyadh-Old	12.5	55.9	16.3	45.1	8.3	1.3	11.9
Sahara	11.1	33.0	14.9	27.2	4.5	2.5	13.7
Noyon	−13.0	55.6	−9.1	41.7	1.3	1.3	17.4
Atacama	5.3	73.0	9.4	55.5	6.5	0.7	12.6
Mojave	3.0	26.3	7.6	17.9	6.8	1.5	14.1
					June/July		
Riyadh-Old	32.6	20.2	36.7	16.5	10.1	4.1	11.3
Sahara	26.9	25.8	30.4	21.2	9.3	3.4	11.5
Noyon	8.6	25.8	13.7	16.4	2.9	3.2	15.0
Atacama	−1.7	33.1	5.1	20.0	1.8	2.5	16.4
Mojave	21.6	26.3	27.5	17.9	6.9	3.3	12.6

The characteristic curve of water adsorption on the MIL-160 demonstrated that the strongest adsorption sites adsorbed water vapor at $\Delta F_{str} \approx 9.0$–$10.0$ kJ/mol (Figure 6), which was lower than $\Delta F_{re} = 11.3$–17.4 kJ/mol at $T_{re} = 80$ °C in all the reference regions (Table 1). Thus, the retained water could be desorbed completely at a quite low temperature of 80 °C, allowing usage of a simple solar collector for the adsorbent regeneration.

The maximum specific water productivity SPmax per cycle can be achieved in the AWHA process if water retained in the adsorbent during the adsorption stage is completely removed and collected during the desorption/condensation stage. The productivity SPmax was calculated from the characteristic curve (Figure 6) as the uptake variation $\Delta w = w_{max} - w_{min} = w(\Delta F_{ad}) - w(\Delta F_{re})$. In all the selected regions, MIL-160 exchanged $\Delta w = 0.31$–0.33 g_{H2O}/g_{ads} at $T_{re} = 80$ °C, which surpassed respective values for various adsorbents, both traditional and novel (Figure 8).

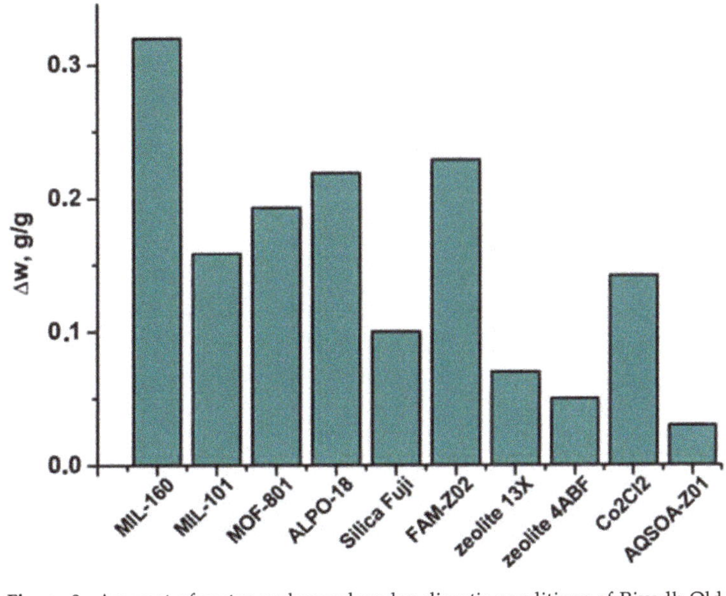

Figure 8. Amount of water exchanged under climatic conditions of Riyadh-Old during the dry season for various adsorbents.

3.3.2. The Fractions of Water Extraction and Collection

Along with the specific water productivity, the fractions δ_{ex} of water extracted from the air during adsorption and δ_{col} of water collected during condensation are also the crucial performance indexes of AWHA. They determine the volume of air to be passed through the adsorber, for harvesting a unit mass of water, and consequently, the energy demand of the system for the air blowing.

The fractions of water extracted δ_{ex} and collected δ_{col} in a simple fixed-bed flowing adsorber (Figure 9a) with MIL-160 as the adsorbent can be evaluated as [13]:

$$\delta_{ex} = (P_{am} - P_{out.ad})/P_{am} = 1 - P_{out.ad}/P_{am} \quad (6)$$

$$\delta_{col} = [P_{out.re} - P_0(T_d)]/P_{out.re} = 1 - P_0(T_d)/P_{out.re} \quad (7)$$

where P_{am}—the partial pressure of water vapor in the ambient air; $P_{out.ad}$—the water vapor pressure in the outlet air during adsorption stage; $P_{out.re}$—the water vapor pressure in the outlet air during regeneration stage; and $P_0(T_d)$—the pressure of saturated water vapor at a day temperature of ambient air T_d.

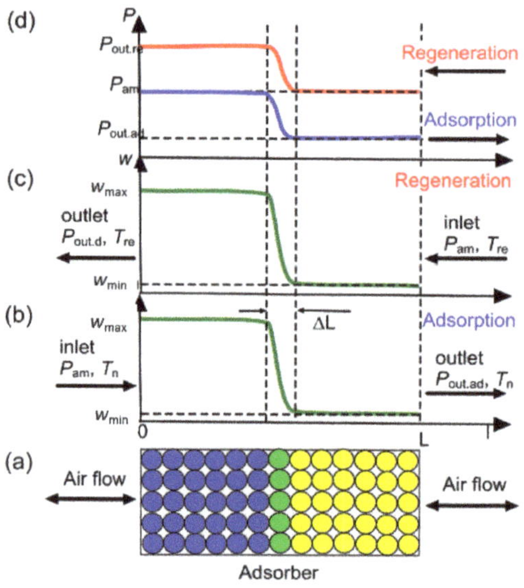

Figure 9. Scheme of a fixed-bed flowing adsorber (**a**) and distribution of water uptake (**b**,**c**) and vapor partial pressure (**d**) along the adsorber.

Here, we assumed that the adsorption front thickness ΔL in a fixed-bed adsorber was negligible in comparison with the adsorber length $L >> \Delta L$ (Figure 9a–c). In this case, the dynamic adsorption capacity w_d of the adsorbent equaled the equilibrium capacity $w(P,T)$. Until the concentration front was inside the bed, the outlet vapor pressure $P_{out.ad}$ during adsorption equaled the equilibrium pressure over the dry adsorbent (Figure 9d). The strongest adsorption centers of the MIL-160 adsorbed water vapor at $\Delta F_{str} \approx 10.0$ kJ/mol (Figure 6). Consequently, the outlet vapor pressure $P_{out.ad}$ during the adsorption stage could be calculated as:

$$P_{out.ad} = P_0(T_n)\exp(-\Delta F_{str}/RT_n) \quad (8)$$

The δ_{ex} values evaluated according to Equations (6) and (8) varied from 0.90 for Riyadh-Old to 0.96 for the Atacama regions during the dry season (Figure 10). During the humid season, δ_{ex} was even higher due to the increase in RH of the ambient air (Figure 10), and reached 0.96–0.98 for the selected regions. Consequently, the affinity of the strongest adsorption sites MIL-160 to water vapor provided effective extraction of water vapor from the ambient air. That promoted a reduced air volume blown through the adsorber to extract a unit volume of water, and consequently, lowering energy consumption for the air purging.

During the regeneration stage, the outlet air was in equilibrium with the adsorbent, saturated with water up to w_{max} right after the adsorption stage (Figure 9c,d). The weakest adsorption sites adsorbed water vapor at $\Delta F_{weak} \approx 3.0$ kJ/mol (Figure 6). Consequently, at $\Delta F > \Delta F_{weak}$, these sites released adsorbed water, and the outlet vapor pressure $P_{out.re}$ during the desorption stage could be calculated as:

$$P_{out.re} = P_0(T_{re})\exp(-\Delta F_{weak}/RT_{re}) \quad (9)$$

It should be noted that for the dry season, the $\Delta F_{ad} > \Delta F_{weak}$ for all regions except Atacama (Table 1). This meant that the weakest adsorption sites remained unsaturated with water during the adsorption stage, and maximum uptake in the cycle equaled $w = w(\Delta F_{ad})$. Accordingly, for the dry season, the pressure $P_{out.re}$ was calculated using the following expression:

$$P_{out.re} = P_0(T_{re})\exp(-\Delta F_{ad}/RT_{re}) \quad (10)$$

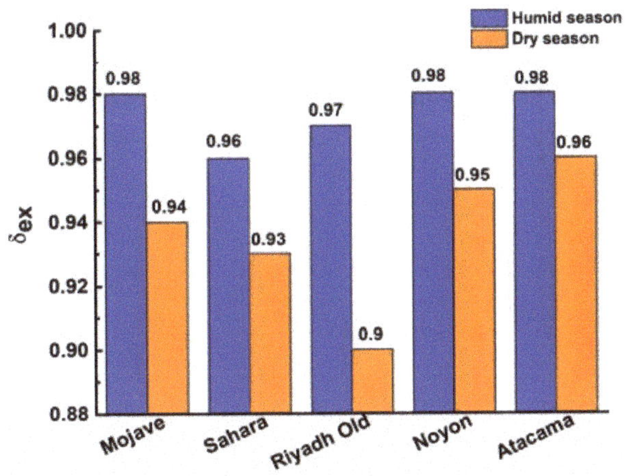

Figure 10. Water-extraction fraction δ_{ex} for the AWHA system based on MIL-160.

Then, the fraction δ_{col} was evaluated as a function of temperature T_{re} using Equation (7). If the condenser was cooled by the ambient air at T_d, δ_{col} values for the selected regions varied in the range 0.48–0.95 and 0.88–0.97 during the dry (July) and humid (January) seasons, respectively, at T_{re} = 80 °C (Figure 11). The increase in the regeneration temperature enhanced the water collection fraction: at T_{re} = 100 °C, the collection fraction rose to 0.77–0.97 and 0.95–0.98 for the dry and humid seasons, respectively. On the contrary, lowering the regeneration temperature to 70 °C resulted in a dramatic reduction of this fraction, particularly for the dry season, down to 0.15–0.9. This showed the necessity of an external heat source for the water desorption, which could be solar or waste heat.

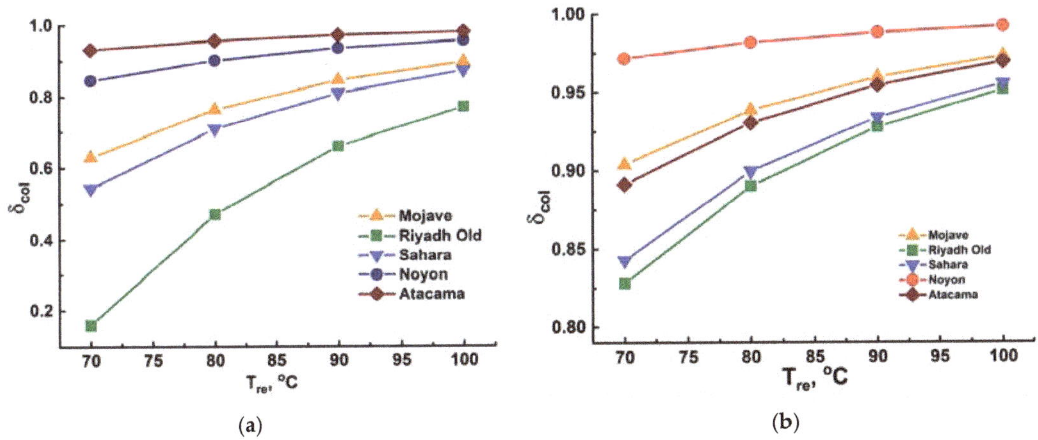

Figure 11. Effect of the regeneration temperature on the water-collection fraction δ_{col} of the AWHA system based on MIL-160 during the dry (**a**) and humid (**b**) seasons, with the condenser cooled by the ambient air.

Another efficient method of increasing the fraction δ_{col} is a decrease in the condensation temperature T_{con}. A tank with a heat exchanger located underground or a condenser connected to a heat pump or adsorptive chiller can be used as a condenser [48,49] instead of that cooled by the ambient air. Figure 12 shows that reducing T_{con} from T_d = 36.1 °C to 10 °C allowed the growth of δ_{col} from 0.48–0.77 to 0.90–0.95 at T_{re} = 80–100 °C under the climatic conditions of Riyadh-Old during the dry season.

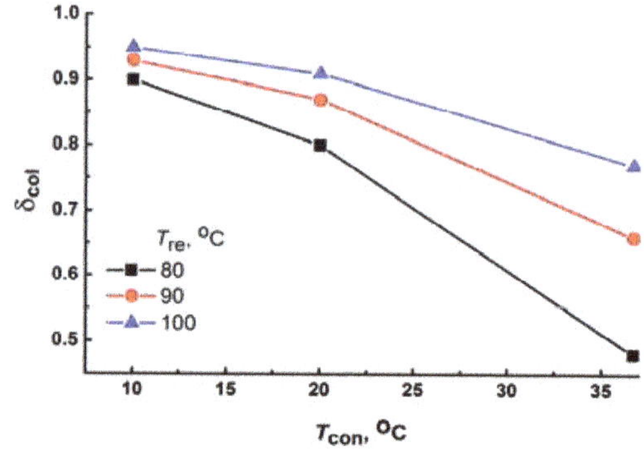

Figure 12. Effect of the condensation temperature on the water-collection fraction δ_{col} of the AWHA system based on MIL-160 during the dry season (July) for the Riyadh-Old region.

Thus, the calculated δ_{ex} and δ_{col} values showed a high performance of the AWHA system employing MIL-160 as an adsorbent. This MOF possessed both strong and weak adsorption sites, exchanged up to 0.33 g_{H2O}/g_{ads} under climatic conditions of the selected arid regions, and provided high fractions δ_{ex} = 0.90–0.98 and δ_{col} = 0.48–0.97 at T_{re} = 80–100 °C.

3.3.3. The Specific Energy Consumption

An important performance index of the AWHA system is the specific consumption of thermal energy SEC for the water production, which can be calculated as:

$$\mathrm{SEC} = Q_{sp.re}/(\Delta w \cdot \delta_{col}) = [\Delta w \cdot Q_{ads} + (C_{pad} + w_{ads} \cdot C_{pH2O}) \cdot (T_{re} - T_{ad})]/(\Delta w \cdot \delta_{col}), \quad (11)$$

where C_{pad} = 1.63 J/(g·K) is the specific heat of MIL-160 at 80 °C [50]), C_{pH2O} = 4.19 J/(g·K) is the specific heat of water at 80 °C, and Q_{ads} = 2.84 kJ/g is the average isosteric heat of adsorption in the range w = 0.05–0.30 g_{H2O}/g_{ads}.

For the selected regions during the dry season, the SEC ranged from 3.7 to 6.8 kJ/g_{H2O} at a regeneration temperature of 80 °C (Figure 13). The SEC was affected by the water-collection fraction δ_{col} (Equation (11)). Accordingly, during the humid season (January), SEC decreased to 3.5–3.8 kJ/g_{H2O} due to a higher water-collection fraction δ_{col}. The decrease in condensation temperature was another way to reduce SEC. Thus, when the condenser was cooled to 10 °C, SEC for the Riyadh-Old region was reduced to 3.6 kJ/g_{H2O} as compared to 6.8 kJ/g_{H2O} for the condenser cooled by the ambient air at temperature T_d (Table 2), owing to an appropriate increase in the water-collection fraction δ_{col}.

Table 2. SEC for water production in the AWHA process employing MIL-160 in the Riyadh-Old region at T_{re} = 80 °C (dry season).

$T_{con} = T_d$ = 33.1 °C	T_{con} = 20 °C	T_{con} = 10 °C
6.8	4.1	3.6

These SEC values were higher than the heat of water vaporization (2.26 kJ/g_{H2O} at 100 °C). Therefore, the energy consumption for AWEA was higher than the energy consumption for conventional water desalination. However, taking into account that in arid regions, solar heat is usually available in abundance and can be used for adsorbent regeneration, the obtained SEC-values for water production were quite acceptable. Accordingly, the AWEA method employing MIL-160 as an adsorbent is promising for arid regions remote from other sources of water.

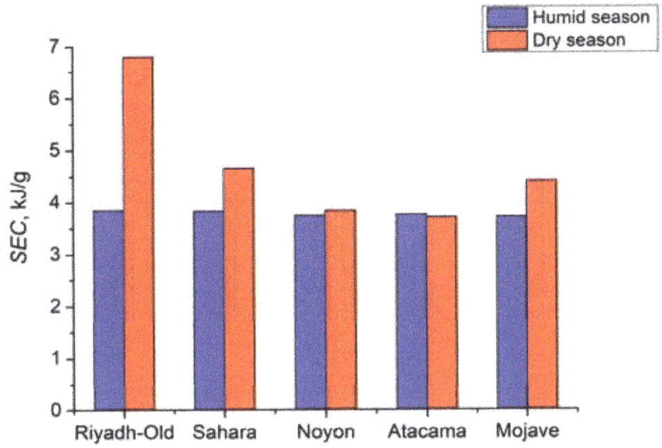

Figure 13. SEC for water production in the AWHA process employing MIL-160 as an adsorbent at T_{re} = 80 °C and $T_{con} = T_d$.

4. Conclusions

In this paper, the water-vapor adsorption on MIL-160 was studied, and the assessment of the feasibility of MIL-160 for the AWHA process in arid climatic regions was carried out. The water-vapor adsorption isotherms were S-shaped curves with a maximum uptake of 0.34 g/g. The isosteric heat of water adsorption equaled 49–52 $\pm$ 1 kJ/mol at a water-uptake range of 0.05 to 0.25 g/g. Under conditions of several arid regions, namely Riyadh-Old (Saudi Arabia), the Sahara Desert, the Mojave Desert, Atacama (Chile), and Noyon (Mongolia), a high maximum specific water productivity of 0.31–0.33 g_{H2O}/g_{ads} per cycle could be achieved with MIL-160. Employing MIL-160 allows a fraction of water extracted δ_{ex} = 0.90–0.98 during the adsorption stage. The fraction of water collected varied in the range of 0.48–0.76 and 0.85–0.97 during the dry and humid seasons, respectively, at T_{re} = 80 °C with natural cooling of the condenser by ambient air. Further increase in the fraction of water collected could be achieved by lowering the condensation temperature to 10 °C and increasing the regeneration temperature to 100 °C. The specific energy consumption for water production varied from 3.5 to 6.8 kJ/g. This is quite acceptable if solar heat, available in abundance in arid regions, can be used to drive the desorption. The AWHA method employing MIL-160 is a promising way to achieve a fresh water supply in arid areas remote from coast-line.

Author Contributions: Conceptualization, L.G. and Y.A.; methodology, L.G. and Y.A.; formal analysis, M.S.; investigation, I.K. and M.S.; resources, L.G; data curation, L.G.; writing—original draft preparation, I.K. and M.S.; writing—review and editing, L.G. and Y.A.; visualization, I.K. and M.S.; funding acquisition, L.G. All authors have read and agreed to the published version of the manuscript.

Funding: This research was funded by the Russian Foundation for Basic Research (grant no. 18-29-04033) and by the Ministry of Science and Higher Education of the Russian Federation within the governmental order for Boreskov Institute of Catalysis (project AAAA-A21-121011390006-0).

Institutional Review Board Statement: Not applicable.

Informed Consent Statement: Not applicable.

Data Availability Statement: The data presented in this study are available on request from the corresponding author.

Acknowledgments: The authors thank T.A. Krieger for XRD analysis of the synthesized MIL-160.

Conflicts of Interest: The authors declare no conflict of interest.

References

1. Zhuang, S.; Qi, H.; Wang, X.; Li, X.; Liu, K.; Liu, J.; Zhang, H. Advances in Solar-Driven Hygroscopic Water Harvesting. *Glob. Chall.* **2021**, *5*, 2000085. [CrossRef]
2. Arnell, N.W. Climate change and global water resources: SRES emissions and socio-economic scenarios. *Glob. Environ. Chang.* **2004**, *14*, 31–52. [CrossRef]
3. Vörösmarty, C.J.; Green, P.; Salisbury, J.; Lammers, R.B. Global water resources: Vulnerability from climate change and population growth. *Science* **2000**, *289*, 284–288. [CrossRef]
4. Kalmutzki, M.J.; Diercks, C.S.; Yaghi, O.M. Metal–Organic Frameworks for Water Harvesting from Air. *Adv. Mater.* **2018**, *30*, 1–26. [CrossRef]
5. Van der Bruggen, B.; Vandecasteele, C. Distillation vs. membrane filtration: Overview of process evolutions in seawater desalination. *Desalination* **2002**, *143*, 207–218. [CrossRef]
6. Beysens, D.; Milimouk, I. The case for alternative fresh water sources. *Sécheresse* **2000**, *11*, 11–16.
7. Tu, Y.D.; Wang, R.Z.; Zhang, Y.N.; Wang, J.Y. Progress and expectation of atmospheric water harvesting. *Joule* **2018**, *2*, 1452–1475. [CrossRef]
8. Al-Farayedhi, A.A.; Ibrahim, N.I.; Gandhidasan, P. Condensate as a water source from vapor compression systems in hot and humid regions. *Desalination* **2014**, *349*, 60–67. [CrossRef]
9. Alayli, Y.; Hadji, N.E.; Leblond, J. A new process for extraction of water from air. *Desalination* **1987**, *67*, 227–229. [CrossRef]
10. Gordeeva, L.G.; Tokarev, M.M.; Parmon, B.N.; Aristov, Y.I. Selective water sorbents for multiple applications: 6. Fresh water production from the atmosphere. *React. Kinet. Catal. Lett.* **1998**, *65*, 153–160. [CrossRef]
11. Talaat, M.A.; Awad, M.M.; Zeidan, E.B.; Hamed, A.M. Solar-powered portable apparatus for extracting water from air using desiccant solution. *Renew. Energy* **2018**, *119*, 662–674. [CrossRef]
12. Hamed, A.M. Absorption-regeneration cycle for production of water from air-theoretical approach. *Renew. Energy* **2000**, *19*, 625–635. [CrossRef]
13. Gordeeva, L.G.; Solovieva, M.V.; Aristov, Y.I. Potable water extraction from the atmosphere: Potential of MOFs. *Renew. Energy* **2020**, *148*, 72–80. [CrossRef]
14. Aristov, Y.I.; Tokarev, M.M.; Gordeeva, L.G.; Snytnikov, V.N.; Parmon, V.N. New composite sorbents for solar-driven technology of fresh water production from the atmosphere. *Sol. Energy* **1999**, *66*, 165–168. [CrossRef]
15. Arakawa, H. Climates of Northern and Eastern Asia. *World Surv. Climatol.* **1969**, *8*, 1–248.
16. Takahashi, K.; Arakawa, H. Climates of Southern and Western Asia. *World Surv. Climatol.* **1981**, *9*, 1–333.
17. Griffiths, J.F. Climates of Africa. *World Surv. Climatol.* **1972**, *10*, 1–604.
18. Gentilli, J. Climates of Australia and New Zealand. *World Surv. Climatol.* **1971**, *9*, 1–405.
19. Groth, W.; Hussmann, P. Process and system for recovering water from the atmosphere. U.S. Patent 4 146 372, 27 March 1979.
20. Bennett, C.E. Heat energized vapor adsorbent pump. U.S. Patent 4,377,398, 22 March 1983.
21. LaPotin, A.; Zhong, Y.; Zhang, L.; Zhao, L.; Leroy, A.; Kim, H.; Rao, S.R.; Wang, E.N. Dual-Stage Atmospheric Water Harvesting Device for Scalable Solar-Driven Water Production. *Joule* **2021**, *5*, 166–182. [CrossRef]
22. Kumar, M.; Yadav, A. Experimental investigation of solar powered water production from atmospheric air by using composite desiccant material "$CaCl_2$/saw wood". *Desalination* **2015**, *367*, 216–222. [CrossRef]
23. Wang, J.Y.; Liu, J.Y.; Wang, R.Z.; Wang, L.W. Experimental investigation on two solar-driven sorption based devices to extract fresh water from atmosphere. *Appl. Therm. Eng.* **2017**, *127*, 1608–1616. [CrossRef]
24. Shimooka, S.; Oshima, K.; Hidaka, H.; Takewaki, T.; Kakkiuchi, H.; Kodama, A.; Kubota, M.; Matsuda, H. The evaluation of direct cooling and heating desiccant devices coated with FAM. *J. Chem. Eng. Jpn.* **2007**, *40*, 1330–1334. [CrossRef]
25. Gordeeva, L.G.; Aristov, Y.I. Composites "salt inside porous matrix" for adsorption heat transformation: A current state of the art and new trends. *Int. J. Low-Carbon Technol.* **2012**, *7*, 288–302. [CrossRef]
26. Férey, G. Hybrid porous solids: Past, present, future. *Chem. Soc. Rev.* **2008**, *37*, 191–214. [CrossRef]
27. Horike, S.; Shimomura, S.; Kitagawa, S. Soft porous crystals. *Nat. Chem.* **2009**, *1*, 695–704. [CrossRef] [PubMed]
28. Yaghi, O.M.; O'Keeffe, M.; Ockwing, N.W.; Chae, H.K.; Eddaoudi, M.; Kim, J. Reticular synthesis and the design of new materials. *Nature* **2003**, *423*, 705–714. [CrossRef]
29. Canivet, J.; Fateeva, A.; Guo, Y.; Coasne, B.; Farrusseng, D. Water adsorption in MOFs: Fundamentals and applications. *Chem. Soc. Rev.* **2014**, *43*, 5594–5617. [CrossRef]
30. De Lange, M.F.; Verouden, K.J.F.M.; Vlugt, T.J.H.; Gascon, J.; Kapteijn, F. Adsorption driven heat pumps—The potential of Metal-Organic Frameworks. *Langmuir* **2015**, *31*, 12783–12796. [CrossRef]
31. Song, D.; Bae, J.; Ji, H.; Kim, M.-B.; Bae, Y.-S.; Park, K.S.; Moon, D.; Jeong, N.C. Coordinative Reduction of Metal Nodes Enhances the Hydrolytic Stability of a Paddlewheel Metal−Organic Framework. *J. Am. Chem. Soc.* **2019**, *141*, 7853–7864. [CrossRef]
32. Bae, J.; Lee, C.Y.; Jeong, N.C. Weak Coordination Bond of Chloromethane: A Unique Way to Activate Metal Node Within an Unstable Metal–Organic Framework DUT-34. *Korean Chem. Soc.* **2021**, *42*, 658–666. [CrossRef]
33. Furukawa, H.; Cordova, K.E.; O'Keeffe, M.; Yaghi, O.M. The chemistry and applications of metal-organic frameworks. *Science* **2013**, *341*. [CrossRef] [PubMed]

34. Seo, Y.-K.; Yoon, J.W.; Lee, J.S.; Hwang, Y.K.; Jun, C.-H.; Chang, J.-S.; Wuttkee, S.; Bazin, P.; Vimont, M.; Daturi, M.; et al. Energy-efficient dehumidification over hierarchically porous metal-organic frameworks as advanced water adsorbents. *Adv. Mater.* **2012**, *24*, 806–810. [CrossRef] [PubMed]
35. LaPotin, A.; Kim, H.; Rao, S.R.; Wang, E.N. Adsorption-based atmospheric water harveisting: Impact of material and component properties on system-level performance. *Acc. Chem. Res.* **2019**, *52*, 1588–1597. [CrossRef]
36. Trapani, F.; Polyzoidis, A.; Loebbecke, S.; Piscopo, C.G. On the general water harvesting capability of metal-organic frameworks under well-defined climatic conditions. *Microporous Mesoporous Mater.* **2016**, *230*, 20–24. [CrossRef]
37. Kim, S.-I.; Yoon, T.-U.; Kim, M.-B.; Lee, S.-J.; Hwang, Y.K.; Chang, J.-S.; Kim, H.-J.; Lee, H.-N.; Lee, U.-H.; Bae, Y.-S. Metal–organic frameworks with high working capacities and cyclic hydrothermal stabilities for fresh water production. *Chem. Eng. J.* **2016**, *286*, 467–475. [CrossRef]
38. Rieth, A.J.; Yang, S.; Wang, E.N.; Dincă, M. Record Atmospheric Fresh Water Capture and Heat Transfer with a Material Operating at the Water Uptake Reversibility Limit. *ACS Cent. Sci.* **2017**, *3*, 668–672. [CrossRef]
39. Kim, H.; Yang, S.; Rao, S.R.; Narayanan, S.; Kapustin, E.A.; Furukawa, H.; Umans, A.S.; Yaghi, O.M.; Wang, E.N. Water harvesting from air with metal-organic frameworks powered by natural sunlight. *Science* **2017**, *356*, 430–434. [CrossRef]
40. Hanikel, N.; Prevot, M.S.; Fathieh, F.; Kapustin, E.A.; Lyu, H.; Wang, H.; Diercks, N.J.; Glover, T.G.; Yaghi, O.M. Rapid cycling and exceptional yield in a metal-organic framework water harvester. *ACS Cent. Sci.* **2019**, *5*, 1699–1706. [CrossRef]
41. Silva, M.P.; Ribeiro, A.M.; Silva, C.G.; Nogueira, I.B.R.; Cho, K.-H.; Lee, U.-H.; Faria, J.L.; Loureiro, J.L.; Chang, J.-S.; Rodrigues, A.E.; et al. MIL-160(Al) MOF's potential in adsorptive water harvesting. *Adsorption* **2021**, *27*, 213–226. [CrossRef]
42. Cadiau, A.; Lee, J.S.; Borges, D.D.; Fabry, P.; Devic, T.; Wharmby, M.T.; Martineau, C.; Foucher, D.; Taulelle, F.; Jun, C.-H.; et al. Design of Hydrophilic Metal-Organic Framework Water Adsorbents for Heat Reallocation. *Adv. Mater.* **2015**, *27*, 4775–4780. [CrossRef]
43. Permyakova, A.; Skrylnyk, O.; Courbon, E.; Affram, M.; Wang, S.; Lee, U.-H.; Valekar, A.H.; Nouar, F.; Mouchaham, G.; Devic, T.; et al. Synthesis Optimization, Shaping, and Heat Reallocation Evaluation of the Hydrophilic Metal–Organic Framework MIL-160(Al). *ChemSusChem* **2017**, *10*, 1419–1426. [CrossRef]
44. Wahiduzzaman, M.; Lenzen, D.; Maurin, G.; Stock, N.; Wharmby, M.T. Rietveld refinement of MIL-160 and its structural flexibility upon H_2O and N_2 adsorption. *Eur. J. Inorg. Chem.* **2018**, *32*, 3626–3632. [CrossRef]
45. Liu, Q.; Chapman, J.; Huang, A.; Williams, K.C.; Wagner, A.; Garapati, N.; Sierros, K.A.; Dinu, C.Z. User-Tailored Metal–Organic Frameworks as Supports for Carbonic Anhydrase. *ACS Appl. Mater. Interfaces* **2018**, *10*, 41326–41337. [CrossRef]
46. Polanyi, M. Theories of the adsorption of gas. *Trans. Faraday Soc.* **1932**, *28*, 316–333. [CrossRef]
47. Food and Agriculture Organisation of the United Nations. Available online: http://www.fao.org/fileadmin/user_upload/newsroom/docs/full-map.png (accessed on 2 April 2021).
48. Bar, E. Extraction of water from air—An alternative solution for water supply. *Desalination* **2004**, *165*, 335. [CrossRef]
49. Habeebullah, B.A. Potential use of evaporator coils for water extraction in hot and humid areas. *Desalination* **2009**, *237*, 330–345. [CrossRef]
50. Cui, S.; Marandi, A.; Lebourleux, G.; Thimon, M.; Bourdon, M.; Chen, C.; Severino, M.I.; Steggles, V.; Nouar, F.; Serre, C. Heat properties of a hydrophilic carboxylate-based MOF for water adsorption applications. *Appl. Therm. Eng.* **2019**, *161*, 114135. [CrossRef]

energies MDPI

Article

Performance Analysis of Variable Mode Adsorption Chiller at Different Recooling Water Temperatures

Ahmad A. Alsarayreh [1,*], Ayman Al-Maaitah [2], Menwer Attarakih [3] and Hans-Jörg Bart [4]

1 Precision Industries, Dubai P.O. Box 37448, United Arab Emirates
2 Wahaj Investment L.L.C., 24B St, Comm-365, Ind 2, Dubai P.O. Box 37448, United Arab Emirates; ayman@wahajsolar.com
3 Department of Chemical Engineering, University of Jordan, Queen Rania St, Amman 11942, Jordan; attarakih67@gmail.com
4 Thermische Verfahrenstechnik, TU Kaiserslautern, 67653 Kaiserslautern, Germany; bart@mv.uni-kl.de
* Correspondence: ahmadsarayrah@ymail.com; Tel.: +971-543-061-810

Abstract: Adsorption cooling can recover waste heat at low temperature levels, thereby saving energy and reducing greenhouse gas emissions. An air-cooled adsorption cooling system reduces water consumption and the technical problems associated with wet-cooling systems; however, it is difficult to maintain a constant recooling water temperature using such a system. To overcome this limitation, a variable mode adsorption chiller concept was introduced and investigated in this study. A prototype adsorption chiller was designed and tested experimentally and numerically using the lumped model. Experimental and numerical results showed good agreement and a similar trend. The adsorbent pairs investigated in this chiller consisted of silicoaluminophosphate (SAPO-34)/water. The experimental isotherm data were fitted to the Dubinin–Astakhov (D–A), Freundlich, Hill, and Sun and Chakraborty (S–C) models. The fitted data exhibited satisfactory agreement with the experimental data except with the Freundlich model. In addition, the adsorption kinetics parameters were calculated using a linear driving force model that was fitted to the experimental data with high correlation coefficients. The results show that the kinetics of the adsorption parameters were dependent on the partial pressure ratio. Four cooling cycle modes were investigated: single stage mode and mass recovery modes with duration times of 25%, 50%, and 75% of the cooling cycle time (denoted as short, medium, and long mass recovery, respectively). The cycle time was optimized based on the maximum cooling capacity. The single stage, short mass recovery, and medium mass recovery modes were found to be the optimum modes at lower (<35 °C), medium (35–44 °C), and high (>44 °C) recooling temperatures. Notably, the recooling water temperature profile is very important for assessing and optimizing the suitable working mode.

Keywords: adsorption; zeolite; SAPO-34; adsorption cooling; mass recovery; variable mode

Citation: Alsarayreh, A.A.; Al-Maaitah, A.; Attarakih, M.; Bart, H.-J. Performance Analysis of Variable Mode Adsorption Chiller at Different Recooling Water Temperatures. *Energies* **2021**, *14*, 3871. https://doi.org/10.3390/en14133871

Academic Editor: Jaroslaw Krzywanski

Received: 18 May 2021
Accepted: 12 June 2021
Published: 27 June 2021

1. Introduction

In recent years, adsorption cooling technology has gained significant attention owing to its ability to utilize low-grade heat sources and its environmental friendliness [1]. Low-grade heat (<100 °C) is easily obtainable with simple solar collectors or as waste heat from power generation and industrial processes. The concept of adsorption is based on the interaction between solid and gas, whereby the physical uptake of the refrigerant (i.e., adsorbate) on the surface of adsorbents, such as zeolite, silica gel, or activated carbon, is facilitated through van der Waals forces or polar bonding [2]. Four main processes are involved in the simple adsorption cooling cycle: isosteric heating, isobaric desorption, isosteric cooling, and isobaric adsorption [3]. Single-stage adsorption cooling systems are the most researched and widely available type of adsorption refrigeration systems [4–6]. Several adsorbent pairs have been studied for cooling applications, such as silica gel/water [7], activated carbon/methanol [8], zeolite (SAPO-34)/water [9], zeolite (13X)/water [10], and MOF aluminum fumarate/water [11].

Silicoaluminophosphate (SAPO-34)/water is one of the most promising adsorbent pairs for refrigeration and cooling applications; a low-grade heat source can drive the SAPO-34/water adsorbent pair, as it has a weak polar framework [12]. As per the IUPAC (International Union of Pure and Applied Chemistry) classification of adsorption isotherms, a weak interaction-based adsorption is characterized by the type V curve [13].

In most studies, the Dubinin–Astakhov (D–A) equation has been used for SAPO-34/water isotherm calculations [9,12]. Sun and Chakraborty [14] developed an uptake equation based on the partition distribution function representing the rigor of each adsorptive site and the adsorption isosteric heat at the zero surface. They compared this equation with the D–A, BET, and Langmuir equations for SAPO-34/water. The results showed a small error of 5%, indicating that the proposed model fit well with the experimental results; by contrast, the Langmuir and BET equations failed to fit to the experimental data. Youssef et al. [15] experimentally and numerically investigated the physical and adsorption-related characteristics of AQSOA-Z02, a commercial variant of SAPO-34, using the Sun and Chakraborty model. The results showed two different values of the constants based on the pressure ratios.

In addition to the adsorbent pairs, researchers have also investigated the effect of operation parameters on the adsorption cooling cycle, such as the cycle time and operation temperature. Wang et al. [16] conducted a study that proved the importance of cooling cycle time in the cooling process, because it affects the cooling capacity and coefficient of performance (COP). Heating fluid, recooling fluid, and chilled fluid temperatures as well as the cycle have a significant impact on the adsorption cooling system COP and capacity. The effects of these parameters were investigated by Ghilen at al. [17], and the study results showed that the highest COP value was obtained at a high heating and evaporation temperature, whereas the lowest value was obtained at a high recooling temperature.

Single-stage adsorption cooling systems are the most researched and widely available type of adsorption refrigeration system [4–6]. Two-stage adsorption cooling systems have also been introduced and investigated by many researchers for operation at higher recooling temperatures and lower heating temperatures compared to single-stage cycle systems [18–21]. Alam et al. [22] proposed a new strategy in the form of a reheating cycle to operate two-stage adsorption chillers and obtain a higher COP compared to the conventional two-stage adsorption cycle. The reheating cycle consists of six main steps: desorption, mass recovery process with heating, pre-cooling, adsorption, mass recovery process with cooling, and desorption. The reheating adsorption cooling cycle performance compared to the single stage cycle at various heating water temperature was investigated by Wirajati et al. [23]; the results showed better COP for short and long reheating cycles at heating temperature lower than 68 °C, while the single stage had a higher COP at heating temperature above 70 °C. The results also showed that the short reheating cycle had a higher cooling capacity at different heating temperatures, and the long reheating cycle had a higher cooling capacity than single stage at heating temperature less than 70 °C.

Since the adsorption cooling cycle has a relatively low COP, the mass-recovery concept was developed to improve adsorption cycle efficiency [24]. This concept involves connecting the high-pressure bed (desorber) to the low-pressure bed (adsorber) at the end of the desorption–adsorption cycle, thereby causing the refrigerant in the desorber to be re-adsorbed in the adsorber because of the pressure difference between the two beds, as shown in Figure 1 [25]. Despite the mass recovery process indicating natural movement of the pressurized refrigerant from the desorber to the adsorber without extra heat input, the mass recovery concept is commonly used in the literature even with continuous heat input, which is similar to the reheating cycle concept, as highlighted above.

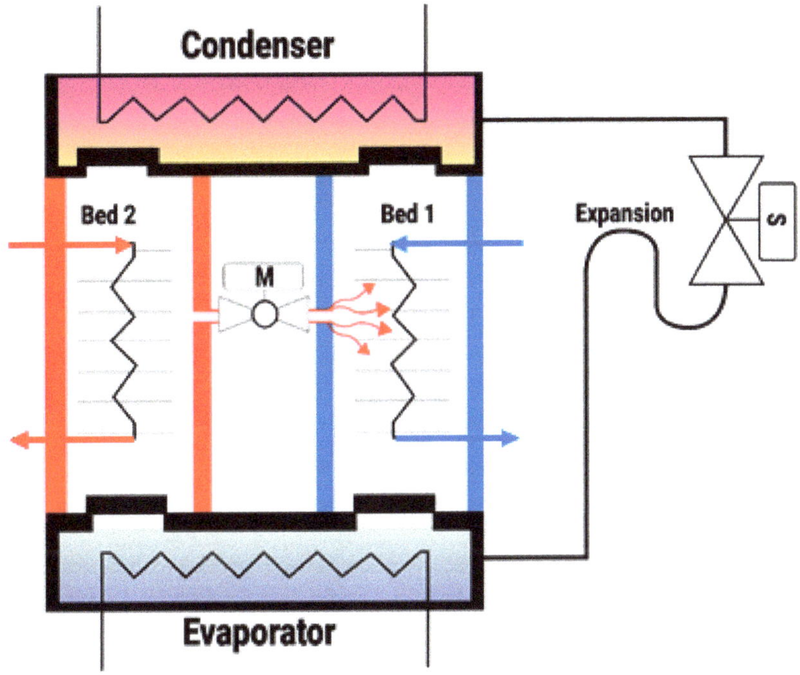

Figure 1. Concept of mass recovery.

The effect of mass recovery without the heating process on the adsorption cooling cycle performance at fixed operation temperatures was investigated by Chan et al. [26]. The simulation results showed that mass recovery improved the COP by about 3.9% with a mass recovery time of 4.7 s. Wang [27] developed and tested an adsorption cooling system at various operation conditions, including mass recovery. Incorporating mass recovery resulted in increased cooling capacity and COP (over 10%). Kabir et al. [28] simulated the effect of mass recovery on the performance of a single-stage adsorption cooling system directly driven by solar energy. The study results showed that the mass recovery process improved the overall performance. Ghilen et al. [17] simulated the effect of mass recovery on a silica gel–water adsorption chiller. The mass recovery was found to improve system performance; however, the improvement level was not quantified. The effect of mass recovery on a three-bed adsorption chiller with various heating temperatures was experimentally investigated by Uyun et al. [29]. The results showed that the COP for the cycle with mass recovery was superior to that of the single-stage cycle for heating temperatures below 75 °C. The effect of mass recovery on the adsorption cooling cycle with different hot water temperatures was investigated by Duong et al. [30] with a transient two-dimensional model. The simulation results showed that mass recovery improved the COP up to 5% at 90 °C. The results also showed that the mass recovery had a greater effect at a lower hot water temperature. This conclusion is in agreement with the result of Uyun et al. [29]. The mass recovery effect on a commercial two-bed adsorption chiller was investigated by Muttakin et al. [31] using a transient lumped model. That study investigated the impact of mass and heat recovery on the chiller performance at various hot water, recooling water, and chilled water temperatures, one with 30 °C and two with 28 °C recooling water, with different mass times up to 25 s. The result showed that there were optimum cycle and mass recovery times for maximum efficiency under specified operating conditions.

There are many studies on the adsorption cooling cycle with mass recovery, but there is a shortage of research on adsorption cooling with recooling water, especially at high temperatures, as is done for heating temperature. This study attempted to bridge this gap by investigating the mass recovery effect at various recooling water temperatures up to 50 °C. The main novelty of the study is that it proposes and investigates a variable mode two-bed adsorption chiller working in single stage mode and short, medium, and long mass recovery modes. In practical situations, it is difficult to maintain a constant recooling water temperature, especially using an air-cooled system on a daily or yearly basis. Moreover, this study provides a methodology to determine the optimum recovery in a dynamic way as a ratio of the cooling cycle duration instead of representing it as time in seconds. In addition, the adsorbent pairs used in this study (SAPO-34/water) were analyzed under different isotherm models to find the optimum isotherm model.

2. System Description

The proposed variable mode adsorption chiller is based on four cycles (steps), as shown in Figure 2. The main concept is to work in a smart way to change the cycle mode based on the ambient or recooling temperature. A patent was filed for this concept by Precision Industries (PI-Dubai) [32]. The working steps can be summarized as follows:

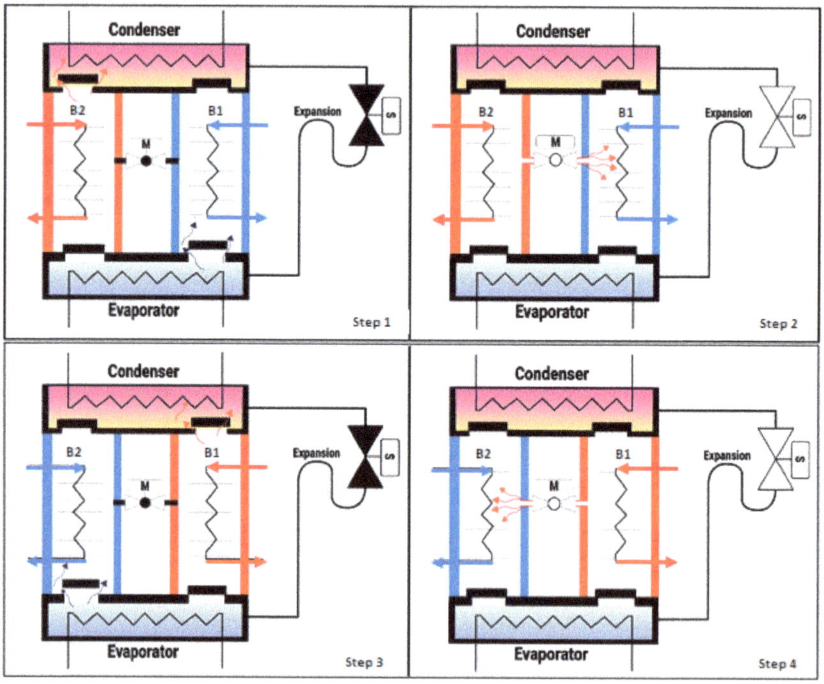

Figure 2. Variable mode adsorption chiller component.

Step 1: hot water flows into the adsorbent bed (B2), which causes desorption of the refrigerant gases by B2; this increases the pressure inside B2. This forces the gases to flow into the condenser through the non-return valve (NRV). The recooling water flows into the condenser, thereby causing the desorbed hot refrigerant gases to condensate at higher relative pressure and accumulate in the condenser. The condenser is connected to the evaporator with a valve (S) to control the flow based on cycle conditions and the expansion mechanism. The recooling water flows into the adsorbent bed (B1), on which the refrigerant gases are adsorbed, thereby reducing the pressure inside it and forcing the refrigerant to flow from the evaporator to B1 through the NRV. The chilled water flows into the evaporator, thus causing the refrigerant gases to evaporate and travel to B1.

Step 2: the motorized valve (M) opens for a certain time, thus allowing the relative high-pressure hot gases to flow from B2 to B1. The time required for this process depends on the recooling water temperature (mass recovery process with heating and cooling).

Step 3: Valve M closes, and the hot water flows into B1, which causes desorption of the refrigerant gases by B1. This increases the pressure inside B1 and forces the gases to flow into the condenser through the NRV. The recooling water flows into the condenser and the desorbed hot refrigerant gases condensate at a high relative pressure to be accumulate in the condenser. The condenser is connected to the evaporator by a valve (S) to control the refrigerant flow based on the cycle and expansion mechanism. The recooling water flows into B2, causing the refrigerant gases to be adsorbed adsorb, thereby reducing the pressure inside it and forcing the refrigerant to flow from the evaporator to B2 through the NRV. Next, the chilled water flows into the evaporator and causes the refrigerant gases to evaporate and travel to B2.

Step 4: M opens for a certain time, which allows the relative high-pressure hot gases to flow from B1 to B2. The duration of this process depends on the recooling water temperature. Therefore, the optimum time increases at higher ambient temperature (mass recovery process with heating and cooling). Then the cycle returns to step 1 to repeat all steps.

Figure 3 shows a Clapeyron diagram of the thermodynamic processes of the proposed system: the first mode (single stage mode, denoted with 1, 2, 3, and 4) and other modes (mass recovery cycle with heating and cooling, denoted with 1, 1′, 2′, 3′, 4′, and 5′). The cooling capacity in the mass recovery mode (represented as the net area in the closed loop) is larger than that during the basic cycle. In addition, the condenser pressure increases, which leads to condensation at a higher temperature. For the first mode, there are four thermodynamics processes: preheating (1–2), desorption (2–3), precooling (3–4), and adsorption (4–1). In the second mode, there are six thermodynamics processes: mass recovery/adsorption with pressurization when the two beds are connected (1–1′), preheating (1′–2′), main desorption process (2′–3′), mass recovery/desorption with depressurization (3′–4′), precooling (4′–5′), and the main adsorption process (5′–1).

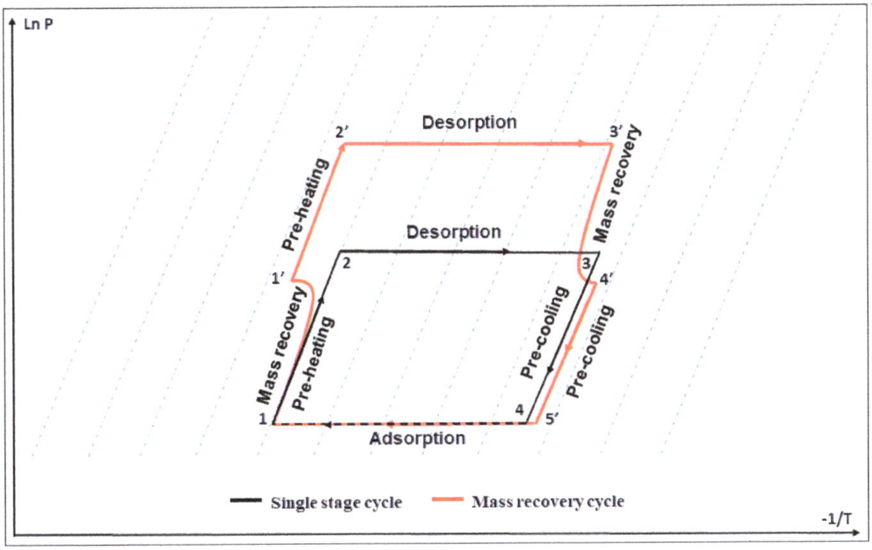

Figure 3. Clapeyron diagram of variable mode adsorption cooling cycle.

3. Materials and Methods

3.1. Adsorbent Material Characterization

In this study, SAPO-34 zeolite was coated on a heat exchanger using epoxy binder [33], and its properties were investigated with regard to adsorption cooling applications. The microstructure of the zeolite crystals was observed via scanning electron microscopy (SEM) using quanta 250 FEG; the zeolite radius was observed to be approximately 5–10 μm, as shown in Figure 4.

Figure 4. SEM image of SAPO-34 zeolite.

3.2. Experimental Determination of Adsorption Isotherms and Adsorption Kinetics

The experimental procedure for determining the adsorption kinetics and isotherms for fitting with various mathematical models are explained in this section. The equilibrium of the water uptake for SAPO-34 was determined using a dynamic vapor sorption (DVS) analyzer (Surface Measurement Systems Ltd., London, UK). A schematic diagram of DVS is shown in Figure 5. Water sorption isotherms of up to 90% P/P^0 at temperatures of 20–70 °C were measured using DVS. The data were recorded in the dynamic mode every 20 s, wherein both the sorbate entry rate, controlled upstream by a mass flow controller, and the sorbate exit rate, controlled downstream by a butterfly valve, were measured. Throughout the experiment, the pressure in the vacuum chamber was controlled (from 1.33 to 133.3 kPa) using a butterfly valve (Baratron by MKS Instruments) that regulated the sorbate exit rate depending on the feedback and adjusted the opening position to maintain the set pressure. A mass flow controller was used to maintain a constant flow rate for the water vapor that was generated at an experimental temperature under thermodynamic equilibrium. Variations in the sample mass were simultaneously measured using an UltraBalance (Surface Measurement Systems Ltd., UK). The sample was kept under constant water-vapor pressure until the mass equilibrium was attained, before the water vapor pressure was increased (adsorption branch) or decreased (desorption branch). The water adsorption isotherms were recorded for 0–90% P/P^0 and for desorption from 90 to 0% P/P^0 at three different temperatures. The samples were outgassed in situ at a temperature of 150 °C under high vacuum (<1.33 × 10^{-3} Pa) and cooled to the desired sorption temperature before each sorption experiment. The water sorption measurements were recorded at 34, 40, and 70 °C.

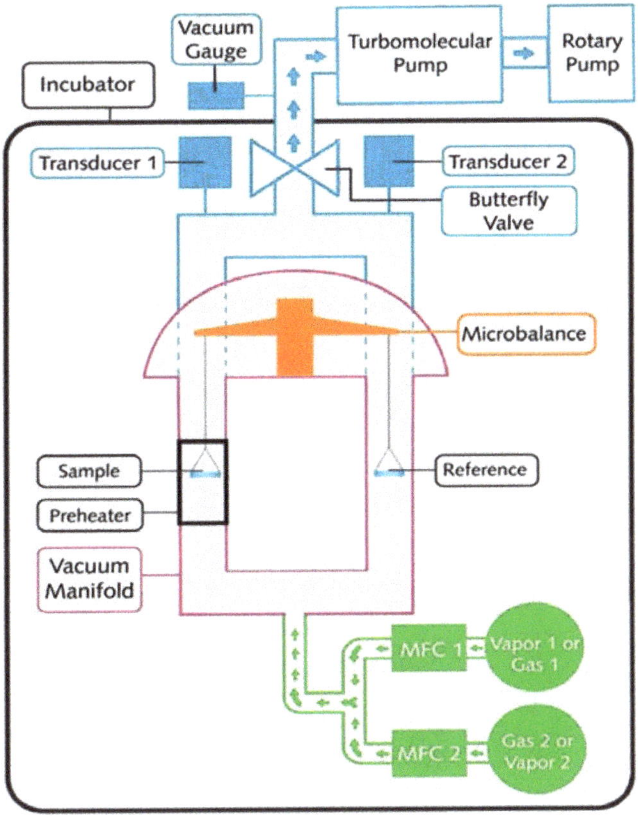

Figure 5. Schematic of DVS vacuum.

3.2.1. Adsorption Isotherms

The isotherm parameters of SAPO-34/water were calculated using several adsorption isotherm models with details provided in Table 1.

Table 1. Adsorption isotherm equations for different models.

Equilibrium Model	Equation	Parameters
Dubinin–Astakhov (D–A)	$\omega = \omega_0\, e^{[-\{(\frac{RT}{E})\ln(\frac{p}{p_0})\}^n]}$	ω: uptake (kg·kg^{-1}) ω_0: maximum uptake (kg·kg^{-1}) E: characteristic energy (kJ·mol^{-1}) n: heterogeneity parameter R: universal gas constant (kJ·mol^{-1}·K^{-1}) T: bed temperature (K) p: equilibrium pressure (kPa) p_0: saturated pressure (kPa)
Freundlich	$\omega = \omega_0 \left(\frac{p}{p_0}\right)^{\frac{1}{n}}$	ω: uptake (kg·kg^{-1}) ω_0: maximum uptake (kg·kg^{-1}) n: heterogeneity factor p: equilibrium pressure (kPa) p_0: saturated pressure (kPa)
Hill	$\omega = \omega_0 \left[\frac{\left(\frac{p}{p_0}\right)^{n_H}}{K_D+\left(\frac{p}{p_0}\right)^{n_H}}\right]$	ω: uptake (kg·kg^{-1}) ω_0: maximum uptake (kg·kg^{-1}) n_H: Hill cooperativity coefficient K_D: Hill constant
Sun and Chakraborty (S–C)	$\omega = \omega_0 \left[\frac{k\left(\frac{p}{p_0}\right)^m}{1+(k-1)\left(\frac{p}{p_0}\right)^m}\right]$ $k = \propto e^{[m(Q_{st}-h_{fg})/RT]}$	ω: uptake (kg·kg^{-1}) ω_0: maximum uptake (kg·kg^{-1}) $\propto$: pre-exponential coefficient m: heterogeneity factor Q_{st}: isosteric heat of adsorption at zero surface coverage (kJ·kg^{-1}) h_{fg}: enthalpy of evaporation (kJ·kg^{-1}) R: universal gas constant (kJ·kg^{-1}·K^{-1}) T: bed temperature (K) p: equilibrium pressure (kPa) p_0: saturated pressure (kPa)

These models were fitted to the experimental results with the least-squared method using the 'nlinfit' function in MATLAB. Root-mean-square error (RMSE) and coefficient of determination (R^2) were used to assess the goodness of fit.

3.2.2. Rate of Adsorption

The linear driving force (LDF) model was used to calculate the adsorption rate (kinetic) parameters. The values of pre-exponential constant D_{so} (m^2·s^{-1}), and activation energy E_a (J·mol^{-1}) were calculated by fitting the DVS data to the LDF model:

$$\frac{d\omega}{dt} = \frac{15D_{so}e^{(\frac{-E_a}{RT})}}{R_p{}^2}\left(\omega_0 - \omega_{(t)}\right) \quad (1)$$

where the terms are defined as follows:
ω: instantaneous water uptake (kg·kg^{-1});
ω_0 : maximum water uptake (kg·kg^{-1});
R: ideal gas constant (J·mol^{-1}·K^{-1});
T: adsorption temperature (K);
R_p: radius of the adsorbent granules (μm);
D_{so}: pre-exponential constant of surface diffusivity (m^2·s^{-1});
E_a: activation energy of diffusion (J·mol^{-1}).

3.3. Adsorption Cooling System Model

In this study, a two-bed adsorption chiller with single stage and mass recovery modes was modelled with the SAPO-34/water pair and numerically and experimentally simulated based on the adsorption isotherms explained in Section 3.2. The lumped model was used to calculate the mass and energy balances, considering the following assumptions:

- The temperature, pressure, and adsorption rates were assumed uniform in the adsorption beds, evaporator, and condenser.
- The water vapor flow from an adsorption bed to the condenser or from the evaporator to an adsorption bed was unrestricted, and the pressure drop was neglected.
- The chiller was well insulated, and there was no heat exchange with the exterior environment.
- The inlet temperatures for hot, recooling, and chilled water were assumed constant.

The energy flow through the main chiller components is illustrated in Figure 6. The amount of heat and mass transfer at each stage was objectively quantified using the lumped model equations (Equations (2)–(24)), as explained subsequently.

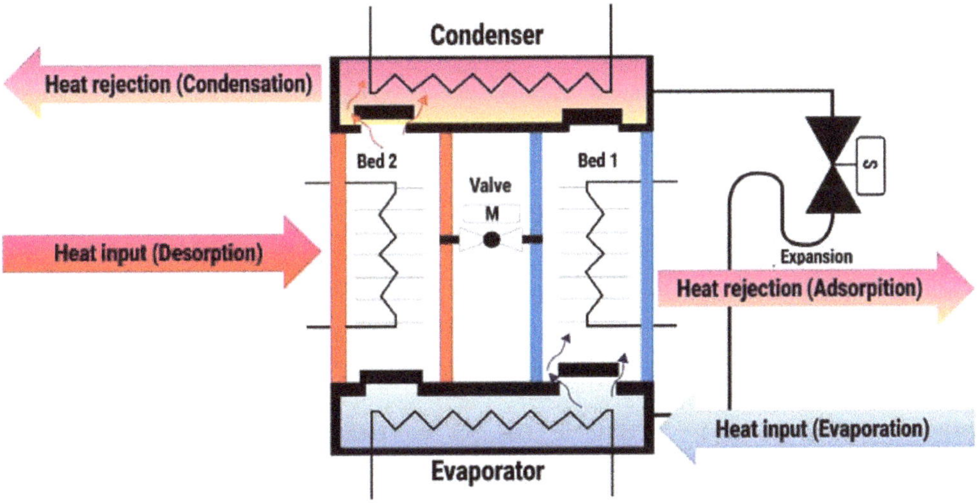

Figure 6. Schematic of chiller with energy flow.

3.3.1. Energy Balance Equation for the Adsorption Bed

The energy balance equation for the adsorption bed is given as:

$$\begin{aligned}&((M_Z \times Cp_Z) + (M_{HX} \times Cp_{HX}) + (M_Z \times Cp_{wv} \times w))\tfrac{dT_{ads}}{dt} \\ &\quad = M_Z \times Q_{st} \times \tfrac{d\omega_{ads}}{dt} - M_Z \times Cp_{wv} \times [\text{Ø} \times (T_{ads} - T_{eva}) + (1-\text{Ø}) \times (T_{ads} - T_{wv})] \times \tfrac{d\omega_{ads}}{dt}] \\ &\quad + \dot{m}_{cw} \times Cp_{cw,ads} \times (T_{cw\_in} - T_{cw\_out\_ads})\end{aligned} \quad (2)$$

where the terms are defined as follows:

M_Z: mass of zeolite in the adsorbent bed (kg);
Cp_Z: specific heat capacity of zeolite (kJ/kg·K);
$M_{HX,ads}$: mass of the adsorbent bed heat exchanger (kg);
$Cp_{HX,ads}$: specific heat capacity of the adsorbent bed heat exchanger (kJ/kg·K);
Cp_{wv}: specific heat capacity of water vapor (kJ/kg·K);
w: amount of water in the bed (kg);
Q_{st}: heat of adsorption (kJ/kg);
T_{ads}: adsorbent bed temperature (K);
T_{wv}: water vapor temperature (K);
T_{eva}: evaporator temperature (K);
$\dot{m}_{cw}$: recooling water mass flow rate (kg·s);
Cp_{cw_ads}: recooling water specific heat capacity (kJ/kg·K);

T_{cw_in}: recooling water inlet temperature (K);
$T_{cw\_out\_ads}$: adsorbent bed recooling water outlet temperature (K).

Here, Ø = 1 if the adsorbent bed is connected with the evaporator and 0 if the adsorbent bed is connected with another bed (mass recovery process). The left-hand side of the equation denotes the required sensible heat transfer of the adsorbent bed (SAPO-34, water, and copper heat exchanger). The first part of the right-hand side of the equation shows the adsorbed heat. The second term shows the sensible heat flow from the evaporator to the adsorbent bed. The third term shows the heat transfer by cooling water.

The cooling water outlet temperature is calculated using

$$T_{cw\_out\_ads} = T_{ads} + (T_{cw\_in} - T_{ads}) \times \exp\left(\frac{-U_{ads} \times A_{ads}}{\dot{m}_{cw\_ads} \times Cp_{cw}}\right) \tag{3}$$

where the terms are defined as follows:

U_{ads}: the overall coefficient of heat-transfer ($kW/m^2{\cdot}K$) for the adsorbent bed;
A_{ads}: the area (m^2) of the adsorbent bed.

The overall heat transfer coefficient of the adsorbent bed is calculated in this study experimentally based on:

$$U_{ads} = \frac{Q_{ads}}{A_{ads} \times LMTD} \tag{4}$$

where Q_{ads} is the adsorption heat capacity and $LMTD$ is the log mean temperature difference being calculated as:

$$Q_{ads} = \dot{m}_{cw} \times Cp_{cw,ads} \times (T_{cw\_in} - T_{cw\_out\_ads}) \tag{5}$$

$$LMTD = \frac{(T_{cw\_in} - T_{ads}) - (T_{cw\_out} - T_{ads})}{\ln\frac{(T_{cw\_in} - T_{ads})}{(T_{cw\_out} - T_{ads})}} \tag{6}$$

3.3.2. Energy Balance Equation for the Desorption Bed

The energy balance for the desorption bed is described as:

$$((M_Z \times Cp_Z) + (M_{HX} \times Cp_{HX}) + (M_Z \times Cp_{wv} \times w))\frac{dT_{des}}{dt} = M_Z \times Q_{st} \times \frac{d\omega_{des}}{dt} + \dot{m}_{hw} \times Cp_{hw} \times (T_{hw\_in} - T_{hw\_out}) \tag{7}$$

where the terms are defined as follows:

$\dot{m}_{hw}$: the mass flow rate (kg/s);
Cp_{hw}: the specific heat capacity (kJ/kg·K);
T_{hw_in}: the inlet temperature (K);
T_{hw_out}: the outlet temperature (K).

The left-hand side of the equation shows the required sensible heat transfer for the adsorbent bed (SAPO-34, water, and copper heat exchanger). The first part on the right-hand side of the equation denotes the desorption heat, and the second term indicates the heat transferred by heating water.

The hot water outlet temperature can be calculated by:

$$T_{hw\_out} = T_{des} + (T_{hw\_in} - T_{des}) \times \exp\left(\frac{-U_{des} \times A_{des}}{\dot{m}_{hw} \times Cp_{hw}}\right) \tag{8}$$

where the terms are defined as follows:

U_{des}: the overall coefficient of heat-transfer ($kW/m^2{\cdot}K$) of the desorber;
A_{des}: the area (m^2) of the desorber.

The overall coefficient of heat-transfer of the desorber is calculated in this study experimentally based on below Equation (14):

$$U_{ads/des} = \frac{Q_{des}}{A_{des} \times LMTD} \tag{9}$$

where Q_{des} is the desorption heat capacity and *LMTD* is the log mean temperature difference:

$$Q_{des} = \dot{m}_{hw} \times Cp_{hw} \times (T_{hw\_in} - T_{hw\_out}) \tag{10}$$

$$LMTD = \frac{(T_{cw\_in/hw\_in} - T_{ads./des}) - (T_{cw\_out/hw\_out} - T_{ads./des})}{\ln \frac{(T_{cw\_in/hw\_in} - T_{ads./des})}{(T_{cw\_out/hw\_out} - T_{ads./des})}} \tag{11}$$

3.3.3. Energy Balance Equation for the Condenser

The energy balance for the condenser is described as:

$$\begin{aligned}&((M_{w\_con} \times Cp_{cw}) + (M_{HX} \times Cp_{HX}))\tfrac{dT_{con}}{dt} \\ &= -M_z \times h_{fg} \times \tfrac{d\omega_{des}}{dt} - M_z \times Cp_z \times (T_{des} - T_{con}) \times \tfrac{d\omega_{des}}{dt} \\ &+ \dot{m}_{cw\_con} \times Cp_{cw} \times (T_{cw\_in} - T_{cw\_out\_con})\end{aligned} \tag{12}$$

where the terms are defined as follows:

M_{w_con}: mass of water inside the condenser (kg);
M_{HX_con}: mass of the condensing heat exchanger (kg);
Cp_{HX_con}: specific heat capacity of the condensing heat exchanger (kJ/kg·K);
h_{fg}: latent heat of evaporation (J/kg);
$\dot{m}_{cw\_con}$: recooling water mass flow rate (kg/s);
T_{con}: condenser temperature (K);
$T_{cw\_out\_con}$: adsorbent bed recooling water outlet temperature (K).

Here, the left-hand side of the equation shows the required sensible heat transfer of the condenser water and the copper heat exchanger. The first part on the right-hand side denotes the latent heat of vaporization for water. The second term indicates the sensible heat transfer from the desorber to the condenser. The third term denotes the heat released by the cooling water.

The cooling water outlet temperature of the condenser water can be calculated by:

$$T_{cw\_out\_con} = T_{con} + (T_{cw\_in} - T_{con}) \times \exp\left(\frac{-U_{con} \times A_{con}}{\dot{m}_{cw\_con} \times Cp_{cw}}\right) \tag{13}$$

where the terms are defined as follows:

T_{con}: the condenser temperature (K);
U_{con}: the overall coefficient of heat-transfer (kW/m^2·K);
A_{con}: the condenser area (m^2).

The overall coefficient of heat-transfer of the condenser is calculated as:

$$U_{ads/des} = \frac{Q_{con}}{A_{con} \times LMTD} \tag{14}$$

where Q_{con} is the adsorption and condensation heat capacity, and *LMTD* is the log mean temperature difference:

$$Q_{con} = \dot{m}_{cw\_con} \times Cp_{cw} \times (T_{cw\_in} - T_{cw\_out\_con}) \tag{15}$$

$$LMTD = \frac{(T_{cw\_in} - T_{con}) - (T_{cw\_out} - T_{con})}{\ln \frac{(T_{cw\_in} - T_{con})}{(T_{cw\_out} - T_{con})}} \tag{16}$$

3.3.4. Energy Balance Equation for the Evaporator

The energy balance for the evaporator is described as

$$((M_{w\_eva} \times Cp_{cw}) + (M_{HX\_eva} \times Cp_{HX\_eva}))\frac{dT_{eva}}{dt} = -M_z \times h_{fg} \times \frac{d\omega_{ads}}{dt} - M_z \times Cp_{wv} \times (T_{con} - T_{eva}) + \dot{m}_{chw} \times Cp_{chw} \times (T_{chw\_in} - T_{chw\_out}) \quad (17)$$

where the terms are defined as follows:

M_{w_eva}: mass of water inside the evaporator (kg);
M_{HX_eva}: mass of the evaporator heat exchanger (kg);
Cp_{HX_eva}: specific heat capacity of the evaporator heat exchanger (kJ/kg·K);
T_{eva}: evaporator temperature (K).

For the chilled water:

$\dot{m}_{chw}$: mass flowrate (kg/s);
Cp_{chw}: specific heat capacity (kJ/kg·K);
T_{chw_in}: inlet temperature (K);
T_{chw_out}: outlet temperature (K).

The left-hand side of the equation represents the required sensible heat transfer of the evaporator (copper heat exchanger and water). The first part of the right-hand side indicates the latent heat of water vaporization. The second term represents the sensible heat required to cool down the incoming condensed water to the condensation temperature. The third term represents the heat transfer by the chilled water.

The chilled water outlet temperature was calculated using:

$$T_{chw\_out} = T_{eva} + (T_{chw\_in} - T_{eva}) \times \exp\left(\frac{-U_{eva} \times A_{eva}}{\dot{m}_{chw} \times Cp_{chw}}\right) \quad (18)$$

where the terms are defined as follows:

U_{eva}: the overall coefficient of heat-transfer (kW/m^2·K);
A_{eva}: the area of the evaporator (m^2).

The overall coefficient of heat-transfer of the evaporator is calculated in this study based on:

$$U_{ads/des} = \frac{Q_{con}}{A_{con} \times LMTD} \quad (19)$$

where Q_{eva} is the adsorption and evaporation heat capacity and $LMTD$ is the log mean temperature difference:

$$Q_{con} = \dot{m}_{chw} \times Cp_{chw} \times (T_{chw\_in} - T_{chw\_out}) \quad (20)$$

$$LMTD = \frac{(T_{chw\_in} - T_{eva}) - (T_{chw\_out} - T_{eva})}{\ln\frac{(T_{chw\_in} - T_{eva})}{(T_{chw\_out} - T_{eva})}} \quad (21)$$

3.3.5. Mass Balance

The system mass balance is described by:

$$\frac{dM_{w\_eva}}{dt} = -M_z \times \left(\frac{d\omega_{ads}}{dt} + \frac{d\omega_{des}}{dt}\right) \quad (22)$$

where the terms are defined as follows:

M_{W_eva}: the mass of water (refrigerant) entering the evaporator (kg);
M_Z: the mass of zeolite in the bed (kg).

3.3.6. System Performance

The cooling capacity Q and the COP of the system were calculated using Equations (28) and (29), respectively. The COP represents the ratio of cooling capacity to the required heat input; the COP of the system is calculated based on the total cycle period (*tcy*), including periods for precooling, adsorption, preheating, and desorption.

$$Q = \frac{\int_0^{tcy} \dot{m}_{chw} \times Cp_{chw} \times (T_{chw\_in} - T_{chw\_out})dt}{tcy} \quad (23)$$

$$COP = \frac{\int_0^{tcy} \dot{m}_{chw} \times Cp_{chw} \times (T_{chw\_in} - T_{chw\_out})dt}{\int_0^{tcy} \dot{m}_{hw} \times Cp_{hw} \times (T_{hw\_in} - T_{hw\_out})dt} \quad (24)$$

The proposed model of energy, mass balance, and performance equations were solved numerically by MATLAB using ODE 45 to calculate the temperatures of the adsorption and desorption beds, condenser and evaporator, as well as system performance.

3.4. Experimental Setup

A lab-scale model of the variable cycle chiller was built at the Precision Industries (PI, Dubai) labs to measure the performance of the chiller experimentally under different modes and various recooling temperatures. The proposed tested adsorption chiller in this study consisted of two adsorbent beds coated with SAPO-34 supplied by Mitsubishi Chemical Corporation (see Figure 7).

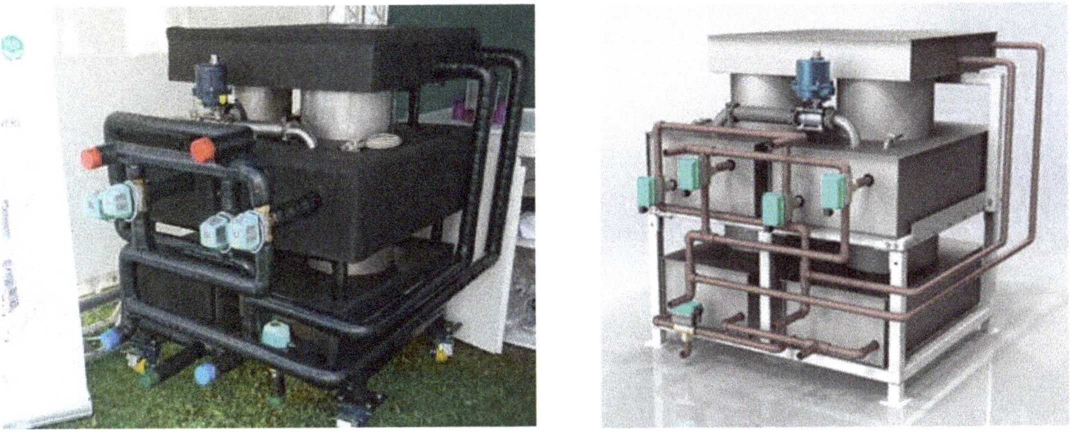

(a) (b)

Figure 7. PI prototype adsorption chiller: (**a**) prototype, (**b**) 3D model.

The chiller performance was analyzed under the following operation conditions:

- The inlet hot water temperature (Th_in) of 90 ± 0.5 °C maintained by an electric water heater with a 500 L buffer tank.
- The inlet recooling water temperature (Tre_in) of 30–50 °C maintained by a dry-cooler with a 500 L buffer tank.
- The inlet chilled water temperature (Tch_in) of 18 ± 0.5 °C maintained by an electrical heater with a 500 L buffer tank, the electrical heater is used as a cooling load in this case.
- Hot, recooling, and chilled water flow rates for were 1.2, 1.2, and 0.71 L/s, respectively.

The following setup of instrumentation was used to measure the experimental performance of the chiller during the mass recovery cycle, as shown in Figure 8. The instrument locations are described as follows:

- Three electromagnetic flow-meters (FM_h, FM_ch, and FM_re) (manufactured by ALIA Inc., Newark, DE, USA) with accuracy of ±0.4% were used to measure the hot water, chilled water and recooling water flow rate, respectively.
- Eight platinum resistance thermometers (PT100 Class A, Pico Technology) integrated with two temperature measuring data loggers (PT-104 is a four-channel logger), a resolution of 0.001 °C, and an accuracy of 0.015 °C were used to measure the inlet and outlet temperatures of the hot, chilled, and recooling water, as well as temperature of the storage tanks.

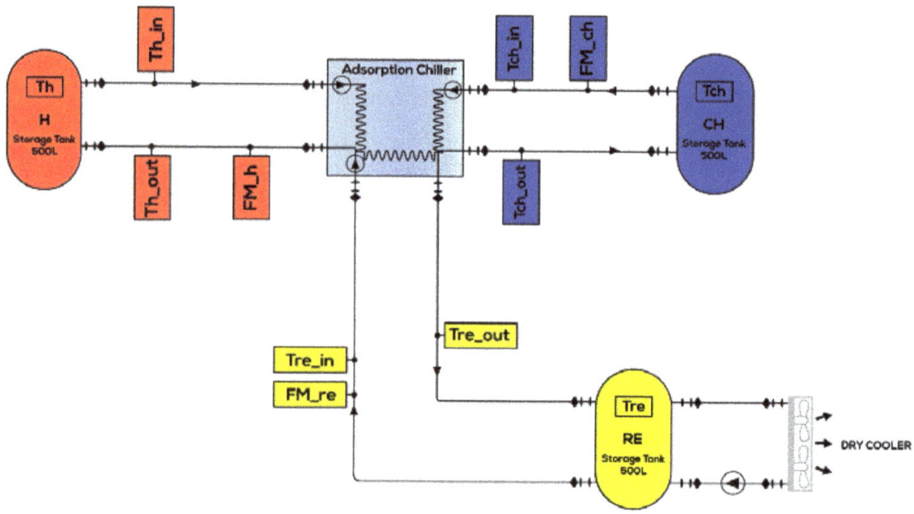

Figure 8. Schematic of prototype adsorption chiller testing equipment.

4. Results and Discussion

Figures 9–11 present the experimental results of water-vapor uptake compared with the calculated values at 34, 40, and 70 °C, respectively. The isotherm parameters for each model are listed in Table 2. As shown in Figures 6–8, the D-A, Hill, and S-C models correlated well with the experimental data, whereas the Freundlich model failed to fit.

Table 2. SAPO-34 isotherm parameters for different models.

Equilibrium Model	Parameters	
D-A	E (kJ·mol^{-1}) 7.1067	n (−) 3.943
Freundlich	n (−) 5.9199	
Hill	n_H (−) 2.2337	K_D (−) 0.0039
S-C	$\propto$ (−) 2.71×10^{-6}	m (−) 2.2307

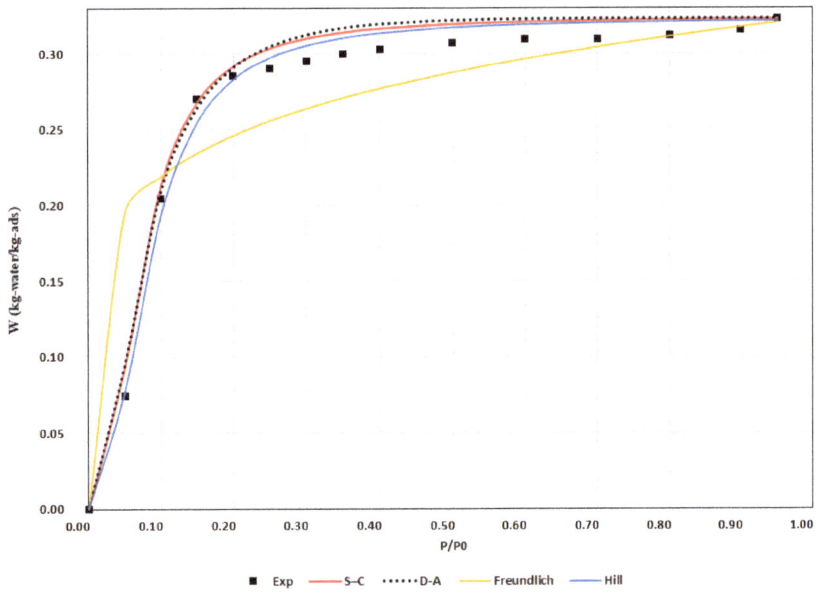

Figure 9. Water uptake comparison between experimental and isotherm model results at 34 °C.

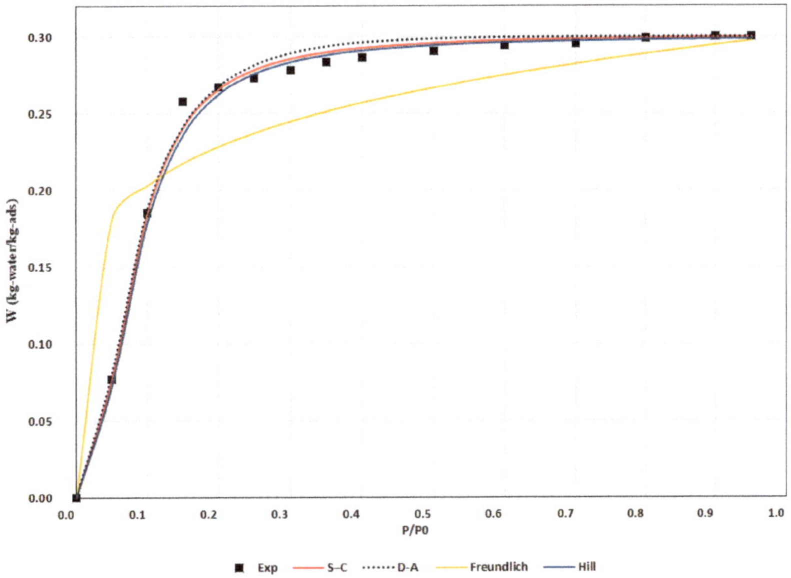

Figure 10. Water uptake comparison between experimental and isotherm model results at 40 °C.

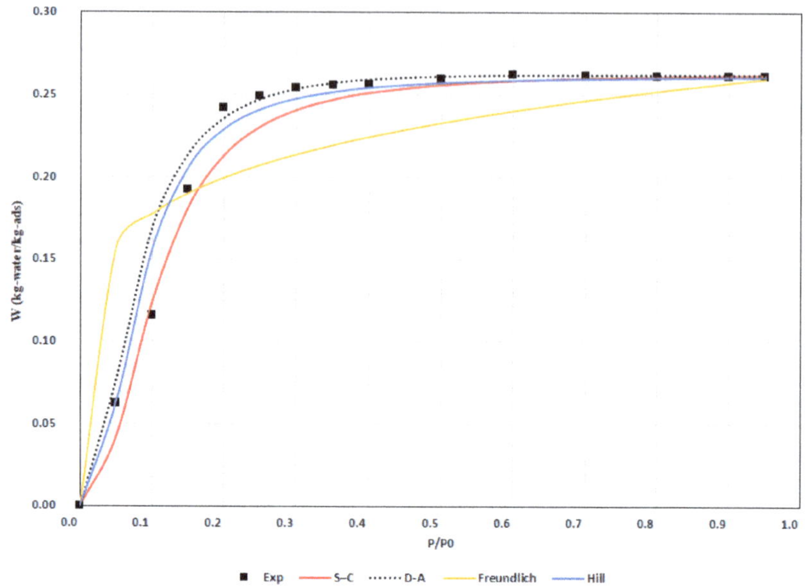

Figure 11. Water uptake comparison between experimental and isotherm model results at 70 °C.

As per the error analysis presented in Table 3, the Hill model exhibited the greatest fitting accuracy at 34 °C, followed by the D–A, S–C, and Freundlich models. At 40 °C, the Hill model achieved the highest fitting accuracy, followed by the S–C, D–A and Freundlich models. The D–A model exhibited the greatest fitting accuracy at 70 °C, followed by the S–C, Hill, and Freundlich models. Overall, the results of the Freundlich model deviated substantially from the experimental results.

Table 3. Errors and statistical analysis of isotherm fitting results.

Temperature (°C)	Model	RMSE	R^2
34	Hill	0.0066	0.9948
	D–A	0.0072	0.9947
	S–C	0.0364	0.8533
	Freundlich	0.0751	0.9926
40	Hill	0.0089	0.9936
	S–C	0.0113	0.9959
	D–A	0.0124	0.9939
	Freundlich	0.038	0.8447
70	D–A	0.0103	0.9848
	S–C	0.0124	0.9817
	Hill	0.0126	0.9774
	Freundlich	0.0384	0.7954

The experimental results of the adsorption kinetics included more than one curve based on the pressure ratios, as shown in Figure 12. The LDF model was fitted to the experimental results for three pressure ratio intervals: $P/P^0 < 5\%$, $5\% < P/P^0 < 15\%$, and $P/P^0 > 15\%$. The fitting results showed that the constants of the adsorption kinetics varied with the partial-pressure intervals; these results are summarized in Table 4. The LDF model was fitted satisfactorily to the experimental data, with an RMSE of 0.0046–0.0119, as shown in Table 4.

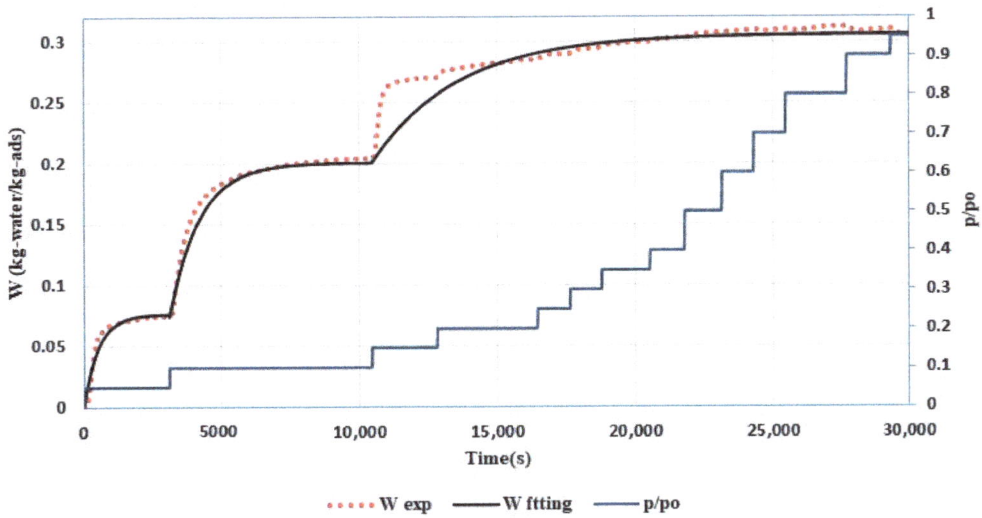

Figure 12. Adsorption kinetics experiment and fitted results at 34 °C.

Table 4. SAPO-34 adsorption kinetic parameters with errors and statistical analysis.

Parameter	E_a (J·mol^{-1})	D_{so} (m^2·s^{-1})	RMSE	R^2
0% < P/P^0 < 5%	2.06×10^4	3.95×10^{-9}	0.0046	0.946
5% < P/P^0 < 15%	2.41×10^4	7.62×10^{-9}	0.0066	0.978
15% < P/P^0	2.974×10^4	2.444×10^{-8}	0.0120	0.871

Figure 13 shows the chiller experimental temperature profile at a total cycle time of 750 s and a mass recovery time of 150 s. Areas A, B, and C indicate the preheating/pre-cooling, adsorption/desorption, and mass recovery with heating and cooling processes, respectively, where both beds were connected with no cooling effect.

Figure 13 shows a cycle period range of 500–1280 s during the preheating of desorption bed 1 (area A). During this process, the temperature increased until it reached a suitable level for desorption, and the pressure increased until the condensation pressure was reached. The condenser exhibited the minimum temperature in this cycle (close to the re-cooling water inlet temperature), and no condensation occurred. Once bed 1 reached a suitable temperature (as shown in area B), the condenser temperature increased, indicating the flow of hot vapor from bed 1 (at condensation pressure or higher) to the condenser. In the pre-cooling phase, the bed 2 temperature decreased until it reached a suitable level for adsorption, and the pressure decreased to less than the evaporator pressure. During this phase, the evaporator exhibited its maximum temperature (close to the chilled water inlet temperature) and no cooling effect occurred. Once bed 2 reached a suitable temperature and pressure for adsorption, the evaporator temperature decreased and the cooling effect occurred (area B). In the mass recovery process (area C), there was a sudden drop in temperature in bed 1 (desorber) and an increase in bed 2 (adsorber). This can be attributed to the increase in adsorption and desorption rates immediately after interconnecting both beds until equilibrium pressure was reached. The desorber temperature then started to increase and the adsorber temperature started to decrease owing to the decrease in adsorption and desorption rates occurring later in the process. Therefore, the heat input became higher than the amount required for desorption, and the rejected heat increased to values higher than the heat of adsorption.

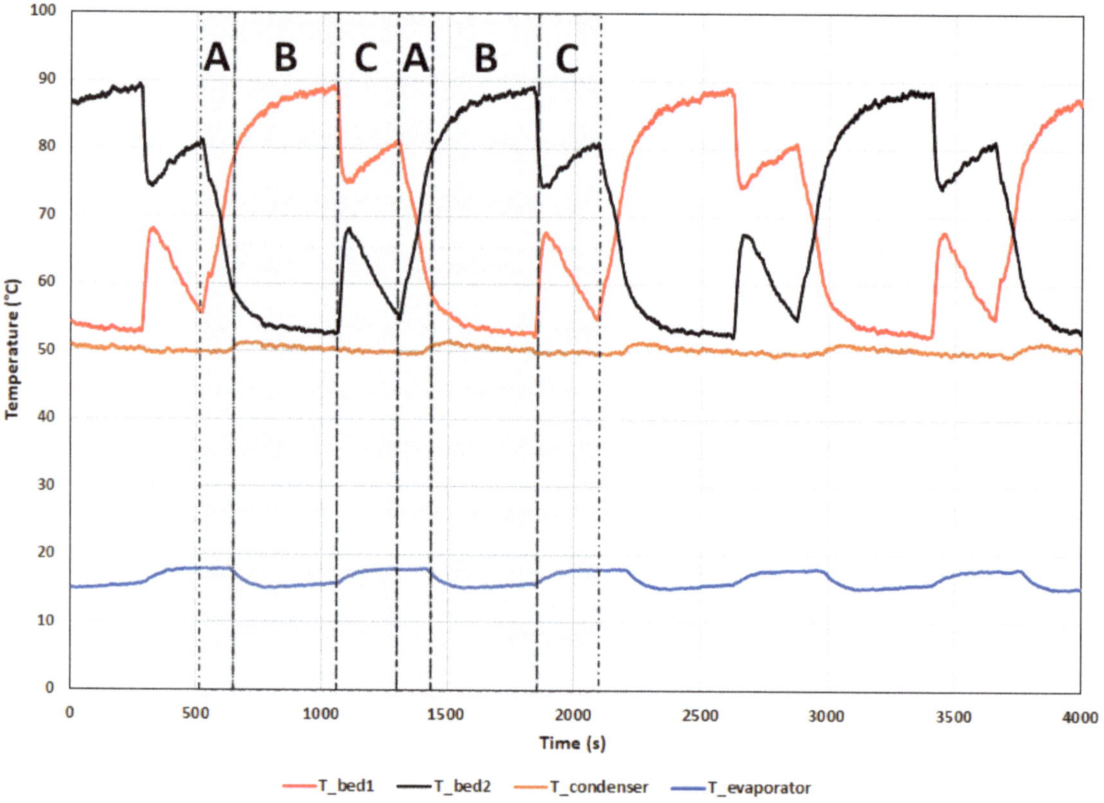

Figure 13. Experimental temperature profile for prototype adsorption chiller with mass recovery with overall cycle time of 750 s. T_{hw_in} = 90 °C, T_{cw_in} = 50 °C, and T_{chw_in} = 18 °C.

The system was investigated at a cooling cycle time of 500 s with different mass recovery times. Figure 14 shows a comparison between the COP and cooling capacity for the simulated model and the experimental measurements, which matched the predicted results and showed the same trends. The mean absolute percent errors for COP and cooling capacity were 9.5% and 6.8%, respectively. Figure 14 indicates that as the mass recovery process time increased, the COP decreased until a saturation trend was observed, where the COP remained constant. In addition, as the mass recovery time increased, the cooling capacity increased until a saturation trend was observed, where the cooling capacity was not affected by the mass recovery process time. In both cases, for COP and cooling capacity, the curve drops for high cycle times.

Figure 15 shows the simulation results of cycle cooling output at different recooling water temperatures, with single stage cycle and two mass recovery durations at a fixed cooling cycle duration of 500 s. The single stage cycle exhibited better cooling output at a recooling temperature below 38 °C compared to higher recooling water temperature. The mass recovery cycle of 250 s exhibited higher cooling output within a recooling water temperature range of 38–49 °C, while the mass recovery cycle of 375 s exhibited a higher output at a recooling water temperature higher than 49 °C.

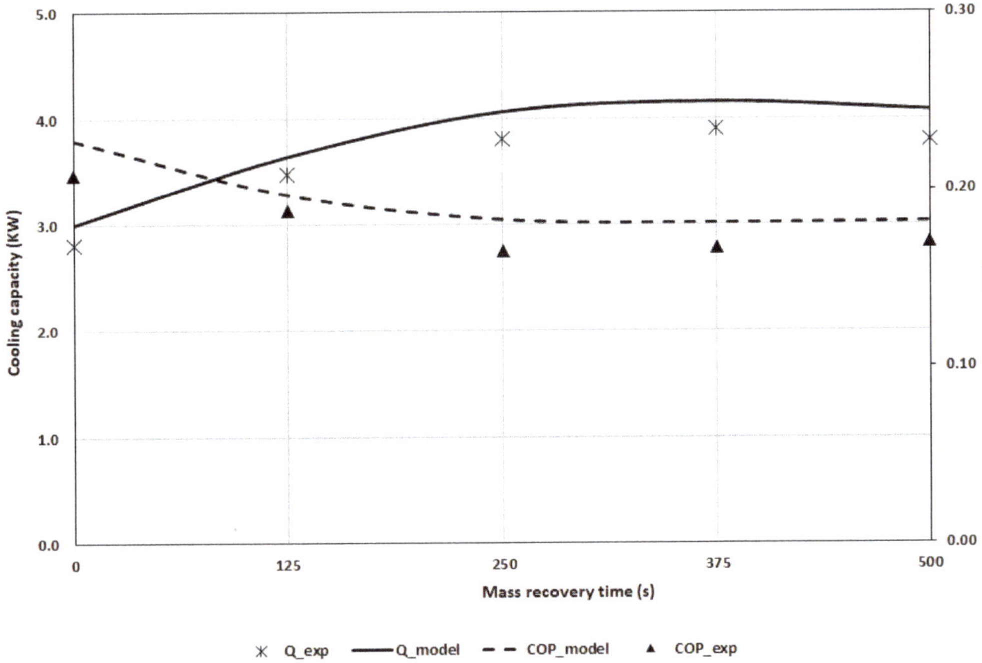

Figure 14. Comparison between experimental and predicted COP and cooling capacity with various mass recovery times at T_{hw_in} of 90 °C, T_{cw_in} of 50 °C, and T_{chw_in} of 18 °C.

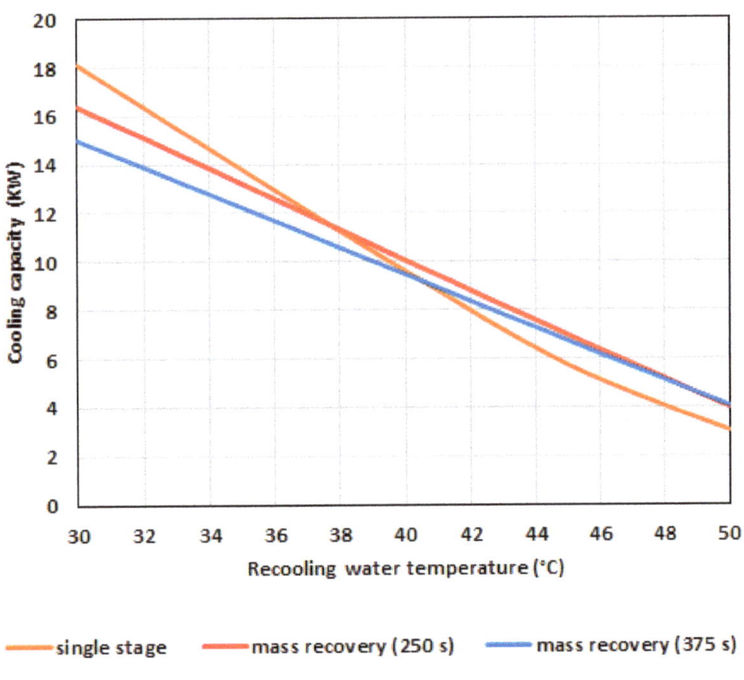

Figure 15. Predicted cooling capacity with various mass recovery times and recooling water temperatures.

The simulated COP results are shown in Figure 16. For a fixed cycle time of 500 s, the simulation shows that the single stage cycle shows the highest COP at various recooling water temperatures. In addition, the COP of the 250 s mass recovery cycle was higher than that of the 375 s mass recovery cycle up to a recooling water temperature of 50 °C, whereas at higher temperatures, similar COP values were seen. These results prove the importance of the variable-cycle mode adsorption cooling system to optimize the chiller performance with changes in recooling temperature.

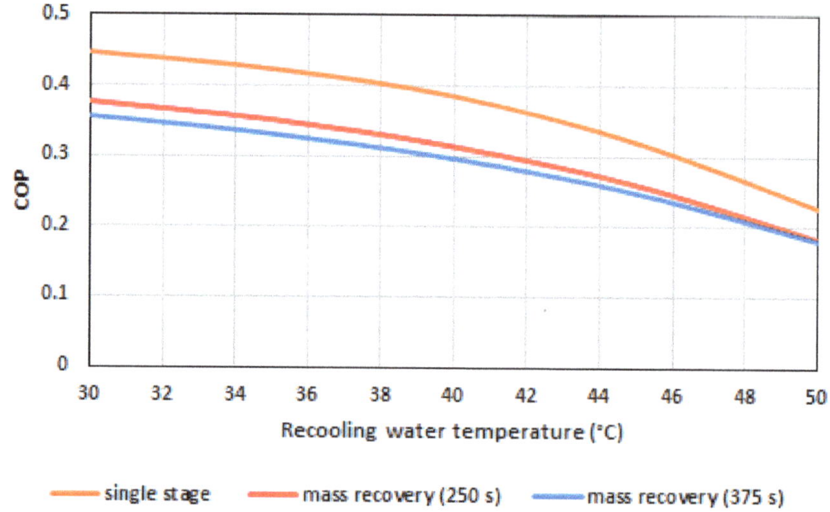

Figure 16. Predicted COP with various cycle stages and recooling water temperatures.

The cycle time has a significant effect on cooling capacity and COP. Therefore, the variable mode cycle was simulated using the model at different cycle times and with various recooling water temperatures to further understand their effects. The mass recovery duration is defined in terms of the ratio of cooling cycle duration, reflecting its dynamic value in a comparable manner. The following four modes were investigated in this study:

- First mode: single stage.
- Second mode: mass recovery with heating and cooling; the duration was 25% of the cooling cycle duration (denoted as short mass recovery mode).
- Third mode: mass recovery with heating and cooling; the duration was 50% of the cooling cycle duration (denoted as medium mass recovery mode).
- Fourth mode: mass recovery with heating and cooling; the duration was 75% of the cooling cycle duration (denoted as long mass recovery mode).

Figure 17 illustrates the modelling results at fixed inlet hot and chilled water temperatures of 90 and 18 °C, respectively. At a low recooling water temperature of 30 °C and a cycle time up to 600 s, the single stage mode exhibited higher capacity than the other modes (see Figure 17). However, at a longer cycle time (more than 600 s), the short mass recovery mode exhibited similar single-stage cooling capacity. With increased recooling water temperature, the system required longer mass recovery time to achieve higher (peak) capacity. Therefore, the uptake of the adsorber should be increased and that of the desorber should be decreased. This can be achieved by increasing the pressure in the condenser and the recooling water temperature or decreasing the adsorption pressure.

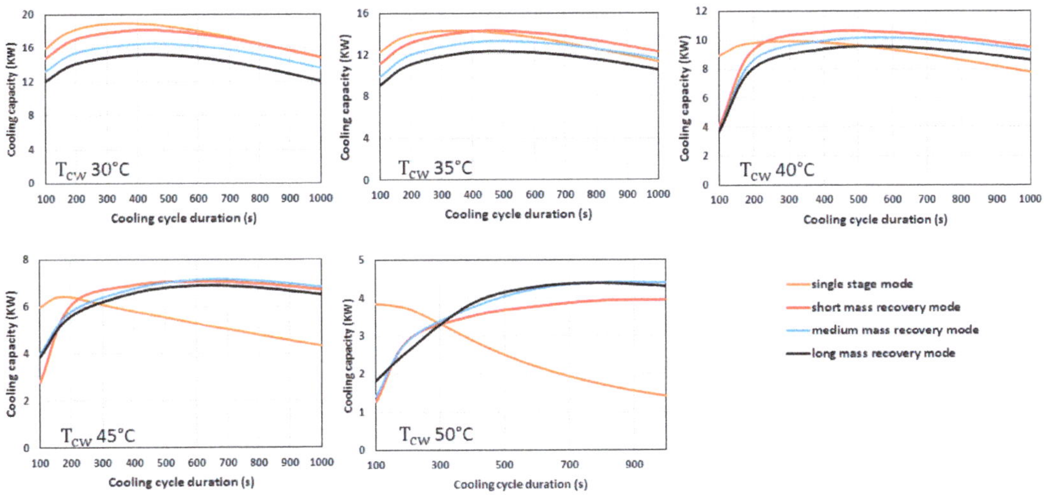

Figure 17. Predicted cooling capacity at various cycle modes and recooling water temperatures.

There is an optimum mode for each recooling water temperature, which is related to the optimum water vapor uptake and the condensation pressure. Figure 17 shows that cooling capacity increased with increased cooling cycle time until a peak point, after which it decreased. The peak cooling capacity was obtained when sufficient time was allowed to reach the optimum water uptake. With a shorter time, the water vapor uptake and cooling capacity were low because the adsorbent bed failed to reach the minimum temperature and pressure to achieve higher adsorption rates. With a longer time, the uptake of the adsorbent bed increased, thereby leading to a lower adsorption rate and cooling capacity. At a fixed recooling temperature, the cooling capacity decreased with an increase in relative mass recovery time because the cooling time became lower than the total cycle time.

Figure 18 illustrates the COP behavior at low and high recooling water temperatures. It can be seen that the COP increased as the cycle time increased because the cooling effect increased more than the heat input during the pre-heating phase. At a low recooling water temperature, the single stage cycle had a higher relative COP compared to the mass recovery cycle at a cooling cycle time less than 900 s and at a longer cycle time the short mass recovery had a higher COP. At a higher recooling water temperature of 50 °C the single stage cycle had a higher cooling cycle time of less than 750 s. While the mass recovery cycle had a higher COP at longer cooling cycle time, the long mass recovery duration had a higher COP followed by medium and short mass recovery duration.

During adsorption cooling, when the heat source is unlimited to waste heat or another free heat source, the cooling capacity value is higher than the COP. Therefore, in this study, the optimum mode cycle time was selected based on the maximum system cooling capacity. Table 5 shows the optimum cooling cycle time (OCCT) for each mode at various recooling water temperatures.

Figure 19 shows the cooling capacity at optimum cycle time for the different modes with different recooling water temperatures. The variable mode curve follows the optimum cooling capacity at different recooling water temperatures. At recooling water temperatures of 30–35, 35–44, and >44 °C, the single stage cycle, short mass recovery mode, and medium mass recovery mode, respectively, are found to be optimum modes.

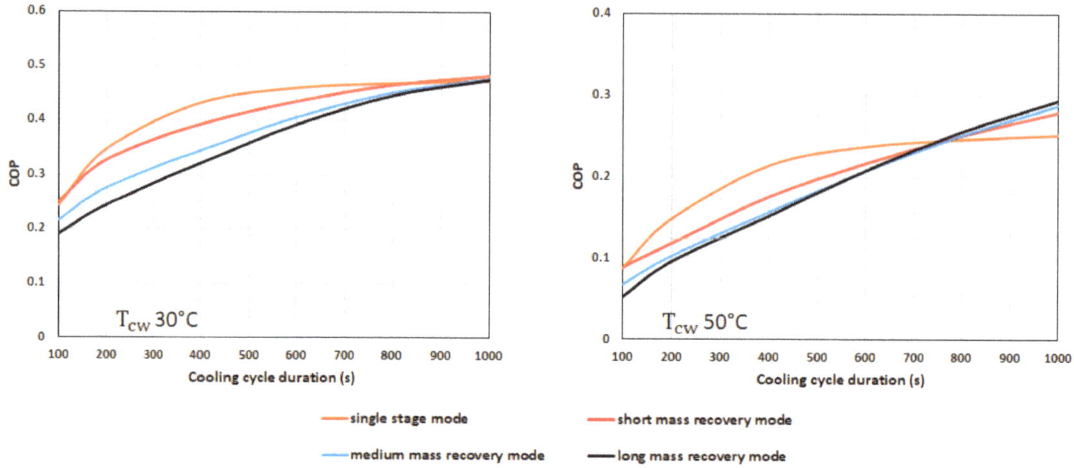

Figure 18. Predicted COP at various cycle mode and recooling water temperatures.

Table 5. Optimum cycle time for each cycle mode.

	Single Stage			Short Mass Recovery			Medium Mass Recovery			Long Mass Recovery		
Recooling Water Temperature (°C)	Optimum Cooling Cycle Time (s)	Qe (KW)	COP	OCCT (s)	Qe (KW)	COP	OCCT (s)	Qe (KW)	COP	OCCT (s)	Qe (KW)	COP
30	350	18.9	0.48	450	18.2	0.41	450	16.5	0.36	450	15.2	0.34
35	350	14.4	0.46	500	14.4	0.39	500	13.4	0.35	500	12.4	0.33
40	300	9.9	0.41	500	10.7	0.35	600	10.2	0.34	600	9.5	0.33
45	200	6.4	0.29	600	7.1	0.31	650	7.2	0.30	650	6.5	0.29
50	100	3.5	0.13	900	3.9	0.27	800	4.4	0.25	800	4.4	0.25

Under the same conditions listed in Table 5 and Figure 19, the COP at the optimum cycle time for the single stage, short mass recovery, medium mass recovery, and variable modes at various recooling water temperatures is shown in Figure 20. As explained previously, the variable mode tracks the optimum mode based on the cooling capacity. At recooling water temperatures of 35–44 °C, the variable mode COP drops to match that of the short mass recovery mode. At recooling water temperatures of 44–50 °C, the COP drops to match the medium mass recovery cycle mode.

This result suggests that the working ambient and recooling water temperatures are very important factors affecting COP cycle optimization. At an ambient temperature where the recooling water temperature is less than 35 °C, the single stage mode is recommended to obtain a higher COP; at higher temperatures, the short mass recovery mode is recommended. If the system changes between the two temperature ranges, a variable mode cycle is recommended.

To quantify and assess the air-cooled variable mode adsorption cycle, it is important to reflect on the ambient temperature profile. Figures 21 and 22 show hourly ambient temperature profiles for two days in Dubai. The figures show hourly simulation of the cooling output and COP for different modes based on the predicted chiller performance curves, assuming the temperature changes on an hourly basis and the recooling water temperature is 5 °C higher than the ambient temperature. On day 1, the variable model gives higher daily cooling capacity and COP of 12% and 3.6% compared to the single stage mode. On day 2, the variable mode gives 3.3% higher cooling capacity and 7.7% lower COP compared to the single stage mode.

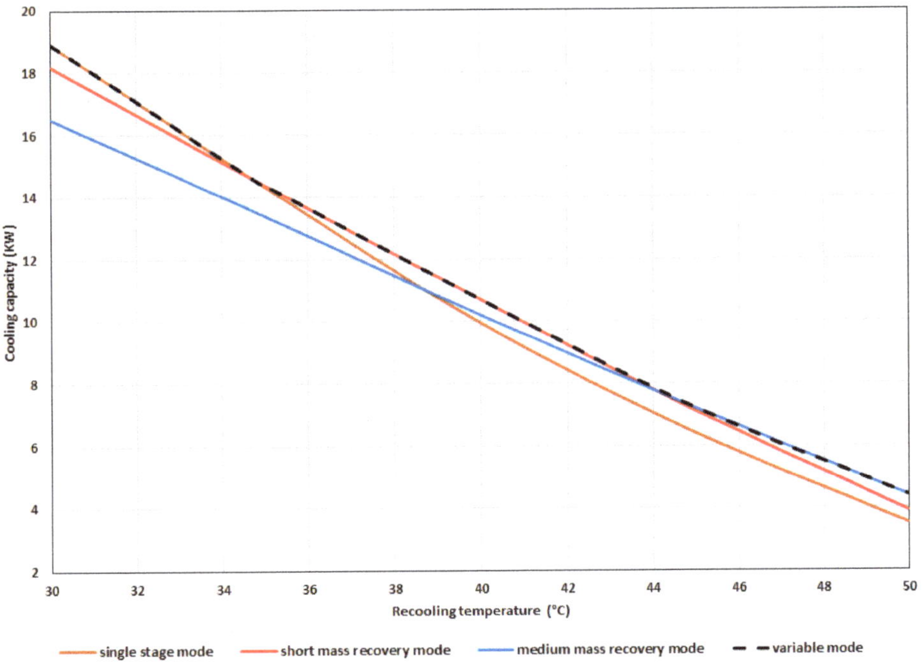

Figure 19. Predicted cooling capacity for various modes and recooling water temperatures.

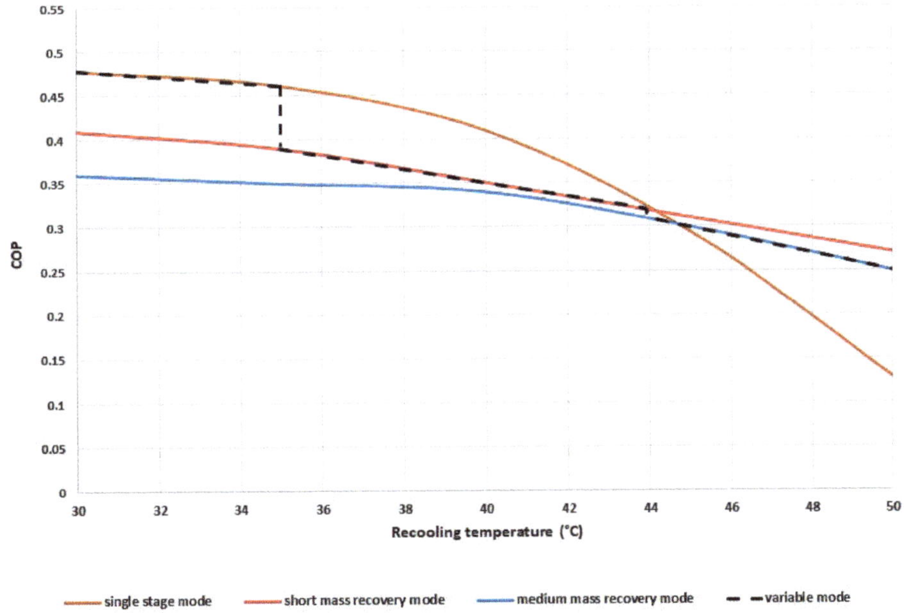

Figure 20. Predicted COP for various modes and recooling water temperatures.

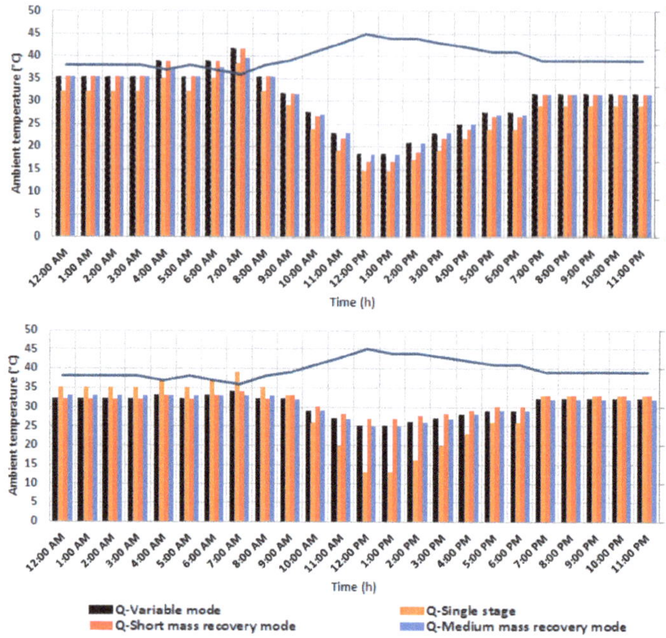

Figure 21. Measured ambient temperature and expected cooling capacity and COP for various modes (day 1).

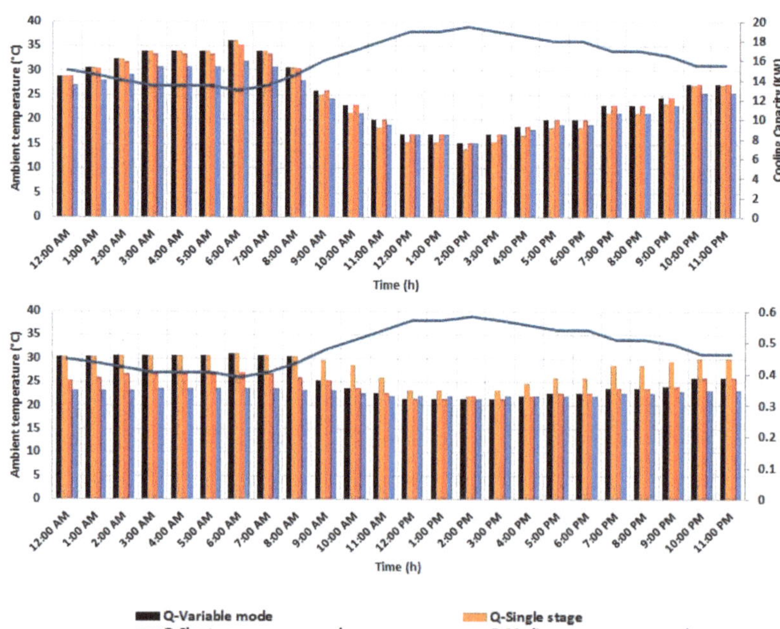

Figure 22. Measured ambient temperature and expected cooling capacity and COP for various modes (day 2).

5. Conclusions

In this study, we experimentally evaluated the adsorption isotherms and kinetics of a SAPO 34-coated heat exchanger. The adsorption isotherms were predicted at three temperatures using the Dubinin–Astakhov, Freundlich, Hill, and Sun and Chakraborty models. The experimental results fit well with the results obtained using these models, except for the Freundlich model. In addition, the adsorption kinetics parameters were calculated using a linear driving force model that was fitted to the experimental data with high correlation coefficients. The results showed that the kinetics of the adsorption parameters were dependent on the partial pressure ratio.

This study proposed and investigated a variable mode adsorption cooling cycle for single stage and different mass recovery durations, which exhibited several advantages compared to other fixed-stage adsorption cooling systems. This system reached optimum cooling capacity and performance at various ambient temperatures in hot and humid areas.

A prototype of the adsorption chiller was designed and experimentally and numerically investigated using the lumped model. The experimental and simulated results showed good agreement and similar trends.

Both experimental and simulation results at a constant cooling cycle time of 500 s indicated that, as the mass recovery time increased, the cooling capacity increased until there was a saturation trend, and as the mass recovery process time increased, the COP decreased until there was a saturation trend.

Four modes were investigated: single stage and three mass recovery modes of short, medium, and long duration. The cycle time was optimized based on the maximum cooling capacity. The single stage, short mass recovery, and medium mass recovery modes were found to be the optimum modes at lower recooling temperatures of less than 35 °C, a medium temperature of 35–44 °C, and higher temperatures of more than 44 °C. Therefore, the variable mode cycle can improve cooling capacity, and the improvement depends on the ambient temperature and recooling water temperature profile.

Author Contributions: Conceptualization, A.A.A.; methodology, A.A.A. and A.A.-M.; software, A.A.-M.; validation, A.A.A. and A.A.-M.; formal analysis, A.A.A.; investigation, A.A.A.; resources, A.A.-M.; data curation, A.A.A.; writing—original draft preparation, A.A.A.; writing—review and editing, A.A.-M.; visualization, A.A.A.; supervision, A.A.-M., M.A. and H.-J.B. All authors have read and agreed to the published version of the manuscript.

Funding: This research received no external funding.

Institutional Review Board Statement: Not applicable.

Informed Consent Statement: Not applicable.

Data Availability Statement: Not applicable.

Acknowledgments: This research was supported by the Precision Industries Specialized Power Generation Company, Dubai, UAE.

Conflicts of Interest: The authors declare no conflict of interest.

References

1. Goyal, P.; Baredar, P.; Mittal, A.; Siddiqui, A.R. Adsorption refrigeration technology—An overview of theory and its solar energy applications. *Renew. Sustain. Energy Rev.* **2016**, *53*, 1389–1410. [CrossRef]
2. De Boer, J.H.; Custers, J.F.H. Adsorption by van der Waals forces and surface structure. *Physica* **1937**, *4*, 1017–1024. [CrossRef]
3. Ambarita, H.; Kawai, H. Experimental study on solar-powered adsorption refrigeration cycle with activated alumina and activated carbon as adsorbent. *Case Stud. Therm. Eng.* **2016**, *7*, 36–46. [CrossRef]
4. Núñez, T.; Mittelbach, W.; Henning, H.-M. Development of an adsorption chiller and heat pump for domestic heating and air-conditioning applications. *Appl. Therm. Eng.* **2007**, *27*, 2205–2212. [CrossRef]
5. Wang, R.Z. Adsorption refrigeration research in Shanghai Jiao Tong University. *Renew. Sustain. Energy Rev.* **2001**, *5*, 1–37. [CrossRef]
6. Restuccia, G.; Freni, A.; Vasta, S.; Aristov, Y. Selective water sorbent for solid sorption chiller: Experimental results and modelling. *Int. J. Refrig.* **2004**, *27*, 284–293. [CrossRef]

7. Xia, Z.Z.; Chen, C.J.; Kiplagat, J.; Wang, R.Z.; Hu, J.Q. Adsorption Equilibrium of Water on Silica Gel. *J. Chem. Eng. Data* **2008**, *53*. [CrossRef]
8. Habib, K.; Saha, B.B.; Koyama, S. Study of various adsorbent-refrigerant pairs for the application of solar driven adsorption cooling in tropical climates. *Appl. Therm. Eng.* **2014**, *72*, 266–274. [CrossRef]
9. Shi, B. Development of an MOF Based Adsorption Air Conditioning System for Automotive Application. Ph.D. Thesis, University of Birmingham, Birmingham, UK, 2015.
10. Sayılgan, Ş.Ç.; Mobedi, M.; Ülkü, S. Effect of regeneration temperature on adsorption equilibria and mass diffusivity of zeolite 13x-water pair. *Microporous Mesoporous Mater.* **2016**, *224*, 9–16. [CrossRef]
11. Elsayed, E.; Al-Dadah, R.; Mahmoud, S.; Elsayed, A.; Anderson, P.A. Aluminium fumarate and CPO-27(Ni) MOFs: Characterization and thermodynamic analysis for adsorption heat pump applications. *Appl. Therm. Eng.* **2016**, *99*, 802–812. [CrossRef]
12. Chen, H.J.; Cui, Q.; Li, Q.G.; Zheng, K.; Yao, H.Q. Adsorption-desorption characteristics of water on silico-alumino-phosphate-34 molecular sieve for cooling and air conditioning. *J. Renew. Sustain. Energy* **2013**, *5*, 053129. [CrossRef]
13. Sing, K.S.W. Reporting physisorption data for gas/solid systems with special reference to the determination of surface area and porosity (Recommendations 1984). *Pure Appl. Chem.* **1985**, *57*, 603–619. [CrossRef]
14. Sun, B.; Chakraborty, A. Thermodynamic formalism of water uptakes on solid porous adsorbents for adsorption cooling applications. *Appl. Phys. Lett.* **2014**, *104*, 201901. [CrossRef]
15. Youssef, P.G.; Mahmoud, S.M.; AL-Dadah, R.K. Performance analysis of four bed adsorption water desalination/refrigeration system, comparison of AQSOA-Z02 to silica-gel. *Desalination* **2015**, *375*, 100–107. [CrossRef]
16. Wang, R.Z.; Wu, J.Y.; Xu, Y.X.; Wang, W. Performance researches and improvements on heat regenerative adsorption refrigerator and heat pump. *Energy Convers. Manag.* **2001**, *42*, 233–249. [CrossRef]
17. Ghilen, N.; Gabsi, S.; Benelmir, R.; El Ganaoui, M. Performance Simulation of Two-Bed Adsorption Refrigeration Chiller with Mass Recovery. *J. Fundam. Renew. Energy Appl.* **2017**, *7*. [CrossRef]
18. Al-Maaitah, A.A.S.; Al-Maaitah, A.A. Two-Stage Low Temperature Air Cooled Adsorption Cooling Unit. U.S. Patent No. 8,479,529, 9 July 2013.
19. Alsarayreh, A.; Albaali, A.; Alkatib, G. Performance Analysis of Solar Air Cooling (Two-Stage Adsorption Chiller) in Jordan. In Proceedings of the 1st International Conference & Exhibition on the Applications of Information Technology in Developing Renewable Energy Processes and Systems, IT-DREPS 2013, Amman, Jordan, 29–31 May 2013.
20. Al-Maaitah, A.A.; Hadi, F.M. Experimental Investigation of Two Stages Adsorption Chiller with Four Generators Utilizing Activated Carbon and Methanol as Working Pair. *Int. J. Curr. Eng. Technol.* **2017**, *7*, 43–54.
21. Baiju, V.; Muraleedharan, C. Investigations on Two Stage Activated Carbon R134a Solar Assisted Adsorption Refrigeration System. In Proceedings of the Internationa Conference on Mechanical, Nanotechnology and Cryogenics Engineering, Kuala Lumpur, Malaysia, 25–26 August 2012; pp. 223–227.
22. Alam, K.C.A.; Khan, M.; Uyun, A.; Hamamoto, Y.; Akisawa, A.; Kashiwagi, T. Experimental study of a low temperature heat driven re-heat two-stage adsorption chiller. *Appl. Therm. Eng.* **2007**, *27*, 1686–1692. [CrossRef]
23. Wirajati, I.; Akisawa, A.; Ueda, Y.; Miyazaki, T. Experimental Investigation Of A Reheating Two-Stage Adsorption Chiller Applying Fixed Chilled Water Outlet Conditions. *Heat Transf. Res.* **2015**, *46*, 293–309. [CrossRef]
24. Akahira, A.; Alam, K.C.A.; Hamamoto, Y.; Akisawa, A.; Kashiwagi, T. Experimental investigation of mass recovery adsorption refrigeration cycle. *Int. J. Refrig.* **2005**, *28*, 565–572. [CrossRef]
25. Wang, K.; Vineyard, E. Vineyard. New opportunities for solar adsorption refrigeration. *ASHRAE* **2011**, *53*, 14–24.
26. Chan, K.C.; Tso, C.Y.; Chao, C.Y.H. Simulation Study of the Heat and Mass Recovery on the Performance of Adsorption Cooling Systems. In Proceedings of the ASME 2014 8th International Conference on Energy Sustainability Collocated with the ASME 2014 12th International Conference on Fuel Cell Science, Engineering and Technology, Boston, MA, USA, 30 June–2 July 2014.
27. Wang, R.Z. Performance improvement of adsorption cooling by heat and mass recovery operation. *Int. J. Refrig.* **2001**, *24*, 602–611. [CrossRef]
28. Kabir, K.M.A.; Alam, K.C.A.; Sarker, M.M.A.; Rouf, R.A.; Saha, B.B. Effect of Mass Recovery on the Performance of Solar Adsorption Cooling System. *Energy Procedia* **2015**, *79*, 67–72. [CrossRef]
29. Uyun, A.S.; Miyazaki, T.; Ueda, Y.; Akisawa, A. Experimental Investigation of a Three-Bed Adsorption Refrigeration Chiller Employing an Advanced Mass Recovery Cycle. *Energies* **2009**, *2*, 531–544. [CrossRef]
30. Duong, X.Q.; Cao, N.V.; Hong, S.W.; Ahn, S.H.; Chung, J.D. Numerical Study on the Combined Heat and Mass Recovery Adsorption Cooling Cycle. *Energy Technol.* **2018**, *6*, 296–305. [CrossRef]
31. Muttakin, M.; Islam, M.A.; Malik, K.S.; Pahwa, D.; Saha, B.B. Study on optimized adsorption chiller employing various heat and mass recovery schemes. *Int. J. Refrig.* **2021**, *126*, 222–237. [CrossRef]
32. Al-Maaitah, A.A.; Alsarayreh, A.A.; Aldaour, M.M.; Precision Industries. Multiple Cycles Smart Adsorption chiller for High Ambient Temperatures. GCC Patent Office 48423, 12 October 2020.
33. Tabata, D.; Okamoto, K.; Taniguchi, K.; Kubokawa, S.; Mitsubishi Chemical Corp. Adsorptive Member and Apparatus Using the Same. U.S. Patent 8,795,418B2, 5 August 2014.

energies

Review

Adsorbents, Working Pairs and Coated Beds for Natural Refrigerants in Adsorption Chillers—State of the Art

Piotr Boruta, Tomasz Bujok, Łukasz Mika and Karol Sztekler *

Department of Thermal and Fluid Flow Machines, Faculty of Energy and Fuels, AGH University of Science and Technology, Mickiewicza 30 Av., 30-059 Cracow, Poland; boruta@agh.edu.pl (P.B.); bujok@agh.edu.pl (T.B.); lmika@agh.edu.pl (Ł.M.)
* Correspondence: sztekler@agh.edu.pl

Abstract: Adsorption refrigeration systems are promising, sustainable solutions for many cooling applications. The operating range and the performance of an adsorption cooling cycle are strongly dependent on the properties of adsorbents, adsorbates, and bed coatings. Therefore, further research and analysis may lead to improved performance of adsorption coolers. In this paper, studies on working pairs using natural refrigerants and the properties of adsorbent coatings were reviewed. The selected working pairs were then thermodynamically characterised and ranked in terms of refrigerant evaporation temperature values. This was found to be a key parameter affecting the applicability of a given adsorbent/adsorbate pair and the value of SCP (Specific Cooling Power), COP (Coefficient of Performance) parameters, which are now commonly used comparison criteria of adsorption chillers. In the analysis of the coating studies, the focus was on the effect of individual parameters on the performance of the cooling system and the effect of using coated beds compared to packed beds. It was found that a fundamental problem in comparing the performance of different cooling systems is the use of different operating conditions during the tests. Therefore, the analysis compares the performance of the systems along with the most important thermodynamic cycle parameters for the latest studies.

Keywords: adsorption chiller; adsorption working pairs; coated beds; comparative analysis; natural refrigerants

Citation: Boruta, P.; Bujok, T.; Mika, Ł.; Sztekler, K. Adsorbents, Working Pairs and Coated Beds for Natural Refrigerants in Adsorption Chillers—State of the Art. *Energies* **2021**, *14*, 4707. https://doi.org/10.3390/en14154707

Academic Editors: Antonio Zuorro and Gabriele Di Giacomo

Received: 3 July 2021
Accepted: 29 July 2021
Published: 3 August 2021

1. Introduction

In 2018, the total global electricity consumption was 24,738.9 TWh, of which about 2075 TWh was the energy demand for cooling [1]. Moreover, the use of cooling in buildings has been increasing rapidly for several years [1]. This is due to the increasing standard of living and architectural trends observed in the building industry, as well as the increase in average and maximum temperatures from year to year [2]. Therefore, energy efficiency has to be increased as a remedial measure for global warming and increasing energy consumption [1]. In addition, it is also important to note the importance of the refrigeration industry for healthcare. Currently, due to the development of healthcare in underdeveloped countries and the need to store many medicines, the demand for refrigeration in the healthcare sector is increasing worldwide.

The issues raised clearly demonstrate the need to look for alternative refrigeration technologies, which include adsorption refrigeration. Hassan et al. [3] divided adsorption systems into open (air conditioning and dehumidification) and closed (freezing, cooling, and air conditioning). This is a rather conventional division, but it draws attention to the different applications of cold to be obtained and thus the different requirements for refrigerant temperature.

Regarding the use of adsorption chillers to cool buildings, it should be seen as an opportunity to reduce the consumption of non-renewable primary energy and carbon footprint. Adsorption chillers are most commonly used in large facilities such as office buildings, hotels, hospitals, and manufacturing plants. Chillers that operate at regeneration temperatures in the range of 45–60 °C can be used in hotels and powered by heat recovered from used hot water, while chillers requiring higher desorption temperatures can be used in industrial chillers where waste heat is often in the 60–80 °C range. Refrigerators powered by waste heat enable operation regardless of climate conditions and can be applied in any place where it is possible to utilize waste heat. In addition, apartment blocks and other buildings that heat their buildings with heat from CHP (Combined Heat and Power) plants can be considered. Connecting to a district heating plant involves paying a fixed charge for the capacity demanded regardless of the low heat consumption in summer. Therefore, in these buildings the use of adsorption chillers for air conditioning of apartments can also make economic sense and enable more efficient operation of district heating plants, which have to cope with excess heat in summer. However, it is also possible to use adsorption chillers in buildings where there are no waste heat sources. In every building where there is a possibility to apply solar collectors, it is possible to apply an adsorption chiller, which will use the heat from the solar radiation to cool the building. Adsorption chillers powered by heat from renewable energy sources (mainly solar energy) operate most efficiently in warm climate zones with high solar radiation supply (tropical, subtropical, temperate climate zones). This solution reduces the load on the energy system in the summer and allows efficient cooling without the need for cooling storage. On the other hand, the further development of global healthcare is very often linked to the provision of adequate storage conditions for drugs and vaccines. Here, special attention should be paid to the medical facilities of underdeveloped countries and developing countries, where the lack of widespread access to the electricity grid very often prevents the use of compressor chillers. Taking into account the fact that most of these countries are in African, Asian, and Central American regions, it can be assumed that adsorption cooling systems powered e.g., by solar energy may be a great opportunity for local communities to improve the quality of medical services provided there.

Adsorption systems reduce electricity consumption in refrigeration and air conditioning by exploiting the thermal compression effect of the refrigerant. Therefore, the main driving energy of the adsorption cycle is heat, which can be low-temperature heat or even waste heat. Therefore, adsorption systems are of paramount importance for sustainable use of energy. Additionally, adsorption systems are characterised by a lack of moving parts, which contributes to their silent operation. On the other hand, it should be noted that these systems have been already investigated in the 1980s of the previous century and, despite many studies, are still characterised by low efficiency coefficients and COP and SCP values. This fact often results in limited applicability due to the size of the equipment and cost-effectiveness of the systems used.

However, the use of COP and SCP as evaluation criteria for adsorption chillers does not allow a reliable evaluation and comparison with other appliances. The COP is defined as the ratio of the heat of evaporation of the refrigerant to the heat of preheating and desorption during the operating cycle of the chiller [4]. This formula does not take into account the fact that the unit is powered by, e.g., waste heat, which is simply irretrievably lost in many industrial processes. Furthermore, considering the fact that compressor units are powered by electricity draws attention to another aspect of this comparison, namely, the differences in power and heat quality. Therefore, an actual comparison of the different systems can be made, for example, based on the basis of an exergetic analysis of the adsorption and compressor cycle operation. SCP, on the other hand, is defined as the ratio of the latent heat of vaporisation of the refrigerant to the mass of the adsorbent [4]. This parameter seems to be a more appropriate comparison criterion than COP, since it allows assessment of the mass of the whole system, which is important in the context of the application of the given solutions. Nevertheless, the COP and SCP parameters take into account the heat of vaporisation of the refrigerant, which is directly related to another parameter characteristic of sorption phenomena, namely, uptake. This parameter is used in publications interchangeably with the concept of adsorption rate, and both can be defined as the mass of adsorbate vapour that has been adsorbed per unit mass of the adsorbent under given conditions. The adsorption rate depends on the temperature, gas pressure and the size of the specific surface area of the adsorbent. Therefore, to compare specific devices, it is necessary to analyse all three of these parameters and pay attention to the desorption temperature and the cold produced. This approach minimises the risk of erroneous conclusions, which can be reached by directly comparing a chiller producing cold for air conditioning purposes with a freezing device. On this basis, attention should be drawn to the need to seek other alternative comparison criteria.

Of course, cooling efficiency is important, especially for the end user who uses the equipment. However, it is necessary to analyse the efficiency of the entire process chains, which consist of several processes linked to different energy carriers. At each stage there are energy or exergy losses, so it is possible to determine the cumulative energy and exergy consumption in the production of a given product/good. In the literature, one can encounter the concept of environmental cost, which refers to the consumption of exergy of non-renewable natural resources but does not consider the impact of CO_2 emissions [5]. Based on this scheme, the approximate efficiency of compressor chillers can be determined, assuming an average EER (Energy Efficiency Ratio) for chillers of 3.5, an average EU electricity generation efficiency of 45%, and an efficiency of energy distribution and transmission of 92% [6]. It can be concluded that compressor chillers have an average primary energy efficiency of 1.5. It should be noted that among the energy sources used in the EU, about 40% of primary energy is generated from fossil fuels, and this rises to more than 70% in some countries. After considering the fact that adsorption chillers can be powered with low-temperature heat, often waste heat or renewable energy sources, a COP of adsorption systems in the range of 0.5 to 0.8 qualifies these units for further research.

When analysing the various comparative parameters of adsorption chillers, it is important to note which component of the system limits the cooling capacity the most. Many components can be mentioned, such as bed heat exchanger, evaporator, condenser, and control system [7], but the first component that affects the cooling cycle is the adsorbent material and its interaction with the adsorbate. Adsorbents, due to their porosity, are characterised by limited heat transport in the bed. It is also worth noting that porosity is a relative concept. From the adsorbent point of view, the pore size is essential (Figure 1), because adsorption occurs most intensively in the micropores and only after they are saturated with adsorbate does it move to mesopores and macropores. Thus, the selection of an optimal adsorbent-adsorbate pair is essential to improve heat and mass transport in the bed [8]. Modifications related to bed design and interference with the thermodynamics of the adsorption process should be implemented for the best materials, according to the authors.

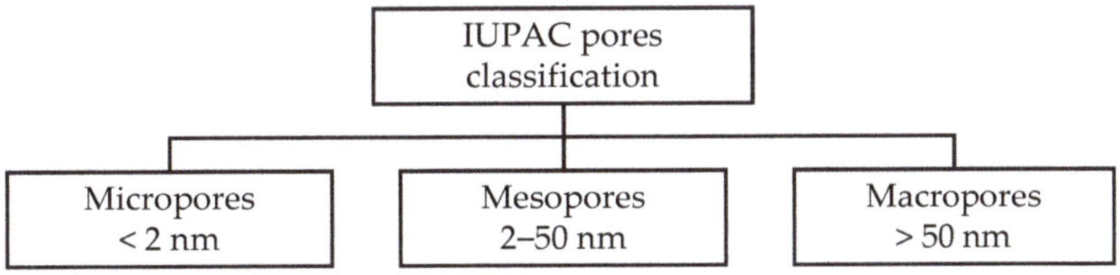

Figure 1. IUPAC pore classification in the characterisation of porous materials [9].

Due to the need to develop alternative cooling technologies and to meet the postulates of sustainable development, this article focuses exclusively on the analysis of adsorption working pairs using natural refrigerants (Table 1). Nevertheless, it should be noted that such a trend was already observed in the 19th century, when ammonia or carbon dioxide were used in refrigeration systems. However, the later development of synthetic refrigerants made natural refrigerants less important. It was not until the current regulations concerning ozone-depleting substances [10] that research into alternative refrigerants was stimulated, because more refrigerants are gradually being phased out, which affects the growing costs of mechanical refrigerators [5].

Table 1. Properties of natural refrigerants [11,12].

Refrigerant	Critical Temperature [°C]	Critical Pressure [kPa]	Boiling Point [1] [°C]	Heat of Evaporation [kJ/kg]	Thermal Conductivity [2] [W/m·K]	ODP	GWP
Water [3]	373.95	22,064	99.97	2256	0.556	0	0
Carbon Dioxide	30.98	7377	−78.46	379.5	0.015	0	1
Methanol	240.20	8220	337.85	1165	0.204	0	-
Ethanol	240.80	6250	78.2	919	0.171	0	-
Propane	96.74	4251	−42.11	428	0.017	0	11
Isobutane	134.66	3629	−11.75	367	0.107	0	3
Ammonia	132.41	11,357	−33.33	1372	0.540	0	0

[1] values are given for the liquid at 1 bar pressure. [2] values are given at temperature 0 °C. [3] water vapour is a greenhouse gas, but it is not considered a cause of man-made global warming.

Water as a refrigerant has good thermodynamic properties, but a pressure in the range of 0.6–1.2 kPa is required to obtain chilled water of 0–10 °C [13], which means that a high tightness of the system must be maintained. The group of low-pressure refrigerants also includes methanol and ethanol. On the other hand, carbon dioxide, propane, isobutane, and ammonia have low boiling points at atmospheric pressure. Ammonia enables operation at pressures of 0.5–5 bar, which allows the production of chilled water at temperatures as low as −50 °C [13]. This fact clearly distinguishes this refrigerant from the others. Nevertheless, researchers point to the toxicity of ammonia vapours as a feature affecting the limited number of studies with this refrigerant [14]. It should also be added that an ammonia mixture of 16–27% by volume with air is explosive. Among the refrigerants listed in Table 1, propane is also characterised by explosive properties. For the safety of the cooling system, it must be considered that isobutane, propane, ammonia, methanol, and ethanol are flammable refrigerants. Analysing the data included in Table 1, it should be noted that water has the highest latent heat of the natural refrigerants listed. On the other hand, ammonia, methanol, and ethanol have latent heats 40%, 50%, and 60% lower than water, respectively. Propane, isobutane, and CO_2 all have a latent heat about 80% lower than water. All the refrigerants listed in Table 1 have no ozone depleting effect and are characterised by extremely low GWP. It must therefore be concluded that all the factors listed have a negligible impact on the environment. Thus, when selecting a specific refrigerant, its thermophysical and thermodynamic parameters should be analysed for the specific refrigeration application. There are different adsorbent-natural refrigerant working pairs and, in each application, the one with the best sorption kinetics should be selected.

The currently leading physical, chemical, and composite adsorbents are discussed and the further possible development of each of these material groups is evaluated. In addition, by analysing a number of different adsorbent/adsorbate pairs, the focus is on the applicable comparative criteria of different studies. In the context of adsorbents, the focus is on evaluating pore volume and size, surface development, particle size, and thermal conductivity. On the other hand, only studies that involve the use of natural refrigerants such as water, methanol, ethanol, ammonia, CO_2, and organic hydrocarbons were selected for the analysis of adsorption working vapours. In addition, the analysis of working adsorption pairs focus on the operating parameters of the refrigeration system. The application of the different working vapours was decided to be divided into three groups: freezing, refrigeration, and air conditioning. Refrigerant evaporation and desorption temperatures have not been analysed as potential comparison criteria of different studies published so far, which affects the difficulty in comparing the different adsorption working pairs. On the other hand, the main aspect of the coating analyses is the influence of the application and design of the adsorbent coatings on the adsorption cycle performance. In this analysis, the basic parameters of the coating structure and the application technology were evaluated with particular focus on the effects on heat and mass transport. Furthermore, possible gaps in previous research are focused on, with the aim of indicating possible directions for further work related to the search for optimal pairs.

The main objective of this paper is to review, discuss, and compare the research conducted mainly between 2015 and 2021 on the materials currently used in adsorption cooling. Adsorbents as well as coated beds have been considered to improve heat and mass transport efficiency in sorption processes. Consequently, the possibility of improving the COP and SCP performance of the adsorption chiller by modifying the adsorbents was evaluated.

2. Adsorbents and Their Influence on a System Performance

As mentioned in the introduction, the adsorption process involves the binding of refrigerant vapours in the pore volume of the adsorbent. Therefore, it should be stated that the adsorbent is the main component of the adsorption refrigeration cycle and is largely responsible for the performance of the refrigeration process. Properties such as sorption kinetics, specific surface area, pore size and pore volume, thermal conductivity, and stability form the basis for evaluating the suitability of a given adsorbent for use in each refrigeration application. In addition, the adsorbent should have the propensity to adsorb large amounts of adsorbate over a narrow temperature range and the ability to desorb rapidly. Moreover, the adsorbent properties should be stable over a wide operating range.

There are many types of adsorbents, both naturally occurring in nature and artificially produced. Generally, adsorbent materials can be divided into three groups:

- physical and novel porous adsorbents;
- chemical adsorbents;
- composite and doped adsorbents [4,15].

Chemical adsorbents are a group of sorption materials that include compounds such as chlorides (e.g., strontium, magnesium, lithium), hydrides (e.g., lithium, calcium), and metal oxides and hydrates of inorganic salts. Chemical adsorbents differ from physical adsorbents because in chemical adsorption there is a strong chemical bond between the adsorbent and the refrigerant. Chemical adsorbents allow higher adsorption capacity, but also require higher desorption temperatures. In addition, chemical adsorbents are also susceptible to swelling and agglomeration phenomena, which generate the problem of clogging of material pores. Therefore, they are rarely used in adsorption systems.

Sharma et al. [16] in their work analysed sorption materials in the form of halide salts such as $CaCl_2$, $SrCl_2$, $MnCl_2$, and $FeCl_2$. Of the salts analysed, only two selected salts operate at desorption temperatures of 80–100 °C. Nevertheless, it is the $CaCl_2$ salt that shows the best adsorption performance among the analysed compounds. It is characterised by relatively high adsorption capacity at 0.5 g/g and low enthalpy of desorption at 56.59 kJ/mol, which affects the relatively high COP of 0.53.

In general, chemical adsorbents do not play a leading role in the reviewed literature, which includes recent research in the field of materials used in adsorption cooling processes. However, this fact does not exclude their use as admixtures with other adsorbents. Chemical compounds such as strontium chloride [17], calcium chloride [18–20], lithium bromide [21], sodium chloride [20,22], and lithium chloride [20] are currently used as additives in composite materials. The combination of the above-mentioned chemical compounds with classical adsorbents such as SG (Silica Gel), AC (Activated Carbon), and zeolites allows in many cases to improve the heat and mass transport in the material and to increase the adsorption size. At the same time, in many cases, the appropriate choice of the mass proportion of the chemical compound allows to eliminate the negative characteristics of chemical adsorbents, which are the common phenomena of swelling, agglomeration, crystallization, which causes clogging of pores and limiting the size of adsorption. Therefore, Section 2.2 of this paper discusses commonly used composite materials, which can be regarded as an attempt to combine the best properties of physical and chemical adsorbents.

2.1. *Physical Adsorbents*

These adsorbents are used in adsorption chillers and depend on van der Waal's forces to retain the adsorbate. The most widely used physical adsorbents in refrigeration and thus the best studied are activated carbons, zeolites, and silica gels. These materials are often referred to as classical by researchers due to their influence on the development of adsorption refrigeration. Nevertheless, further development of this scientific discipline has led to the identification of novel porous adsorbents such as aluminium phosphate (AlPO), silicoaluminophosphate (SAPO) and metal-organic frameworks (MOFs) [15,23].

2.1.1. Activated Carbons

Activated carbons, despite being well known and widely studied materials, are still the basis for further research and this trend is expected to continue. This is due to the fact that, despite the significant development of other adsorbent groups, activated carbons (ACs) provide high sorption dynamics and are characterised by low manufacturing cost. Allouhi et al. [13] pointed out that activated carbons can be produced from various raw materials such as wood, coconut, any nutshells, or coal. When properly treated, these materials exhibit high degree of surface development, the highest among the adsorption materials used. The HDACFs (high-density activated carbon fibres) analysed in Kumit's study [24] are characterised by a surface development of 3263 m^2/g, which is the best among the materials summarized in Table 2. In general, the selected ACs are characterised by good surface development, but vary in a wide range of 800–3000 m^2/g. Such a high surface development of ACs results in a significant pore volume contribution to the adsorbent mass, which is in the range of 0.43–1.85 cm^3/g, and the best porosity is characterised by PR KOH4. Analysing the properties of selected activated carbons collected in Table 2, it should be noted that with the increase of surface area, the pore volume in the adsorbent increases. Nevertheless, these two parameters are not sufficient to evaluate the quality of the adsorbent material.

The porosity of the adsorbent is an essential parameter, but the dynamics of the adsorption process is mainly determined by the size of these pores. Physical adsorption occurs first in the micro pores, because they have the strongest interaction with the adsorbate. Only after saturation of the smallest pores, the adsorbate starts to settle in the mesopores. Therefore, the average pore size is an important parameter because it realistically affects the adsorption rate and allows the shape of the sorption curve to be predicted. Brancato et al. [21] analysed the characteristics of five activated carbons: SRD 1352/3, FR20, AP4-60, ATO and COC-L1200, which are characterised by different origins. They found that the pore width of the adsorbent is an important parameter characterising the adsorption capacity of an adsorbent. Therefore, materials such as HDACF [24], Maxsorb III (H2) [18,25,26], and PR KOH4 [27] characterised by very large pore volume (1.7–1.85 cm^3/g) will not necessarily allow high adsorption size, precisely because of the large average pore size. In general, by analysing the average pore size of the materials summarized in Table 2, it should be concluded that ACs are characterised by small pore size, on the order of 0.5–0.6 nm. Nevertheless, treatments that modify activated carbons, such as increasing thermal conductivity by pelletisation, can have a negative effect on pore size. Another treatment is pressing, which can also lead to a reduction in micropores' volume, such as for HDACF material.

A parameter related to porosity is the mentioned thermal conductivity of the adsorbent. Adsorption cooling characteristics require heating of the adsorbent in order to regenerate the bed. Therefore, porosity is important from a heat transport point of view because it slows down the heating of the adsorbent, which is reflected in the very low thermal conductivity values of adsorbent materials. However, this parameter is rarely compiled by researchers e.g., the ACs conductivity of 0.066 $W/(m{\cdot}K)$ for Maxsorb III (H2) is largely responsible for the low heat transfer efficiency of the bed. On the other hand, HDACF after the granulation process has a much higher thermal conductivity of about 0.2 $W/(m{\cdot}K)$;

nevertheless, as already mentioned, granulation reduces the volume of micropores in the adsorbent.

Table 2. Physical properties of activated carbons.

Ref.	Adsorbent	Pore Volume [cm^3/g]	Thermal Conductivity [W/(m·K)]	BET Surface Area [m^2/g]	Adsorbent Size [mm]	Average Pore Size [nm]
[21,28]	SRD 1352/3	0.65	-	2613	0.5–2.0	0.56
[21]	FR20	0.75	-	2180	0.01	0.59
[21]	AP4-60	0.47	-	1428	>4.0	0.64
[21]	ATO	0.64	-	1745	0.25–0.6	0.59
[21]	COC-L1200	0.49	-	1412	0.42–1.0	0.59
[22]	AC	0.435	-	1237	-	-
[29]	SRD 1352/3	0.65	-	-	0.71–1.18	-
[30,31]	CSAC	0.43	-	804	0.00022	1.76
[24]	HDACF	1.70	0.162–0.205	3263	0.018	2.10
[18,25,26]	Maxsorb III (H2)	1.70	0.066	3045	-	1.12
[26]	PR KOH4	1.85	-	3060	-	1.25
[32]	ACM-35.4	0.69	-	1200	-	2.30

The last parameter in Table 2 is the adsorbent particle size. Analysing the given materials, it should be noted that ACs can be characterised by a size of a few micrometers or millimeters. Brancato et al. [29] in their work investigated the adsorption dynamics for different particle sizes of SRD 1352/3 activated carbon. Their study allows us to conclude that the adsorption capacity decreases as the adsorbent particle size increases. At the same time, the adsorption time increases with decreasing adsorbent particle size. This tendency is due to the fact that intermolecular mass transfer resistances limit the sorption dynamics. On the other hand, for adsorbent grains of increasing size, the sorption kinetics is limited by intramolecular heat and mass transfer. This observation points to the need to analyse the size of the material used, since the use of the same material but with different particle sizes is associated with a change in bed performance over a wide range. Brancato analysed activated carbon particles of 0.21–1.18 mm, which resulted in SCP values in the range of 1.5–2.41 kW/kg [29]. Such values are high, and unprecedented in other publications.

In turn, the fact that activated carbons can have such different particle sizes makes it possible to conduct the analysis of the behaviour of the blends consisting of different particle sizes. In their study, Hamrahi et al. [33] focused on comparing the performance of adsorption cooling based on activated carbon and a mixture of activated carbons. The mixture was obtained by adding different amounts of nano-activated carbon to micro-activated carbon. It was calculated that a 5% addition of nano-AC to the adsorbent improves the adsorption volume and system performance by about 10% with reference to micro-activated carbon. In turn, about 20% proportion of nano-AC can improve the system performance by about 30%. Moreover, the said adsorbent works with regenerative heat, on the order of 70–80 °C.

2.1.2. Silica Gels

Referring to the previous studies, it is possible to outline the thesis that silica gels (SGs) are currently a less frequently studied material than, e.g., activated carbons or composite adsorbents. Nevertheless, this material is still widely used in adsorption chillers operating based on solar regenerative heat. This is due to the fact that SGs allow the chiller to operate at low desorption temperatures. Furthermore, as noted by Yaici et al. [34], silica gels as adsorbents are cost-effective and widely available.

Analysing the data in Table 3 [35–41], it should be noted that SGs are characterised by a surface development in the range of 650–1000 m^2/g and a pore volume in the range of 0.3–0.5 cm^3/g. Additionally, the silicas listed in Table 3 have pores with an average size of 0.9–3.2 nm. The parameters listed are characterised by lower values than those described by activated carbons. Nevertheless, pore volume and pore size directly affect the heat and mass transfer in the adsorption bed. Therefore, the lower porosity of SGs, relative to other adsorbents, affects the higher value of the thermal conductivity of silica gels. The thermal conductivity of SGs [38] is 0.72 W/(m·K) and for SG type RD is 0.2 W/(m·K). These values are several times higher than for ACs, so lower temperature heat can be used to regenerate silica gels than to regenerate activated carbon.

Table 3. Physical properties of silica gels.

Ref.	Adsorbent	Pore Volume [cm^3/g]	Thermal Conductivity [W/(m·K)]	BET Surface Area [m^2/g]	Adsorbent Size [mm]	Average Pore Size [nm]
[35]	Silica Gel	-	0.72	-	0.26	-
[36]	RD	0.462	-	827.5	0.7–1.0	3.24
[36,37]	RD-2060	0.335	-	686.3	0.3–0.7	3.19
[38,39]	RD	-	0.198	-	0.2	-
[40]	A-type SG	0.491	-	997	-	0.90
[41]	Silica Gel	0.375	-	650–800	2.0–4.0	-

Analysing recent studies conducted on silica gels, it should be noted that they focus on the evaluation of the effect of SG particle size on sorption processes. Vodiannitskaia et al. [35] compared different grain sizes of silica gel and addressed the determination of the optimal particle size of the material. It was found that decreasing the particle size affects the porosity of the adsorbent, due to the fact that the micropores become clogged during the material crushing process. The best performance of the system was determined for a particle size 0.5 mm and no significant change in performance was observed with further grinding of the material. Nevertheless, as the particle size increases from 0.5 to 2 mm, SCP decreases by about 10% and reaches a value of 70 W/kg. The same change in adsorbent particle diameter results in a 5% decrease in COP, which is about 0.55. In contrast, Radu et al. [42] investigated the sorption dynamics for two silica gel sizes (0.45 and 0.85 mm) as a function of the number of sorbent layers. The authors found that smaller particles showed better sorption kinetics in monolayer configurations. Moreover, the more layers, the slower the adsorption and desorption processes. The study also determined a negligible effect of increasing the effective diffusivity inside the adsorbent particles on the SCP value. In contrast, decreasing diffusivity significantly decreases the system performance, especially for large adsorbent particle sizes. A similar relationship of decreasing sorption size with increasing particle size was observed by Yaici et al. [34], who investigated the effect of silica gel particle size on adsorption. Their study also confirmed that the use of a silica gel with a smaller size reduces the cycle time. Moreover, the analysis of adsorption curves reveals a relationship that for larger particle sizes the adsorption rate becomes increasingly linear over time.

2.1.3. Zeolites

As already mentioned in this chapter, zeolites belong to the group of physical adsorbents that, besides SGs and ACs, can be called classical. Analysing the materials summarised in Table 4, it can be observed that zeolites are characterised by lower pore volume and surface area parameters than the previously described activated carbons and silica gels. The surface area of zeolites is 30–720 m^2/g and their pore volume is 0.1–0.35 cm^3/g. However, with a pore size of 0.6–1.2 nm on average and thermal conductivity of 0.02–0.11 W/(m·K), it can be concluded that zeolites are ranked between active carbons and silica gels in terms of sorption properties. It is worth mentioning that zeolites are minerals and there are about 40 naturally occurring zeolites and about four times as many artificially produced ones [13]. This observation influences the fact that these materials continue to be analysed in further studies. In their work, Kayal et al. [43] addressed the study of AQSOA zeolites (Z01 and Z02) using different experimental approaches. It should be noted that these materials are characterised by a significant difference in specific surface area and pore volume parameters. Zeolite Z01 is characterised by a surface area development of 132 m^2/g and a pore volume of 0.087 cm^3/g. On the other hand, zeolite Z02 is characterised by four times higher values of the given parameters. The authors attributed such significant differences in properties to the difference in ionic radii of Fe^{3+} and Si^{4+} in the lattice structure of AQSOA-Z01 and AQSOA-Z02, respectively. Furthermore, based on the study, zeolite Z01 is suitable to operate at desorption temperatures below 65 °C, while zeolite Z02 operates more efficiently at 80 °C, which is related to the adsorption curves of these materials. The AQSOA family of zeolites shows very good thermal stability, adequate adsorption charge and high desorption rate.

Similar to SGs, for zeolites also studies are being conducted to evaluate the effect of adsorbent particle size on sorption kinetics. Girnik et al. [44] investigated the study of different particle sizes of AQSOA™-FAM-Z02 material. They found that as the zeolite grain size increases, the SCP of the cycle decreases, which is directly related to the increase in adsorption and desorption time for the material with larger particle size. Nevertheless, it should be noted that the adsorption curves for different particle sizes are very similar in shape. Relating these observations to the studies of Yaici et al. [34] and Brancato et al. [29], it should be concluded that SGs, ACs, and zeolites show higher adsorption dynamics for smaller particle sizes. On the other hand, the effect of adsorbent particle size on the shape of the adsorption curve cannot be clearly determined. Girnik et al. [45] in another publication also studied the effect of silica gel grain size on the adsorption size of adsorbate. It is worth mentioning that the authors conducted their study on a monolayer of adsorbent, which marginalises the problem related to vapour diffusion of adsorbate, and mainly draws attention to the role of heat transfer. The sorption processes occurring in the adsorbent monolayer cannot be compared with those occurring in a bed consisting of several layers of material, due to the significant difference in the conditions of the process occurrence. Hence, another doubt arises regarding the comparison of different test results. Due to the complexity of the phenomenon in question, individual papers should be carefully collated to avoid drawing incorrect conclusions.

Table 4. Physical properties of zeolites.

Ref.	Adsorbent	Pore Volume [cm^3/g]	Thermal Conductivity [W/(m·K)]	BET Surface Area [m^2/g]	Adsorbent Size [mm]	Average Pore Size [nm]
[43,46]	AQSOA-Z01	0.071–0.087	-	132–190	0.005–0.008	1.18
[43,46]	AQSOA-Z02	0.269–0.277	-	590–718	0.005	1.18
[46]	AQSOA-Z05	0.07	-	187	-	1.18
[45]	AQSOA-FAMZ02	-	0.019	-	0.20–0.25	-
[38]	FAM-Z01	-	0.113	-	0.2	-
[47]	AlPO4	0.33	-	642	-	0.66
[47]	SAPO4	0.34	-	659	-	0.66

As for other adsorbents, the adsorption mechanism for zeolites also occurs primarily in micropores. This phenomenon is also of interest to researchers because further understanding of this phenomenon may enable significant development of adsorbents. Fan et al. [48] analysed the interaction of water molecules with the pores of zeolites. The authors found that the interaction potential between the adsorbate and adsorbent can be determined by the isosteric heat of adsorption as a function of AQSOA pore width. Based on their study, they concluded that the interaction between adsorbate molecules and zeolite channels is strongest in the centre of the zeolite channel. The developed relationships indicate the validity of pore size analysis in the adsorbent and can be used to design new types of adsorbents with high kinetics and adsorption size.

However, even the analysis of several parameters does not always allow for unambiguous comparison of adsorbent parameters. Analysing the data collected in Table 4, a conclusion arises that in studies of adsorbents the thermal conductivity of the material is relatively rarely determined, while it is a very significant parameter, as it allows evaluation of the heat transfer in the material. As a proof of this statement, analyses conducted by Kim's team [47] were quoted for the analysis of the application potential of a new AlPO4 adsorbent, which was compared with another, already known zeolite SAPO4. The AlPO4 zeolite is characterised by very similar parameters to SAPO4. The good adsorption capacity and favourable micro-pore structure are worth mentioning. Despite the very similar sorption properties of these materials, there is a difference in the sorption processes occurring with these sorbents. For AlPO4, the desorption process ends at 75 °C, while SAPO4 continues desorption under these conditions. For SAPO4, raising the desorption temperature to 110 °C is most beneficial. This observation highlights the advantage of AlPO4, which requires less heat of regeneration, which is directly related to the difference in performance of systems based on these zeolites. On the other hand, it is also necessary to analyse the behaviour of a given material at variable refrigerant regeneration temperatures, as these characterise real refrigeration systems. Mohammed et al. [37], based on their experimental studies, found that silica gel RD-2060 is superior to zeolite AQSOA-Z02 in terms of SCP achieved over the entire range of refrigerant evaporation temperatures analysed. On the other hand, the zeolite in question allows obtaining stable cooling performance over a wide range of adsorption and desorption times. This draws attention to the fact that the values of the COP and SCP parameters are instantaneous and the cooling processes are often characterised by dynamic changes of the individual parameters.

2.1.4. Metal-Organic Frameworks

MOFs are porous materials with a crystalline structure consisting of inorganic metallic nodes and organic ligands. MOFs usually show a steeper adsorption curve than other sorption materials [49]. These materials show very different properties, as the surface area of NU-1000 [49] is less than 2400 m^2/g, while for $(CH_3)_2$-MOF-801 it is just over 750 m^2/g. Similar disparities are observed for the pore volume, being in the range of 0.3–1.5 cm^3/g. As for the average pore size, MOFs are characterised by pore sizes in the range of 0.8–3.5 nm (Table 5).

Table 5. Physical properties of MOFs.

Ref.	Adsorbent	Pore Volume [cm^3/g]	Thermal Conductivity [W/(m·K)]	BET Surface Area [m^2/g]	Adsorbent Size [mm]	Average Pore Size [nm]
[50,51]	MOF-801	0.37–0.44	-	820–864	0.4–0.5	1.08
[52]	MIL-101-3	1.12	-	2047	0.42–0.85	2.54
[53]	NH_2-MIL-125	0.57	-	1305	0.4–1.8	-
[51]	$(CH_3)_2$-MOF-801	0.30	-	756	-	1.10
[49]	UiO-66	0.98	-	1508	-	1.06
[49]	NU-1000	1.49	-	2362	-	1.43–3.54
[49]	DUT-67	0.43	-	936	-	0.75

Due to the increasing popularity of these materials, they are the subject of numerous studies as well as review articles. Gordeeva et al. [54] developed a review of the characteristics of adsorbents of MOFs type. In this work, attention was paid to the stability of the materials, i.e., the need for cyclic, long-term operation without degradation of the sorption properties of the material. In addition, an important aspect of the study is the analysis of the stability of MOFs under water contact conditions. The introduction of water into the metal-ligand bond results in the formation of a released cation and a free ligand, that is, it destabilises the material. On this basis, the authors developed a division of MOFs into: thermodynamically stable, high and low kinetic stability and unstable. In turn, Karmakar et al. [55] pointed out that MOFs make it possible to create a material with properties selected specifically for given applications. Recent developments in the field of MOFs are reviewed and the important factors required for the generation of stable MOFs are discussed. Furthermore, the advantages and disadvantages of these materials in relation to classical adsorbents are discussed. It was concluded, as in the work of Gordeeva [54], that MOFs should be characterised by long-term stability, high thermal conductivity and sorption kinetics, and high adsorption in a narrow temperature range.

With the literature review conducted, newly developed MOFs were highlighted and analysed for adsorption cooling. Han et al. [51] compared classical MOF-801 and MOF-801 with the addition of methyl functional groups. The addition of $(CH_3)_2$ improves the micro porosity of the material and its stability. Consequently, the new material shows up to two times faster adsorption kinetics than conventional MOF, and this is reflected in the obtained SCP and COP cycle parameters. Youssef et al. [56] analysed a MOF named Al.-Fumarate in their study. The most important conclusion of the study is that this material is insensitive to the reduction of the desorption temperature from 85 to 65 °C. A similar change in desorption temperature for silica gel and AQSOA-Z02 zeolite results in a decrease in SCP by about 90%, while for the MOF analysed it decreases by only about 15%. This fact is particularly important in the context of the application of a given adsorbent in a real system, where the driving heat used can be characterised by varying temperatures. Nevertheless, theAQSOA-Z02 zeolite has an advantage over the other investigated adsorbents in the form of a constant SCP value at a variable evaporation temperature of the medium. This fact, in turn, is important in the aspect of cooling systems, which are characterised by variability of chilled water temperature. As can be seen, the values of the mentioned adsorption, desorption, and refrigerant temperatures are important criteria for comparing different adsorbents. Ma et al. [52] undertook the analysis of MIL-101-3 as a new type of adsorbent for refrigeration applications. Based on the study, it was found that MOF achieves complete desorption at a lower temperature than activated carbons. Moreover, this material has high adsorption capacity and a low desorption temperature of 80–100 °C. In addition, after analysis of 60 cycles of operation, the material was found to be completely stable, based on the lack of degradation of pore volume and specific surface area. Furthermore, Solovyeva et al. [53] analysed a new material from the MOF group, which is NH_2-MIL-125. This adsorbent is characterised by desorption temperature (70–90 °C), work stability, and high surface area.

The last parameter that allows the analysis of adsorbents is the grain size of the material. In their work, Solovyeva et al. [57] addressed the study of MOF-801 in the context of evaluating the effect of material grain size on the kinetics of sorption processes. Increasing the particle size of the adsorbent from 0.2 to 0.8 mm results in an increase in adsorption and desorption times by about 300%, for a material arranged as a monolayer. Doubling the number of layers generates an additional increase in sorption time of about 100%. These observations are consistent with the adsorption dynamics of other physical adsorbents. On the other hand, Solovyeva et al. [53] investigating NH_2-MIL-125 found that the adsorption size of the adsorbate for grain sizes of 0.2–1.8 mm does not depend on the material size. This is a particularly significant observation as it stands in opposition to the studies described in publications [31,37,48]. On this basis, it should be concluded that the i of adsorbent particle size on the system performance must be analysed each time.

2.2. Composite and Doped Adsorbents

Composite adsorbent materials were developed to optimise the advantages of physical and chemical adsorbents. By design, composite materials are characterised by high adsorption capacity and efficient heat and mass transport. Additives used in composite or doped materials, such as expanded graphite, significantly improve heat and mass transport in the material. It should also be added that the use of additives in adsorbents very often requires the use of a binder to bond the adsorbent and the additive. On the other hand, Seol et al. [40] stated that composite adsorbents are somehow porous matrices impregnated with inorganic salts, so that they show high theoretical adsorption capacity compared to ordinary adsorbents. However, depending on the additives used, there is a risk of eliminating some of the micropores, such that the actual adsorption capacity is not satisfactory. Additionally, the rate of the sorption processes should also be analysed, as composite materials are often characterised by dynamic changes in this aspect. Therefore, the study of composite materials focuses on both heat and mass transport intensification.

Analysing the data on composite adsorbents collected in Table 6, it should be concluded that this is currently the most studied group of materials in the context of application in adsorption chillers. Since composites are formed by doping one material with another, it can be concluded that there is an unlimited number of new materials that can be created in this way. Based on the literature review conducted, several composite materials have been selected and are currently under investigation.

Younes et al. [58] analysed silica gel-based composites with the addition of EG500 and EG100 (Expanded Graphite) as a material to improve thermal conductivity, which reaches 1.55 W/(mK) and using PVP (polyvinyl pyrrolidone) as a binder. This thermal conductivity value is very high but is associated with a very low COP of the cycle, which is 0.058, so it can be concluded that the proportion of doped graphite in the adsorbent is inappropriate. The authors found that the addition of 2% PVP has a negligible effect on clogging the pores of the adsorbent and thus limiting the adsorption pore volume up to 0.9 cm^3/g. Nevertheless, this 2% binder is required to prepare a stable composite adsorbent. As the graphite addition increases, the thermal conductivity of the adsorbent increases. Each time 10 wt% graphite is added to SGP (silica gel powder) it generates a 200% increase in the thermal conductivity of the material. However, this effect is accompanied by a decrease in pore volume and composite surface area by about 10%. Therefore, too much graphite addition will decrease the adsorption pore volume in the bed.

Yagnamurthy et al. [59] analysed the thermophysical properties and adsorption characteristics of new composite materials based on Maxsorb III activated carbon. The composite is made by adding natural graphite nanoplatelets and using PVA (polyvinyl alcohol) as a binder. The addition of graphite improves thermal properties while not overly limiting mass transport. In their study, the authors considered different mass proportions of graphite and the results are in line with the study of Younes [58]. As the proportion of graphite in the material increases, the specific surface area and pore volume of the composite decreases. The addition of 40 wt% graphite generates a decrease in the pore volume and specific surface area of the material by about 70% compared to the base AC, while the thermal conductivity increased 64 times, from 0.066 W/(m·K) for the AC to 4.33 W/(m·K) for the composite. The composites proposed by the research team show an inverse relationship between the increase in material conductivity and the amount of refrigerant adsorption. Combining these relationships, it should be concluded that the composites analysed have similar sorption kinetics, which is determined by water vapour diffusion and heat transfer. Thus, the selection of a particular material depends on the specific design of the bed and the given application.

On the other hand, Pal et al. [60] dealt with a composite material based on Maxsorb III activated carbon (similarly to Yagnamurthy et al. [59]), to which graphene nanoplatelets (GNPs) were added and the whole material was consolidated using polyvinyl alcohol (PVA) as a binder. These composites are characterised by a very high surface area of more than 2000 m^2/g and a large pore volume of the order of 1.0–1.3 cm^3/g. Moreover, depending on the amount of graphene addition, the developed composite can have up to 23 times better thermal conductivity than the base AC. Based on these studies, one should observe a large variation in the properties of the composite materials depending on the composition. This clearly indicates that further research on composite adsorbents is warranted, as application-specific dedicated materials can be developed. The possibility to work out the structure of the material with specific dominant features such as pore size or volume, thermal conductivity and diffusivity, or surface development allows for concluding that that further development of composite adsorbents will allow an increase in the efficiency of adsorption chillers.

Kumita et al. [24] analysed a composite adsorbent based on activated carbon fibre (ACF) using polytetrafluoroethylene (PTFE) as a binder. The authors found that by consolidating the fibres into larger particles, cooling performance could be improved. The adsorption and desorption behaviour of the composite HDACF was investigated. The effect of adsorbent density on the adsorption rate was determined. It was observed that the adsorption and desorption rates increase with increasing density of the composite reaching a maximum value for a density of 380 kg/m^3. This situation is due to the fact that increasing the density of the composite improves the heat transport in the material. However, further densification of the activated carbon reduces the mass exchange. Thus, it should be concluded that for each material there is a certain limiting density where the balance between heat transfer dynamics and mass exchange in the material allows the maximum adsorption/desorption efficiency to be achieved.

Dzigbor et al. [22] studied composites formed from activated carbon doped with NaCl. They found that doping with NaCl at 30 wt% improves the thermal conductivity of the sorbent by about 25 times. On the other hand, such doping reduces absorption by about 35%. The author speculates that sodium chloride blocked access to the AC micropores, which limits the amount of refrigerant adsorption.

Chen et al. [61] analysed a composite adsorbent based on 13X zeolite and doped with $CaCl_2$. The authors found a reduction in adsorption capacity when trying to improve the thermal conductivity of the material. The answer is proposed by Chen's team [62], who found in an earlier study that such a doped zeolite has a larger pore volume, a better adsorption capacity, and that the increasing proportion of $CaCl_2$ is responsible for the increase in pore volume and the development of the material surface area. In general, the study of composite adsorbents impregnated with hygroscopic salts is interesting due to their high affinity for water.

Table 6. Physical properties of composite adsorbents.

Ref.	Adsorbent	Pore Volume [cm^3/g]	Thermal Conductivity [W/(m·K)]	BET Surface Area [m^2/g]	Adsorbent Size [mm]	Average Pore Size [nm]
[22]	AC + 20% NaCl	0.391	0.0023	1120	-	-
[22]	AC + 25% NaCl	0.365	0.005	1069	-	-
[22]	AC + 30% NaCl	0.331	0.0054	793	-	-
[61]	13X/$CaCl_2$	0.34–0.54	0.2	608–622	0.002	-
[40]	WSS + 20 wt% LiCl	0.368	-	149	-	9.5
[58]	S2-EG500	0.257	0.34	548	0.026–0.23	-
[60]	AC + 40 wt% GNPs	0.958	1.55	1935	-	2.2
[58]	SGP	0.284	0.12	601	0.1–2.0	-
[58]	S3-EG100	0.246	0.28	520	0.026–0.23	-
[63]	S1-PVP	0.274	0.16	572	-	-
[64]	MWCNT/MIL-100(Fe)	0.7	-	-	0.005	-

Grekova et al. [20] addressed the development of a new composite based on AA (anodic alumina) that is impregnated with hygroscopic salts ($CaCl_2$ and LiCl). The significant adsorption dynamics at the beginning of the cycle (about 150s) should be noted, as well as the subsequent stagnation of adsorption. The reason for this shape of the adsorption curve lies in the structure of the adsorbent. As the process continues, the pores of the material become blocked by the hydrated salt. It should also be added that AA can be characterized by different parameters of pore size and volume, depending on the environment in which it was produced.

Analysing the latest data on adsorbents, it should be concluded that despite many ongoing studies, there is still a great potential for their development. The optimal selection of adsorption materials for a certain refrigeration application allows for the intensification of adsorption and desorption processes.

When comparing different adsorbents' parameters, special attention should be paid to the grain size and average pore size. Analysing the grain size of different adsorbent materials for most of them, a deterioration of sorption dynamics with increasing grain size can be observed. Nevertheless, Mitra et al. [25] analysed the effect of adsorbent particle size on adsorption dynamics. Based on their analysis, they concluded that it is difficult to unambiguously determine the effect of particle size on the adsorption rate because it is specifically correlated with the bed structure.

In general, all material properties are related to pore size and the amount of energy required to remove adsorbate from the micropores. This is why sorption curves have such different shapes, because there are certain energy levels that must be exceeded to allow adsorption/desorption in a given pore group. Moreover, adsorption always starts in micropores, and only after they are filled do adsorbate molecules move to mesopores, which are characterised by a lower affinity for adsorption of refrigerant, which affects the inhibition of adsorption process [43]. Therefore, how the doping of different materials affects the pore volume should be evaluated each time. Considering the characteristics of the adsorption process, it should be noted that a drastic reduction in the micropore volume significantly reduces the adsorption volume as well as its kinetics. Adsorption occuring in mesopores is characterised by much slower kinetics than that taking place in the smallest spaces of the sorbent. However, this observation does not only apply to composite adsorbents. All procedures involving granulation or fragmentation of adsorbent materials carry the risk of clogging the micropores; therefore, the above-mentioned processes require special attention so that the target effect of improving a given parameter of sorption properties of the material does not have a negative impact on the other properties.

This analysis of sorption materials focuses on comparing the thermodynamic and physical properties of the materials, but other factors such as price and availability of a given adsorbent are also important. Comparing the prices of different sorbents, it can be concluded that ACFs are the cheapest. The next group are zeolites, which are on average two times more expensive than activated carbons. On the other hand, silica gels are about several times more expensive than ACFs [13]. In contrast, the price of MOFs is about $2300 per 100 g, so this limits its commercial applications [54].

Therefore, the ACs, SGs, zeolites, named as classical materials, despite inferior physical parameters compared to MOFs and composite materials, have lower prices and better availability. It should be noted that ACs can be produced from plant waste, while zeolites or silica gels occur naturally in nature. Therefore, these materials fit better into the concept of sustainable cooling, which should work with natural refrigerants and adsorbents. On the other hand, further such intensive development of MOFs and composite materials may allow overcoming some technological barriers regarding the low efficiency of adsorption refrigeration systems, and this would allow real competition of adsorption chillers with compressor chillers.

3. Impact of Coated-Bed Application

Following the selection of appropriate material properties, another factor affecting the efficiency and performance of adsorption cooling systems is the configuration of adsorbent deposition in the bed. In the adsorption cooling process, it is crucial to provide the best possible heat and mass transport to achieve high adsorption and desorption kinetics. Heat transport can be improved essentially in two ways. The first is based on improving the thermal conductivity of the adsorbent and heat exchanger material. The second way is to increase the exchange area by developing the surface area of the heat exchanger elements or modifying the bed structure to reduce the contact thermal resistance. Among the structural configurations of the adsorbent bed, the following bed types can be distinguished: packed beds, consolidated (bonded) beds, coated beds, and hybrid (coated and packed) beds. Currently, packed beds, in which the adsorbent granules are placed in the free space between elements of the heat exchanger, are the most commonly used. This type of bed is characterised by good mass transport, but heat transport is limited by contact thermal resistance and the presence of voids in the boundary layer. The solution to the problem of poor thermal conductivity of packed beds is the application of a binder, which is used in consolidated beds. The material in the consolidated beds is placed in the voids after mixing with the binder, forming a continuous structure after curing. By using the binder, voids are filled and point thermal contact is replaced by surface contact. A significant disadvantage of this type of bed is that vapour diffusion and thus mass transport is significantly reduced. Therefore, additional steam channels are used. An extension of the consolidated bed concept is coated beds, where the adsorbent material is applied to the heat exchanger elements, forming a thin coat. This solution allows for the positive effects of forming a consolidated boundary layer while maintaining sufficient mass transport. The disadvantage of coated beds is the reduction of the mass content of the adsorbent material in the bed. The last type of beds is hybrid beds, which combine the advantages of coated and packed beds. The concept of building this type of bed is based on the formation of a coating on the exchanger elements and placing granulated adsorbent material in the remaining volume of the bed. The use of coatings improves heat transport and the filling of free spaces with adsorbent has a beneficial effect on the HEX to adsorbent mass ratio.

Table 7 presents recent research related to coated and hybrid beds. Regarding the improvement of heat and mass transport in coated and hybrid beds, special attention should be paid to the coating technique and the basic coating parameters, which include thickness, particle size, structure, and the binder used. These aspects are discussed in the following chapters.

3.1. Adsorbent Coating Technologies

Currently, several technologies have been developed to create and apply adsorbent coatings. Wang et al. [15] proposed to classify the techniques for obtaining coatings into ex situ and in situ methods. In ex situ methods, coating formation is based on the application of appropriately selected adsorb ent particles with a binder to the surface of the heat exchanger. The in situ method is based on direct crystallisation of adsorbent particles on the heat exchanger surface. When comparing these methods, three main aspects should be considered: heat and mass transport, coating strength, and technological applicability.

In the case of the in situ method, due to the possibility of resigning from adding a binder, the heat transfer resistance between the heat exchanger and the adsorbent is negligible. The lack of binder also has a positive effect on the mass transport, as the binder does not close the adsorbent pores and it is possible to obtain a thinner coating, minimising the resistance to mass transfer to deeper layers of the bed. However, the use of a thinner coating results in the disadvantage of potentially lower chiller performance due to the unfavourable adsorbent to heat exchanger mass ratio.

Coatings made by the ex situ method have better strength properties. The use of a binder increases the flexibility of the coating, which affects the cracking intensity of the coating. Palomba et al. [27] investigated SAPO-34 zeolite coatings obtained by direct crystallisation and dip coating methods for strength. Pull-off test results showed that the coating samples for the in situ method had a mechanical resistance (average) of 0.78 MPa, while the dip-coated samples had a mechanical resistance of 0.82 MPa. In addition, the choice of coating manufacturing method was found to have a greater influence on the fracture mechanism of the coating layers than on the mechanical resistance of the coating. However, it should be noted that the fracture of the coating itself has a significant impact on the fatigue strength and therefore on the durability of the coating.

Ex situ methods are technologically better mastered and easier to scale up for industrial applications. The main advantage of these methods is their wide applicability, i.e., coatings with a binder can be applied to any substrate, only the appropriate selection of the binder is required. Additionally, compared to in situ methods, they usually do not require high temperatures during coating and drying.

Table 7. Research on adsorption coated bed.

Ref.	Coating Technology	Working Pair	Coating Thickness [μm]	Particle Size [μm]	Binder	Presented Tests	Characteristics
[27,65]	dip-coating	SAPO-34/water in situ	tens of microns 100–200	-	-	MacBain test (uptake); pull-off test; CFD simulation	conceptual studies of a coated exchanger based on graphite plates; in situ coatings have lower mechanical strength; in situ coatings have better adsorption properties
[65]	-	polymer (super desiccant polymer)/water	100–300	-	-	simulations based on analytical model	a simulation study of the effect of geometric parameters on the performance of an adsorbent-coated bed chiller; there is an optimal adsorbent coating thickness for a given COP value and cycle length; SCP is always larger for smaller coating thicknesses;
[66]	coated with Zehntner applicator	SAPO-34/water	60–460	-	2-hydroxyethyl ether	isochoric temperature step; heat of adsorption (calculated); isochoric equilibrium sorption curve; water uptake(calculated)	experimental study on the effect of mass transport on system performance; mass transfer limitation is independent of coating thickness; binder contribution and particle size affect inter-molecular mass transfer
[67]	-	SG/water	1500	-	epoxy resin	LFM method (thermal diffusivity); CFD simulation	the numerical analysis of heat transfer of coated and packed adsorption bed; the use of adsorbent coating intensifies heat transfer
[68]	dip-coating	SAPO-34/water	100	-	N-propyl trimethoxy-silane	SEM (coating morphology and surface coverage grade); peel, pull-off, impact and microhardness tests (mechanical characterization of the coating); thermogravimetric method (uptake, adsorption kinetic);	experimental testing of SAPO-34 coating; Comparison of coated and packed bed; SCP increases after coating application; Coating has a life time of at least 600 cycles
[61]	-	13X/$CaCl_2$/water	-	-	-	tests on lab-scale adsorption chiller	performance studies of a compact dual cooling system with a coated adsorption bed; the coating significantly improves heat transfer; SCP increased by 256%
[69]	in situ	SG/water	1000–3000	70–149; 149–250; 250–400;	polyvinyl alcohol (PVA)	breakthrough curves	the effect of coating thickness and particle size on heat and mass transport was investigated; a thinner coating with larger particles can achieve better mass transfer performance;

Table 7. *Cont.*

Ref.	Coating Technology	Working Pair	Coating Thickness [μm]	Particle Size [μm]	Binder	Presented Tests	Characteristics
[70]	dip-coating	Y zeolite/methanol	10,000	up to 35	bentonite	tests on lab-scale adsorption chiller	experimental studies on improving the performance of the adsorption cooling system; the use of coating allows to reduce cycle time and improve SCP
[71]	-	zeolite/water	4000	-	alumina gel precipitated in situ	simulations based on analytical model	simulation studies on the effect of adsorbent coating application; coating application improved heat and mass transport properties; effect of coating thermophysical parameters on SCP was investigated
[72]	spray- coating	WSS/water SG (A)/water	up to 800 up to 1000	-	10% LDM6680, Celvolit	LFD + gravimetric method (internal and interfacial mass transfer, sorption dynamic)	adsorption properties of coatings with WSS and silica gel were investigated experimentally; a method of determining internal and inter-facial mass transport was proposed; A-type silica gel showed lower total mass transfer coefficient than WSS
[73]	Combination of sol-gel and electrophoretic deposition techniques.	SiO_2/Al composite/water	60,000–150,000	0.1–0.83	-	SEM (coating morphology); adsorption uptake (volumetric experiments)	experimental study of adsorption properties of a new type of coating obtained by direct deposition of silica on aluminium plates; the influence of technological parameters on the properties of the coating was investigated; an increase in SDS concentration resulted in an increase in particle size
[74]	-	SG(3A)/water SG(RD)/water	1300 700	-	epoxy; PVA; corn-flour; HEC; gelatin; bentonite; sepiolite	AUTOSORB-1 analyzer (BET surface); HYDROSORB analyzer (adsorption isotherms); FESEM (coating morphology)	experimental study of adsorption properties of silica gel coatings; analysis of the effect of binder selection; improved heat and mass transport allows reduce half- cycle time and simplified exchanger geometry
[75]	Mitsubishi Plastics Method	AQSOA/water	-	-	-	tests on lab-scale adsorption chiller	experimental study of a new coated-bed adsorption cooling system;

Table 7. *Cont.*

Ref.	Coating Technology	Working Pair	Coating Thickness [μm]	Particle Size [μm]	Binder	Presented Tests	Characteristics
[44]	-	ACM-35.4/methanol	800–14,400	800–1800	PVA	Volumetric large temperature jump method (adsorption uptake, adsorption isotherms); SCP (calculated)	experimental study of the effect of adsorbent grain consolidation on the ad-/desorption process; the effects of grain size and film thickness were investigated; the use of a binder accelerates ad-/desorption
[76]	-	Y zeolite/water SAPO-34/water	102–135 230–290	4.8 ± 0.3 3.69 ± 0.01	SilRes MP50E	3D LSM (size of particles, height profile); thermobalance manufactured by Rubotherm (equilibrium water adsorption characteristics); thermogravimetric method (adsorption of pure powder); LPJ method (kinetic adsorption characteristics)	experimental study on Y zeolite and SAPO-34 based coatings; effect of binder on adsorption process was investigated; implementation of coatings improves performance; hydrothermal stability was studied
[77]	-	MOF (CPO-27(Ni)/water AQSOATM FAM-Z02/water	100 -	-	-	CFD simulation; simulations based on analytical model	simulation study of tubular AHEx units possessing adsorbent/ copper foam composites; performance for the composite is inferior to that obtained for AQSOATM FAM-Z02 under similar conditions
[78]	-	SG/water	100	1	silica binder	simulations based on analytical model	simulation studies of adsorption bed based on adsorbent-coated microchannels; 77% improvement in COP over conventional systems using adsorbent-coated microchannels
[79]	-	Aluminium fumarate/water	300; 400; 500	-	-	simulations based on analytical model	simulation studies of coated and packed bed; the effects of geometrical and thermophysical parameters were studied; the use of a binder improves thermal contact, thus increasing the effective thermal conductivity
[80]	bar coating	FAM—Z01	50–800	10–300	epoxy YDPN	simulations based on analytical model	simulation studies of the effects of cycle time, binder, particle size, and coating thickness on the adsorption cooling process; experimental validation of the model was carried out; coating application reduces cycle time; there is an optimal coating thickness for SCP;

Table 7. *Cont.*

Ref.	Coating Technology	Working Pair	Coating Thickness [μm]	Particle Size [μm]	Binder	Presented Tests	Characteristics
[81]	manual coating	SG/water	-	<200; 1000	PVA; Hydroxyethylcellulose; 2-Hydroxyethylcellulose	optical microscope (coating morphology); dynamic mass measurement during desorption (uptake)	experimental studies of different binder and adsorbent particle sizes configurations; the choice of binder affects the adsorption dynamics
[82]	Mitsubishi Plastics Method	FAM-Z02/water	330	-	silicon dioxide based binder	tests on lab-scale adsorption chiller	comparative experimental study of coated and packed adsorption bed; an increase in SCP and COP was observed for the coated bed compared to the packed bed
[83]	-	SG/water	-	-	-	simulations based on analytical model	simulation studies of hybrid (coated+ packed) adsorption bed; application of coated first adsorbent layer and filling the remaining space with granulate eliminates thermal contact resistance and improves SCP and COP
[84]	PST	SAPO-34/water (Aluminum fiber composite)	-	-	silane-based binder	tests on lab-scale adsorption chiller	experimental studies of a new type of bed based on coated aluminium fibres; a new UA index was developed to compare adsorbers for different boundary conditions
[85]	dip- coating	SG + FAM-Z02/water	-	-	silane-based binder	tests on lab-scale and full-scale adsorption chiller	experimental study of a hybrid adsorption bed system (coated + packed); coated bed, packed bed, and two hybrid bed configurations were compared; increase in coating thickness implies a decrease in adsorption dynamics
[86]	in situ	SG/water	15–150	-	-	thermogravimetric method (uptake); SEM (coating morphology); simulations based on analytical model	experimental and simulation studies of the system with a coated adsorption bed, the effect of coating thickness was examined; COP and SPC were compared to the system with a packed bed
[87]	in situ	zeolite X/water	up to 100	-	-	SEM (coating morphology); volumetric method (adsorption isotherm); XRD (crystallinity and phase identification)	experimental studies of adsorbent coating obtained by direct synthesis from solution under microwave heating; the applied coating technique significantly increases the crystallisation index;

The ex situ method of adsorbent application includes several techniques. The simplest technique, which can be used practically only in laboratory conditions when preparing samples for testing, is the two-step bonding technique. First, a binder is applied to the surface and then the adsorbent particles are evenly distributed on the surface. Another approach for application to heat exchangers is the use of an adsorbent slurry containing a binder, a solvent (usually water), which is applied to the exchanger surface and forms a coating when dry. Capri et al. [82] distinguish several slurry application techniques: dip coating, spin coating, spray coating, and droplet coating.

The most common adsorbent coating technique used in research is the dip coating technique. The process of applying the slurry itself involves complete immersion of the substrate (heat exchanger elements) in the slurry vessel. This technique makes it possible to obtain coatings of uniform thickness on components with complex geometry. The disadvantages of this technique include the limitation of the coating thickness and high adsorbent consumption. Another technique for applying coatings to elements with complex geometries is the spray coating technique. This technique is based on the use of a spray head, airbrush or paint gun. This technique requires less slurry compared to the dipping technique, but does raise the problem of penetrating the substrate cavities adequately.

Droplet coating involves the mutual movement of the applicator and substrate, where the applicator delivers successive portions of the suspension in the form of droplets. In this way, material consumption is reduced and, additionally, as in the case of the dipping technique, relatively simple automation of the process is possible.

A similar technique is spin coating. An applicator placed over the substrate feeds the material and the substrate is set in motion by a vortex. As a result of the inertial force, the slurry is properly distributed on the surface. This technique is designed for less complex geometries.

Wang et al. [15] classified in situ methods into reagent-assisted and inert-assistedmethods. When using the first type, the reactants for crystallisation of the zeolite layer are also taken up either by partial dissolution of the support or by extraction from an inert support matrix. In contrast, the second type is based on building the zeolite layer from gels or solutions containing all reactants. The in situ method is mainly used to produce coatings of such adsorbents as mesoporous aluminosilicate molecular sieves, zeolites and metal-organic structures (MOFs) [88].

3.2. Relevant Coating Parameters

The important parameters describing the structure of the coating include the following: coating thickness, adsorbent grain size, its structure, and the binder used. The primary purpose of introducing coatings is to intensify heat and mass transport, so the above-mentioned parameters were analysed for their influence on these phenomena.

The thickness of the adsorbent coating is a parameter that has received much attention in research [24,39,50,65–67,70,74,76,79,83,84,89,90]. The main factor that determines the coating thickness is the technology used. In the case of direct crystallisation, coatings with thicknesses of tens of microns are usually used. In contrast, methods using binders allow for larger coating thicknesses, i.e., 100–200 μm [27]. The coating thickness has been found to affect the SCP [65,79,80,91]. As the thickness of the adsorption coating decreases, the specific power increases. This is due to the reduction in mass transfer resistance [71], relative thermal resistance, and relative thermal inertia of the heat exchanger [91], leading to increased heat and mass transfer [80]. It should be noted that decreasing coating thickness is also associated with a decrease in adsorbent mass and an increase in the metal-adsorbent mass ratio [79]. This results in a decrease in chiller efficiency. However, because of improvements of the heat transfer for thinner coatings, it is possible to reduce the cycle time, resulting in improved COP. It was found that there is an optimum thickness and it depends on the objective function, i.e., to achieve maximum COP or SCP, under the given conditions, i.e., adsorbent used and exchanger design [76].

The effect of particle size on the adsorption/desorption process in coated beds has not been as widely studied as the coating thickness. However, based on a review of works [66,69,80], relatively large particle sizes are recommended for coated beds. Ammann et al. [66] conducted an experimental analysis of mass and heat transfer during water sorption on SAPO-34 coatings. The main limiting factor was mass transport. A particular mass transport limitation was found when using adsorbent particles with a diameter of less than 10 μm. This is attributed to an increase in the interparticle resistance to steam flow. Similar results were obtained by Duong et al. [80], where the use of particles with diameters between 10 μm and 300 μm was analysed. It was found that decreasing the particle size resulted in an increase in steam flow resistance. It was also observed that smaller adsorbent particles have a positive effect on the COP value. As a result of using adsorbent particles with a diameter of 150 μm instead of 300 μm, a 30% increase in COP value was obtained. The authors relate this phenomenon to an increase in intraparticle mass transport at smaller particle sizes.

The use of binders was determined by the requirement to reduce the contact thermal resistance by filling the free interparticle space and thus increasing the contact area of adsorbent particles with the exchanger elements. It should be noted, however, that the introduction of the coating also leads to a reduction in the mass transport of water vapour, which adversely affects the efficiency of the system. An important issue is also to ensure adequate strength of the resulting coating. On this basis, it is possible to define several criteria that a suitable binder should meet. Li et al. [74] reviewed adsorbent-binder pairs and defined a good binder as one that meets the following criteria: (1) improves heat transport along the exchanger-adsorbent surface path, (2) does not affect or improves vapour uptake, (3) allows the formation of sufficiently strong adhesive bonds between the adsorbent and the metal base, (4) has high mechanical and thermal strength under varying humidity conditions, and (5) is chemically inert with respect to the adsorbent-adsorbate pair. Similar criteria were presented in an analysis by Sztekler et al. [81]. Duong et al. [80] investigated the effect of binder weight percentage in the coating on the adsorption process. The experiment was conducted for 0%, 5%, 10%, and 15% binder content. Increasing the binder content resulted in a decrease in the adsorption rate to 53% for 15% binder content compared to the sample without binder. COP and SCP also decreased with increasing binder content. For samples with 15% binder content there was an up to 20% decrease for COP and up to 55% decrease for SCP. As can be seen, the binder has a negative effect on the performance of the adsorption cooling system, but the authors emphasise that the use of binders is necessary; hence a maximum low binder content should be used. It should be noted that no attention has been paid to in situ methods that do not require the use of a binder, but as described earlier they are technologically more complicated compared to ex situ methods. The binders considered in the study include, among others, organic binders such as epoxy resin [67,74,80], gelatin [74], hydroxyethylcellulose (HEC) [74,81], polyvinyl alcohol (PVA) [44,74,81], corn flour [74] and inorganic ones: bentonite [70,74,90], silane-based binders [68,85], sepiolite [74].

The last parameter considered is the structure of the adsorbent bed, more specifically the proportion and distribution of porosity. High porosity improves uptake but also increases heat transfer resistance. In contrast, low porosity improves heat transfer but reduces adsorption. Therefore, it is necessary to optimise the porosity proportion to ensure adequate heat and mass transport. Li et al. [92] propose a novel approach to the porosity problem i.e., introducing a non-uniform porosity distribution. A simulation study was carried out to analyse the distribution variation along the x, y, z directions, where the x, y directions are parallel to the exchanger surface and the z direction is perpendicular. It was found that the change in porosity along x, y direction leads to deterioration of adsorption and SCP, while a change along the z direction with an appropriate distribution allows for an improvement in adsorption and SCP. Linear, quadratic and step distributions were investigated. All types of distributions were considered in two variants with increasing and decreasing porosity from the cooling side to vapour side. Stimulation results indicate that a distribution with increasing porosity from the cooling side to the vapour side is preferred. Additionally, it is desirable that the porosity gradient increases from the cooling side to the steam side. A suitable gradient was obtained for a quadratic porosity distribution. Using a quadratic porosity distribution, an increase in SCP of 9.5% was found compared to a uniform distribution [92].

3.3. *Impact of Coated-Bed Application—Research Results*

In order to properly analyse the effect of using coated beds on the effectiveness and efficiency of the adsorption cooling process, it is necessary to consider cases under identical or comparable conditions, which consist of adsorbent/adsorbate pair, chiller design, exchanger design, and thermodynamic cycle parameters, i.e., the temperature of the chilled water in the evaporator (T_{eva}), cooling water temperature (T_{chill}), heating water temperature (T_{heat}), and cycle time. On this basis, the work was selected, subjected to further analysis, the results of which are summarised in Table 8.

The introduction of an adsorbent coating causes an increase in the specific cooling power. This is due to the intensification of heat transfer by reducing the thermal contact resistance. The reduction of thermal contact resistance is based on increasing the thermal contact between the adsorbent particles and the exchanger elements and reducing the proportion of free space and porosity in the bed volume [79]. The effect of improved heat transport in the bed was found, among others, in the study of Grabowska et al. [67], where the application of the coating allowed for an increase in the average temperature during desorption by 6% compared to a packed bed. Improved heat transfer allows for an acceleration of desorption and adsorption process, resulting in a shorter cycle and thus an increase in power.

For the effect on efficiency and COP, two opposing research results were observed. Some studies showed a decrease in COP after coating [58,71], while some of the review works showed an increase [19,70,82,86]. The reason for the decrease in performance has been linked to the clogging of pores by the binder, thus reducing adsorption capacity and increasing in the metal to adsorbent mass ratio, leading to an increase in losses associated with heating and cooling of the exchanger [79]. Additionally, the use of a binder causes a reduction in vapour flow paths, leading to a reduction in mass exchange. On the other hand, the authors of the studies in which a positive effect of the coating on COP has been reported identify an increase in efficiency due to a shorter chiller cycle and a reduction in the amount of heat extracted during desorption. Therefore, it is necessary to further investigate the influence of coated beds on the adsorption cooling process, especially in terms of their effect on COP.

Table 8. Effect of adsorbent coating on adsorption process performance.

Ref.	Working Pair	Bed Type	SCP [W/kg]	COP	SCP Change [2]	COP Change [3]	Cycle Time [s]	$T_{ads}/T_{des}/T_{eva}$ [°C]	Coating Technique	Type of Research
[68]	SAPO-34/water	coated packed	675 498	0.24 0.40	35.5%	−40%	300	28/90/15	dip coating	experiment
[19]	13X/$CaCl_2$/water	coated packed	377 106	0.27 0.16	256%	68%	3060 1060	28/85/14 22/85/14	-	experiment
[70]	Y zeolite/methanol	coated packed	30–60 10	0.1–0.12 0.08–0.1	200–500%	20–25%	15–20 60–120	40/90/10	dip coating	experiment
[79]	Aluminium fumarate/water	coated packed	825 290	- -	184%	-	400	22/90/20	-	simulation
[80]	FAM-Z01/water	coated packed	485 354	0.454 0.504	37%	−10%	480 840	30/80/13	dip coating	simulation
[82]	FAM-Z02/water	coated packed	456 118	0.27 0.21	286%	28%	600	30/90/15	Mitsubishi Plastics Method	experiment
[83]	SG/water	Hybrid [1] packed	- -	0.70 0.67	-	4.5%	-	-	-	simulation
[86]	Aluminium oxide/water	coated packed	- -	- -	50%	40–70%	30–600	-	in situ	simulation

[1] coated bed in combination with packed bed. [2] change in SCP value of coated bed in comparison with packed bed. [3] change in COP value of coated bed in comparison with packed bed.

4. Adsorption Working Pairs for Different Refrigerant Evaporation Temperatures

The working pair in the adsorption/desorption cycle is a key component of the adsorption cooling system. To ensure high cooling efficiency, the appropriate adsorption working pair should be selected according to the heat source temperature. Then, the appropriate adsorption refrigeration cycle parameters should be considered and selected them according to the application. The application areas and characteristics are different for different working pairs of adsorption refrigeration systems.

In his analysis of the adsorption cycle working pairs, Shmroukh et al. [50] focused on the experimental approach; he determined whether the given studies were experimental, theoretical or simulation. He identified zeolite-water, silica gel-water, and AC-methanol pairs as classical pairs for which the maximum adsorption capacity is about 0.26 g/g; however, recent studies of these classical material pairs have obtained uptakes of 0.5–0.6 g/g [24,36]. On this basis, it can be concluded that further development of classical adsorption pairs is possible. Nevertheless, new adsorption pairs using composite adsorbents and MOFs allow for higher adsorption capacity.

Younes et al. [93] in a compiled review of adsorption material pairs focused on comparing COP and SCP parameters. Based on their review, they found that SCP and COP are mainly determined for ideal refrigeration cycles, so it is difficult to relate them to real systems. Nevertheless, even laboratory values of COP and SCP obtained for specific operating conditions allow for the assessment of the suitability of a given pair in the context of a specific application.

Goyal et al. [94] in their work collected information on solar-powered adsorption systems. They found that solar systems are still not competitive. Manufacturers of such systems struggle to keep their solutions in the market due to technological and economic constraints.

Shabir et al. [95] developed a review of materials used in adsorption chillers, where he compared the physical properties of adsorbents, adsorption equilibrium, and uptake, but nothing was mentioned about the refrigerant evaporation temperature, which significantly affects the COP and SCP of the system. On the other hand, Papakokkinos et al. [96] in their study pointed out the crucial importance of the values of the system operating temperatures. The authors presented software to predict the minimum desorption temperature for adsorption isotherms classified according to IUPAC. They found that the minimum desorption temperature depends on the shape of the adsorption isotherm.

According to the main objective of the paper, adsorption working pairs using natural refrigerants and desorption heat of less than 100 °C were analysed. In addition, according to the authors, different adsorption working pairs should be compared considering a set of three temperatures: refrigerant evaporation, adsorption and desorption. Only the combination of the mentioned temperatures with the COP and SCP parameters allows a reliable evaluation of the given working pairs. Therefore, in Table 9, selected adsorption working pairs with characteristic parameters of the cycle, arranged according to increasing temperature of refrigerant evaporation, are presented to enable comparison of working pairs operating under similar conditions. Nevertheless, the data summarised in Table 9 are characterised by very high variability and even their ordering does not completely solve the mentioned problem. In addition, Palomba et al. [97] in their publication pointed out the need for a simulation framework for their research. In the mentioned paper [97], a model for conducting calculations of adsorption systems was presented. Unfortunately, it is not fully functional since different components of systems are used by different research teams. This affects the difficulty of comparing different research results and, at the same time, constitutes a problem to be solved: how to interpret research conducted under extremely different conditions.

Table 9. Properties of selected adsorption working pairs.

Ref.	Adsorbent	Adsorbate	Evaporation Temperature [1] [°C]	Adsorption Temperature [°C]	Desorption Temperature [°C]	Adsorption Pressure [kPa]	Desorption Pressure [kPa]	Uptake [g/g]	SCP [kW/kg]	COP [-]
[98]	AC/ENG-TSA	Ammonia	−15	20	80	-	-	0.06	-	0.23
[17]	$SRCl_2$	Ammonia	−5	-	130	-	-	-	-	-
[99]	Maxsorb	Ethanol	−5	-	70	-	-	1.129	-	-
[99]	ATO	Ethanol	−5	-	70	-	-	0.437	-	-
[100]	MIL-101	Isobutane	−5	30	85	12.5	-	02	0.022	-
[100]	AC	Isobutane	−5	30	85	125	-	01	0.006	-
[52]	MIL-101-3	Ethanol	−5	25	80	0.5	-	0.304	-	-
[52]	MIL-101-3	Ethanol	−5	25	100	1.1	-	0.45	-	-
[98]	AC/ENG-TSA	Ammonia	−5	30	124	-	-	0.124	-	0.262
[29]	SRD 1352/3	Ethanol	−3	25	90	1.266	7.845	0.25	1.49	-
[28]	SRD 1352/3	Ethanol	−2	30	90	-	-	-	0.057	0.075
[21]	SRD 1352/3	Ethanol	−2	30	90	-	-	0.151	-	0.55
[21]	FR20	Ethanol	−2	30	90	-	-	0.103	-	0.47
[21]	AP4-60	Ethanol	−2	30	90	-	-	0.086	-	0.49
[21]	ATO	Ethanol	−2	30	90	-	-	0.103	-	0.48
[21]	COC-L1200	Ethanol	−2	30	90	-	-	0.26	-	0.39
[21]	SG/LiBr	Ethanol	−2	30	90	-	-	0.162	-	0.64
[101]	AC35	Methanol	−2	35	100	-	-	-	0.66	0.329
[18]	$SG/CaCl_2$	Water	5	50	135	-	-	-	0.058	0.58
[18]	Maxsorb III (H_2)	Ethanol	5	50	130	-	-	-	0.083	0.53
[18]	CAU-3	Ethanol	5	50	115	-	-	-	0.056	0.54
[57]	MOF-801	Water	5	30	85	0.9	4.3	0.21	1.5	0.67
[30]	CSAC	CO2	5	25	80	3450	-	0.52	-	0.09

Table 9. *Cont.*

Ref.	Adsorbent	Adsorbate	Evaporation Temperature [1] [°C]	Adsorption Temperature [°C]	Desorption Temperature [°C]	Adsorption Pressure [kPa]	Desorption Pressure [kPa]	Uptake [g/g]	SCP [kW/kg]	COP [-]
[98]	AC/ENG-TSA	Ammonia	5	20	80	-	-	0.19	-	0.43
[98]	AC/ENG-TSA	Ammonia	5	30	80	-	-	0.09	-	0.31
[60]	AC + 40 wt% GNPs	CO2	5	30	80	4000	7000	0.6	-	0.06
[28]	SRD 1352/3	Ethanol	7	30	90	-	-	-	0.095	0.1
[21]	SRD 1352/3	Ethanol	7	30	90	-	-	0.235	-	0.63
[21]	FR20	Ethanol	7	30	90	-	-	0.143	-	0.53
[21]	AP4-60	Ethanol	7	30	90	-	-	0.122	-	0.58
[21]	ATO	Ethanol	7	30	90	-	-	0.152	-	0.55
[21]	COC-L1200	Ethanol	7	30	90	-	-	0.086	-	0.45
[21]	SG/LiBr	Ethanol	7	30	90	-	-	-	-	0.72
[102]	AQSOA™-FAM-Z02	Water	7	35	90	1.23	4.24	-	1.8	-
[41]	SG	Water	7	30	80	1	10	-	0.0082	0.258
[16]	$CaCl_2$	Ammonia	10	25	100	616.3	1903	0.68	-	0.38
[16]	$SrCl_2$	Ammonia	10	25	100	616.3	2142	0.136	-	0.33
[52]	MIL-101-3	Ethanol	10	25	100	3.15	-	0.7	-	-
[29]	SRD 1352/3	Ethanol	10	25	90	3.106	10.412	0.25	1.51	-
[53]	NH_2-MIL-125	Water	10	30	90	1.24	4.28	0.35	2.2	-
[45]	AQSOA-FAM-Z02	Water	10	30	90	1.23	4.24	0.22	1.3	-
[64]	MWCNT/MIL-100(Fe)	Water	10	35	-	-	-	0.65	0.455	-
[102]	AQSOA™-FAM-Z02	Water	10	30	90	1.23	4.24		2.9	-
[38]	FAM-Z01	Water	10	30	70	-	-	0.1368	0.2795	0.59
[38]	Type-RD silica gel	Water	10	30	70	-	-	0.134	0.207	0.6128
[26]	Maxsorb III (H2)	Ethanol	10	30	70	3.17	10.55	0.37	-	0.8–0.9

Table 9. *Cont.*

Ref.	Adsorbent	Adsorbate	Evaporation Temperature [1] [°C]	Adsorption Temperature [°C]	Desorption Temperature [°C]	Adsorption Pressure [kPa]	Desorption Pressure [kPa]	Uptake [g/g]	SCP [kW/kg]	COP [-]
[26]	PR_KOH4	Ethanol	10	30	70	3.17	1055	0.45	-	0.8–0.9
[26]	H2-treated Maxsorb III	Ethanol	10	30	70	3.17	10.55	0.39	-	0.8–0.9
[37]	AQSOA-Z02	Water	10	30	85	-	-	0.37	-	0.16
[37]	RD-2060	Water	10	30	85	-	-	0.31	-	0.4
[25]	Maxsorb III (H2)	Ethanol	10	30	70	3.2	10.6	0.6	0.4–0.7	-
[103]	Silica Gel	Ethanol	14	30	85	1.01	4.2	0.39	0.268	0.47
[103]	ACF/BCS	Ammonia	14	30	85	103.4	1334.8	0.62	0.245	0.92
[103]	ACF/NCS	Ammonia	14	30	85	284	1303	0.59	0.245	0.58
[104]	13X/$CaCl_2$	Water	14	31	85	-	-	-	0.06	0.15
[104]	Silica Gel	Water	14	31	85	-	-	-	0.045	0.28
[19]	Zeolite 13X/$CaCl_2$	Water	14	28	85	-	-	~0.6	0.319	0.22
[105]	Silica Gel	Water	14	25	75–80	-	-	0.38	0.075	0.42
[19]	Zeolite 13X/$CaCl_2$	Water	14	28	85	-	-	-	0.319	0.22
[46]	AQSOA-Z01	Water	14.8	30	55	-	-	0.215	0.11	0.37
[46]	AQSOA-Z01	Water	14.8	30	80	-	-	0.215	0.19	0.34
[46]	AQSOA-Z02	Water	14.8	30	55	-	-	0.29	0.06	0.22
[46]	AQSOA-Z02	Water	14.8	30	80	-	-	0.29	0.26	0.36
[46]	AQSOA-Z05	Water	14.8	30	55	-	-	0.22	0.025	0.26
[46]	AQSOA-Z05	Water	14.8	30	80	-	-	0.22	0.028	0.17
[51]	MOF-801	Water	14.8	30	80	1	5	0.15	0.54	0.47
[51]	$(CH_3)_2$-MOF-801	Water	14.8	30	80	1	5	0.19	0.79	0.63
[17]	$SRCl_2$	Ammonia	15		180	-	-	-	-	-
[30]	CSAC	CO2	15	25	80	3450	-	0.52	-	0.1
[38]	FAM-Z01	Water	15	30	80	-	-	0.1368	0.397	0.5623

Table 9. *Cont.*

Ref.	Adsorbent	Adsorbate	Evaporation Temperature [1] [°C]	Adsorption Temperature [°C]	Desorption Temperature [°C]	Adsorption Pressure [kPa]	Desorption Pressure [kPa]	Uptake [g/g]	SCP [kW/kg]	COP [-]
[38]	Type-RD silica gel	Water	15	30	80	-	-	0.134	0.382	0.6572
[35]	Silica Gel	Water	15	30	80	1.5	5	0.21	0.068	0.53
[35]	Silica Gel	Water	15	30	80	1.5	5	-	0.08	0.56
[31,40]	CSAC	CO2	15	25	80	3450	-	0.56	-	0.06
[40]	WSS + 20 wt% LiCl	Water	15	30	100	1.57	-	0.35	0.47	0.52
[40]	A-type silica gel	Water	15	30	-	1.57	-	0.14	0.4	0.46
[106]	Silica gel 127B	Water	15	30	80	2	7.5	-	0.215	0.587
[39]	Silica gel RD	Water	15	25	85	-	-	0.16	0.16	0.7
[105]	Silica Gel	Water	18	25	75	-	-	0.38	0.074	0.39
[33]	AC	Methanol	24	30	75	20	-	-	0.028	0.11
[47]	AlPO4	Water	27	35	80	1.6	5.6	0.3	0.523	-
[47]	SAPO4	Water	27	35	80	1.6	5.6	0.3	0.423	-
[22]	AC	Ethanol 99,7%	-	24	95	5	-	0.7	0.085	0.1
[22]	AC + 20% NaCl	Ethanol 99,7%	-	24	95	5	-	0.6	0.074	0.082
[22]	AC + 25% NaCl	Ethanol 99,7%	-	24	95	5	-	0.55	0.076	0.08
[22]	AC + 30% NaCl	Ethanol 99,7%	-	24	95	5	-	0.45	0.055	0.076
[22]	AC	Ethanol 60%	-	24	95	5	-	-	0.075	0.09
[22]	AC + 20% NaCl	Ethanol 60%	-	24	95	5	-	-	0.123	0.121
[22]	AC + 25% NaCl	Ethanol 60%	-	24	95	5	-	-	0.15	0.16
[22]	AC + 30% NaCl	Ethanol 60%	-	24	95	5	-	-	0.113	0.146
[51]	AC	Ethanol	-	-	80	-	-	0.14	-	-
[51]	AC	Ethanol	-	25	120	-	-	0.189	-	-
[107]	Zeolite 13X/$CaCl_2$	Water	-	40	75	0.873	12.352	0.4	0.018	0.76

[1] evaporation temperature given by the adsorption pressure.

Studies on Different Adsorption Working Pairs

Quadir et al. [108] performed research using an adsorption chiller driven by solar heat. The paper points out that many different studies related to the use of solar energy for bed regeneration assume a constant adsorption/desorption cycle time, while the solar input is variable. Therefore, fixed and variable cycle times of the device were compared, knowing that optimal adsorption/desorption times are related to achieving equilibrium adsorbate uptake in the adsorbent. The proposed adaptive cycle adjusts the duration of the sorption processes to the difference between the current and equilibrium uptake taking into account the magnitude of the pressure gradient in the bed. The authors noted that the fixed cycle time is suitable for laboratory conditions, where it allows for high specific cooling power. In real conditions, however, cooling is a dynamic process and cycle times need to be modified. The research conducted allowed for stating that adaptive cycle times improve COP and SCP of the system.

Li et al. [39] evaluated the working steam behaviour of SG RD-type/water at a desorption temperature of 85 °C. They found that reducing the cycle time leads to insufficient heating/cooling of the bed, and this entails a more imbalanced heat transfer. Furthermore, the COP cooling efficiency increases almost linearly with increasing desorption temperature. The authors also concluded that bed porosity analysis should be performed in addition to adsorbent porosity analysis. Different packing of the adsorbent in the adsorption bed leads to porosity grading and changes in cycle efficiency.

Papakokkinos et al. [106] conducted a study of the SG 127B/water working pair, where the desorption energy is supplied by a solar collector system. In this paper, the results of dynamic simulations of the entire adsorption system were analysed. It was found that dynamically varying the cycle time allows for improving the efficiency of the system, and the combination with solar collectors is a way to significantly reduce CO_2 emissions. The analysed system allows for high SCP and COP values of 215 W/kg and 0.59, respectively, which are among the higher values obtained by solar-powered systems.

Seol et al. [40] found that the COP and SCP tend to decrease with increasing adsorption time due to decreasing rate of sorption kinetics. On the other hand, the COP value increases with the cycle time, so it is necessary to optimise the cycle time. From the study, it can also be observed that the WSS + 20 wt% LiCl composite has a COP that is about 5–15% better than the type A silica gel. Similar aspects of the analysis of the appropriate adsorption termination time were analysed by the Velte team [109], who used an experimental approach involving a combination of experiment and computer simulations. An interesting observation is the dependence of uptake on adsorption time. The difference in adsorption time for which the uptake is 80% of the maximum capacity and 90% of the maximum capacity is very significant, about 40–60%, which evidently affects the SCP parameter. In such a system, it is necessary to analyse the process carefully and decide the appropriate time to stop the adsorption process.

Liu's publication [41] analysed similarly the adsorption cycle regenerated by solar heat. An interesting aspect of the study is the focus on the duration of bed heating and pre-cooling. In the analysed system, the pre-cooling takes as long as 120 min. Furthermore, it was found that in the range of adsorption time of 30 min, the desorption time is similar. Nevertheless, the more optimal adsorption time is 45 min, which corresponds to a desorption time of about 28 min. From the point of view of the performance of the whole system, the duration of the whole cycle is important, i.e., the adsorption and desorption times, but also the heating and pre-cooling of the bed. SG was found to be more susceptible to cycle time variations than SAPO-34. This is a very important finding because most often studies focus on evaluating COP and SCP at a specific point. On the other hand, from the point of view of the actual system, a wider range of operation should be analysed, since efficient operation over a wide range of operating parameters is a very important criterion necessary for the commercialisation of adsorption chillers.

The cited studies indicate the need to analyse the refrigeration cycle time. Therefore, when selecting a given adsorbent/adsorbate pair, it is necessary to pay attention to the properties of these materials, as they have a direct impact on the cycle time. Sah et al. [110] in their study also analysed solutions using only low temperature heat i.e., such systems where the driving energy is solar heat or waste heat. Their study shows that the adsorption rate is proportional to the bed pressure and the desorption rate depends on the size of the pore area occupied by the adsorbate and the activation energy.

Radu et al. [42] pointed out that in general the desorption process occurs more intensively than the adsorption process due to the faster diffusion of the adsorbate vapour at higher temperatures. Similar conclusions are drawn from the work of Chan [19], who also notes that desorption is faster than adsorption due to the fact that the vapour moves faster at higher temperatures and pressures (by virtue of kinetic theory of gases).

Elsheniti et al. [105] focused on the analysis of SG/water pair performance. Based on their analysis of adsorption kinetics, they found that higher evaporation pressure enhances adsorption mass transfer mechanisms in the adsorbent, which affects the achievement of higher uptake values for higher adsorption pressures. The present study also highlighted the need to select the application of a particular device, to determine the temperature of the heat source, which affects the selection of a suitable adsorption working pair for a particular application. It was also found that increasing the chilled water temperature by 4 °C allows for a 25% reduction in cycle time, while maintaining the baseline COP and SCP values.

Since higher adsorption pressure has a positive effect on the adsorption kinetics of refrigerant vapours in the adsorbent, attention should be paid to ammonia as a refrigerant. As mentioned in the introduction, ammonia allows operation at pressures close to or even higher than atmospheric pressure, which seems to be a very advantageous feature of this refrigerant. The study by Xu [98] evaluated the possibility of using ammonia as a refrigerant in deep-freezing processes, with a temperature of −25 °C, a desorption temperature of the order of 75 °C and an adsorption temperature of 40 °C. This system has a very low uptake of 0.02 g/g for −25 °C freezing temperature and 0.28 g/g for 15 °C cooling temperature, respectively. Nevertheless, the desire to achieve a freezing effect with such a low temperature requires a desorption heat of 150 °C. These studies simultaneously highlight the problems associated with the use of ammonia in an adsorption refrigeration system, but also provide insight of the potential applications of ammonia in freezing, which requires further research.

Sinha et al. [103] in their study compared classical SG/water pair and pair using ACF/BCS or ACF/NCS composite material along with ammonia. The systems with ammonia allow better energy conversion, which is important for this system that is driven by heat from solar panels.

On the other hand, Boman et al. [18] in their study focused on evaluating vapour applications from the perspective of heating and cooling. They found that the ammonia/AC vapour is suitable for both heating and cooling, while the ethanol/AC vapour performs very well only in cooling applications. The ethanol/MOF pair analysed shows good performance in a cooling context. This research highlights the fact that there are many variants of adsorbent/adsorbate combinations, and the behaviour of a given working pair within changing operating conditions is highly variable.

Dzigbor et al. [22] analysed the AC + NaCl pair working with ethanol. They found that the AC/ethanol pair had better uptake than the pair using the composite adsorbent, which is also reflected in the COP and SCP parameter values. However, the authors noted that the addition of water to ethanol at 40% of the adsorbate volume generates better heat and mass transport in the adsorbent, resulting in an improvement in SCP for the composite steam of about 100% compared to pure ethanol. This information also indicates an interesting research direction related to the mixing of different natural refrigerants and their effect on adsorption performance.

Tso et al. [104] compared systems operating with different working pairs of SG/water and 13X + $CaCl_2$/water. The COP of the system using the composite material is 60% lower than that for the steam with SG. In contrast, under the same conditions, the SCP for the pair with composite is about 30% higher. This is due to the fact that the analysed composite has worse adsorption kinetics than for silica gel. In addition, the composite component is $CaCl_2$, and chemical adsorbents require more heat for desorption than physical adsorbents.

Han et al. [51] analysed MOF-801 and the same material doped with $(CH_3)_2$ ligand. The doped MOF exhibits better COP and SCP parameters than the classical MOF-801 when interacting with water. Nevertheless, both pairs dominate in terms of characteristic parameters over SG type A/water, SG type 3A/water and AQSOA Z01 and ZO2 pairs also with water as adsorbate. The uptake for the new MOF is about 25% better than that of the regular MOF. This draws attention to the need for further research related to modifications of already existing very good MOF adsorbents.

In his work [38], Hong studied the sorption dynamics of another zeolite, FAM-Z01, which works with water as an adsorbate. This pair is characterised by a low desorption temperature of about 80 °C and fast sorption kinetics. Compared to SG type RD/water vapour, zeolite vapour has better SCP with similar COP, which allows the construction of more compact chillers. Moreover, zeolite pair after lowering the refrigerant evaporation temperature has a significant advantage over SG pair in terms of 65% better uptake value and 35% higher SCP value, i.e., it can be concluded that zeolite water pair outperforms SG/water pair in terms of the range of optimum operation.

Brancato et al. [21] evaluated the characteristics of different adsorption materials, specifically 5 different activated carbons: SRD 1352/3, FR20, AP4-60, ATO and COC-L1200, which are characterised by very different origins and grain sizes. The pore width of the adsorbent was found to be an important parameter characterising the adsorption capacity of the adsorbent. Based on the comparison of the mentioned activated carbons with the SG/LiBr composite material, it was observed that the value of the regeneration temperature has less influence on the performance of the systems with the composite material than in the case of AC-based systems. It was found that lowering the desorption temperature by 30 °C, generates a decrease in COP for AC by about 65%, and for the composite by about 50%. Furthermore, lowering the ethanol evaporation temperature from 7 to −2 °C generates a decrease in COP of about 10–15% for all analysed operating pairs. The selection of particular cycle temperatures is, next to the already analysed role of cycle time, a very important element in of chiller operation optimisation.

Allouhi et al. [13] in their study were concerned with the search for an optimal working pair for a refrigeration application. They analysed different classical working pairs at different evaporation temperatures of refrigerant. On this basis, it should be concluded that lowering the evaporation temperature by 5 °C generates a decrease in COP by about 15–20% and a reduction in evaporation temperature by 15 °C implies a decrease in COP value by 43–50%. On this basis it can be concluded that for the analysed working pairs AC/methanol, AC/ammonia, AC/ethanol, zeolite/ethanol, and zeolite/water/SG/water, the decrease in COP is directly proportional to the decrease in evaporation temperature of the refrigerant.

Solovyeva et al. [57] estimated the adsorption dynamics of MOF-801 with water as a working pair. The maximum uptake of this pair is reaching about 0.4 g/g. Nevertheless, the tests carried out at a water evaporation temperature of 5 °C allow the adsorption capacity to obtain 0.2 g/g. It was also observed that lowering the desorption temperature from 90 to 75 °C increases the adsorption time by about 80%. The same author in publication [53] analysed the MOF NH_2-MIL-125 in cooperation with water. This pair achieved an SCP value of 2.2 kW/kg with a water evaporation temperature of 10 °C. In addition, increasing the desorption temperature from 90 to 110 °C reduces the desorption time by half.

Rogala et al. [111] analysed the SG/water pair in their study. They pointed out the possibility of lowering the desorption temperature from 80 to 60 °C, or even less. However, this treatment lowers SCP and COP by about 65 and 50%, respectively, compared to the higher desorption temperature. Thus, if very low temperature heat is to be used, the temperature in adsorption process should be lowered, preferably below 25 °C.

Similar observations come from the study of Singh's team [30], who presented an analysis of the adsorption kinetics of CSAC/CO_2 vapour. It is worth mentioning that CSAC is derived from coconut shells, i.e., it is a waste material. The authors found that the CO_2 adsorption decreases with increasing adsorption temperature, which is related to the decrease in the bond strength between adsorbate and adsorbent during adsorption. Therefore, to maintain sufficient CO_2 uptake, the adsorption bed should be cooled. In addition, a 15 °C increase in evaporating medium temperature generates a 25% increase in COP and about 50% increase in SCP. The opposite trend is observed for increasing adsorption temperature, since an increase of 15 °C implies a decrease in COP and SCP of about 75%.

Jribi et al. [112] presented an experimental approach, where the first step is to perform experimental measurements that are used to validate the developed CFD (Computational Fluid Dynamics) model. In his study, he addressed the validation of ethanol adsorption model on Maxsorb III. He pointed out the necessity to remove the heat of adsorption because the lack of cooling during adsorption generates a significant decrease in process efficiency.

5. Discussion

The results of studies by various authors presented in this paper show a wide variety of working pairs both in terms of materials and working parameters as well as COP and SCP ratios obtained. Some working pairs in adsorption devices may be characterised by low values of COP and SCP and such working parameters that practically exclude their commercial applications.

The figure (Figure 2) shows the dependence of the COP on the SCP, for selected adsorption working pairs listed in Table 9. It should be noted that, according to the criterion imposed when preparing the list, all pairs are characterized by a heat of regeneration not exceeding 100 °C, but they differ in the temperature of evaporation of the refrigerant. Unfortunately, the addition of the third axis on the graph (evaporating temperature of the refrigerant) significantly reduces its readability. Therefore, it was decided to focus on the analysis of the dependence of these two parameters only and to indicate the possible application of the working pair. The working pairs analysed were divided into 3 groups of applications depending on the refrigerant temperature of evaporation (T_{eva}), freezing (T_{eva} < 0 °C), refrigeration (0 °C < T_{eva} < 15 °C) and air conditioning (T_{eva} > 15 °C).

Conducting an analysis of the advantages and disadvantages of each adsorption working pair is not easy, due to the diversity of applications. A given pair may have extremely different operating parameters depending on the values of operating temperatures. Regarding the working pairs with the highest evaporating temperatures that are classified as suitable for air conditioning systems, they have an SCP in the range of 0.4–0.7 kW/kg, with a much lower COP, which is less than 0.5. It is also worth noting that the pairs with the highest T_{eva} value are combinations of different adsorbents with water as the adsorbate. This is directly related to the good performance of water as a refrigerant and the lack of need to keep the bed pressure as low is needed as for lower evaporation temperatures. As can be easily seen, the best COP and SCP parameters are characterised by MOF/water pairs, because they have a COP of over 0.6 and SCP of over 0.8 kW/kg. It is also worth mentioning that the MOF-801/water pair [57] operates at an evaporation temperature of 5 °C (for a pressure of 900 Pa) and at a regeneration temperature of 85 °C. These performance and temperature parameters allow for speaking about the validity of using adsorption chillers for cooling purposes. However, the fact that the best COP and SCP values are related to the use of MOFs, which are several to a dozen times more expensive than classical adsorbents, makes these working pairs uncompetitive. The study of different working pairs at T_{eva} corresponding to freezing processes is associated with obtaining low values of COP and SCP at the level of a few watts per kilogram. Nevertheless, attention should be paid to the AC35/methanol pair [101], which allows obtaining a COP of 0.33 and an SCP of 0.66, at an evaporation temperature of −2 °C and at a regeneration temperature of 100 °C.

These values indicate great potential for the development of adsorption chillers operating at negative refrigerant evaporation temperatures. Freezing applications, however, involve the use of adsorbates other than water, and currently the most extensively studied working pairs are those using methanol or ethanol.

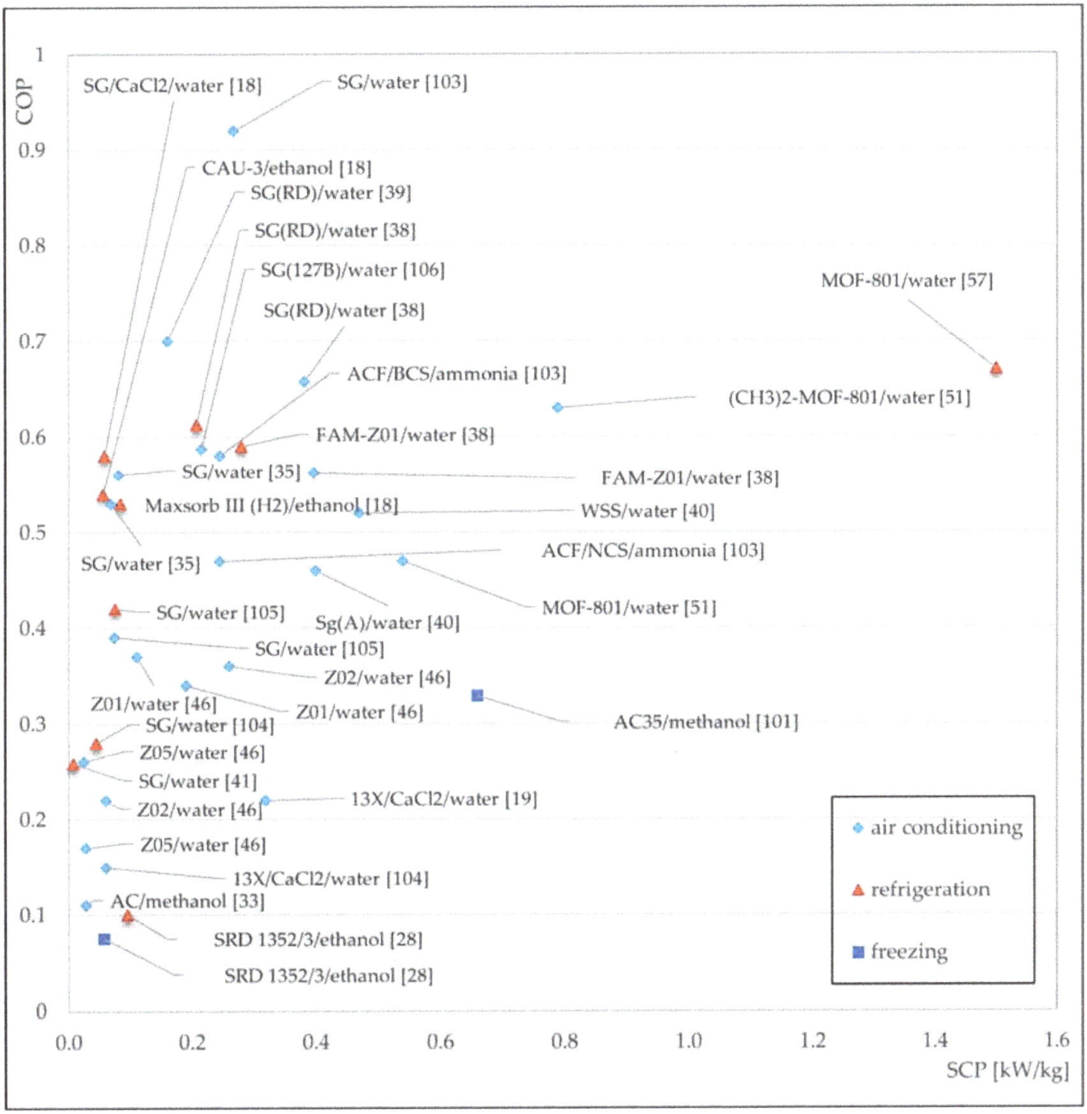

Figure 2. Correlation of the COP parameter on the SCP for different working pairs.

Referring to the data in Table 9 and the conclusions of the selected studies, it should be stated that the selection of the optimal adsorbent/adsorbate pair allows for a significant improvement of the adsorption cycle. Almost every paper has written about the dependence of the increase in COP with increasing adsorption time, which in turn is accompanied by a decrease in SCP. Another element common to all studies is the fact that the values of COP and SCP coefficients increase with increasing the evaporation temperature of the refrigerant, while the values of these characteristic parameters decrease with decreasing desorption temperature. It can be concluded that the tabular statement in the shape of Table 9 and the figure (Figure 2) allow comparing the individual adsorption working pairs. Nevertheless, the attention should also be paid to the differences in the construction of the individual systems. The systems studied in the laboratory are very often characterised using a monolayer of adsorbent, whereas the real ones are packed beds. Therefore, in the case of adsorption beds with several layers of adsorbent, both adsorption and desorption take longer, and this is another criterion that, besides the heat transfer coefficient between adsorbent and bed, affects the performance of the system and requires several analyses.

The discussed studies included analyses of the duration and values of the individual temperatures of the adsorption chiller operating cycle. It should be stated that the common point of the discussed studies is the need for further research in real operating conditions. Modi et al. [101] in their study analysed solar heat as an energy source for adsorbent bed regeneration. The system studied was operated with AC35/methanol vapour, and the researchers determined the effect of actual ambient conditions on cooling performance. It was found that a temperature of 100 °C is the value for which the efficiency of the operating process is the highest. In addition, it was found that actual operating conditions generate a decrease in ice production efficiency and COP of about 30%. This value is related to the conditions in India, which also allows analysing the profitability of the investment. On this basis, the payback time of the 9.25 kW adsorption system was determined to be 3.5 years, with an annual reduction in CO_2 emissions of almost 13 tonnes. However, it is not easy to estimate the overall annual cost of operation of individual solutions, due to the variety of cooler design options. The choice of a specific adsorption working pair influences the amount of heat necessary for bed regeneration and the amount of energy consumed by the vacuum pump that ensures the right bed pressure for the application. Therefore, the annual operating cost of an adsorption chiller is based on the amount of electricity consumed by the vacuum pump, circulating pump, and control systems, and thus the annual operating cost is based on the price of electricity in the region. Of course, it is also necessary to analyse the demand for cooling, to assess whether this demand is constant or variable throughout the day, because all this affects the operating costs. Obviously, for a given chiller and a particular adsorption working pair, the highest operating costs will be for freezing and the lowest for air conditioning. The final annual cost of operating an adsorption chiller depends on the location of the application, electricity prices, the amount of refrigeration demand, the specific application, the source of the bed regeneration heat, and the chiller design in general.

Analysing publications on adsorption systems driven by solar heat, it can be stated that they are mostly based on classical working pairs and are characterised by low SCP values. Nevertheless, most publications on real adsorption systems are based on solar heat as the driving energy. Furthermore, studies focusing on the optimisation of adsorption and desorption times also mostly refer to these systems. This has to do with the characteristics of solar radiation, which is time-varying, affecting the temperature fluctuations of the desorption heat. In addition, the cooling load is also characterised by a variation in demand throughout the day. On this basis, it can be concluded that solar-powered adaptive temporary adsorption chillers will continue to be the subject of further research, as they are already in use in India, for example, and further modifications can only accelerate their wider application, especially in the African, Asian, and Central American regions. It should also be noted that among the solutions discussed, adsorption working pairs operating with a refrigerant evaporation temperature of 0–10 °C predominate, which allows them to be classified in the group of refrigeration or air conditioning solutions. Thus, it can be concluded that the use of adsorption cycle in freezing requires much research and analysis, as it has not received much attention in contemporary research.

Generally, current research focuses on the use of composite adsorbents and MOFs. MOFs are now a very widely researched group of materials because they allow the production of materials with specific parameters that are beneficial for a given application, although the disadvantages of these materials are the high price and the need to analyse the stability and durability of the newly developed MOFs. Nevertheless, further development of classical adsorbent/adsorbate pairs is also observed. Analyzing solutions based on physical adsorbents, it was found that their main advantages are easy availability, relatively low price, and the ability to operate at desorption temperatures of 40–50 °C. On the other hand, these adsorbents are already very well studied and in their original form prevent further improvement of the COP and SCP performance of the chiller. However, this group of materials provides a basis for the development of another one, which are composite and doped adsorbents. Their main advantage is the ability to shape the properties of the adsorbent formed. The use of different additives allows for improving heat and mass transport in the adsorption bed. On the other hand, the disadvantage of this group of adsorbents is the need to analyse the effect of doping on the clogging of pores in the adsorbent.

The most important conclusion from the analysis of the adsorption coating studies is that the use of an adsorption coating compared to a packed bed results in an intensification of heat and mass transport in the bed. The cited results indicate an increase of up to 500% in SCP values for the coated bed compared to the packed bed. In the case of COP, two trends of change were found, i.e., an increase up to 70% and a decrease up to 60%. However, it should be noted that the number of tests carried out for different configurations of the bed structure under comparable conditions is small, so further studies of coated beds are needed to verify contradictions in the results obtained.

Most studies of adsorbent coatings consider SG/water working pairs' application. However, comparative studies of coatings consisting of different adsorbents are lacking. These studies would give an idea of how the different materials interact with the binders.

Among the studies cited on the preparation of adsorption coatings, the immersion technique was most commonly used, due to the simplicity of obtaining the appropriate adsorbent/binder mixture. In contrast, the number of studies of coatings obtained by in situ methods is small. The main reasons for this are the high technological requirements, including high process temperature and pressure, and the small difference in performance improvement of the adsorption cooling system compared to ex situ methods.

An analysis of the influence of basic coating design parameters on heat and mass transport was carried out. A decrease in the SCP value of the system was observed with increasing coating thickness. The occurrence of optimum thickness depending on conditions and COP values was found. From the point of view of the adsorption process, it is more advantageous to use larger adsorbent particles as a coating component and to aim to minimise the proportion of binder. Potential research within the framework of parameters selection for the construction of adsorption coatings should be particularly directed towards the development of technology for obtaining coatings with variable proportion and gradient of porosity, as appropriate adjustment of these parameters allows improvement of SCP values.

The main challenge of the following publications is that there are few studies of different adsorption working pairs conducted under the same conditions (temperature values and heat exchanger type). The simulation studies also require careful analysis, due to the fact that CFD simulations conducted very often use as input data experimental measurements of other researchers, which were conducted under very specific conditions. Using the results data under even slightly changed operating conditions and developing a series of new simulation measurements can be misleading and lead to error propagation.

Analysing the objectives of individual papers, it can be concluded that they boil down to increasing the energy efficiency of the adsorption cooling cycle, which can be achieved by developing better adsorbents, selecting an appropriate adsorption working pair and using adsorbent to coat elements of the bed structure. However, presentation of COP and SCP parameters for a given adsorption working pair as a conclusion from the research carried out, without specifying the conditions under which these parameters were obtained, makes it much more difficult to conduct further analyses involving research in real operating conditions.

The results presented in this paper from different authors show a great diversity in experimental approach, adsorbents, adsorbates, binders, and operating parameters used. Therefore, there are still many different combinations of the above mentioned to be investigated, which will contribute to the further development of adsorption refrigeration.

6. Conclusions

Analysing recent works, many different experimental approaches in the field of adsorption cooling can be observed. This fact makes it difficult to compare the results presented by different authors. Therefore, it has to be concluded that the knowledge on the applications of different adsorption working pairs and coatings is incomplete.

Adsorption chillers are mostly studied for adsorbate evaporation temperatures in the range of 0–15 °C with water as refrigerant. The number of works considering the application of adsorption chillers in freezing is negligible, and the available studies describe solutions using methanol and ethanol as adsorbates. Among the works analysed, the application of most of the working pairs allows for obtaining a SCP < 500 W/kg and COP < 0.6. Only MOFs/water pairs achieve the desired high COP > 0.6 at SCP > 800 W/kg. For many pairs, a high COP is obtained with relatively low SCP < 300 W/kg, leading to increased bed dimensions and thus application problems. MOFs are up to 15 times better than classical working pairs with activated carbons or silica gels, which compensates for the difference in price of these adsorbents. For zeolite/methanol pairs, the application of bed coating allows up to a 5-times increase in SCP, with a 25% increase in COP, which allows to state that further development of classical adsorbents is possible. In the case of MOF/water pairs, the coating allows to increase SCP by 2.5 times and COP by up to 30%. Most of the discussed works focus on coatings thicker than 100 μm, which is connected to dip coating technique. In order to obtain thinner coatings with thicknesses of several microns, the in situ method should be used.

Summarising the collected observations, it should be stated that the modifications related to the adsorption working pair allow for a significant intensification of the heat and mass transfer in the bed, thus improving the COP and SCP coefficients. When considering potential trends for further research, according to the authors, the possibility of using mixtures of multicomponent adsorbents and mixtures of various natural refrigerants should be analysed. Further development of in-situ coatings in combination with MOF adsorbents could be an important step towards improving the efficiency of adsorption systems.

The analysis of the various adsorbates, adsorbents, adsorbent coatings, and operating temperatures used in refrigerators carried out in this work allows us to conclude that there are still opportunities for development in this field. In addition, it should be noted that each combination of the above-mentioned elements and parameters corresponds to an individual sorption curve. On this basis, the authors decided to draw attention to the need for detailed guidelines for testing adsorption chillers, analogous to those for heat pumps. Therefore, each adsorption working pair should be characterised by a set of COP and SCP parameters, determined for characteristic (e.g., 18) measurement points: $T_{ads} = 30$ °C, $T_{des} = \{60; 90; 120$ °C$\}$ and $T_{eva} = \{+12; +7; +2; -2; -7; -15$ °C$\}$. The use of this system objectifies the evaluation of adsorption chillers, allows comparison of bed materials, and is an opportunity for significant progress in the development of the adsorption chiller industry.

Furthermore, tests conducted under real conditions, including heat losses and possible desorption temperature fluctuations (e.g., ±10 °C), are necessary for proper evaluation of adsorption chillers. On the basis of the presented test results, it can be concluded that it is possible both to use environmentally friendly working vapours in the device and to use waste energy or renewable energy sources for its powering (low bed regeneration temperature), which will obtain relatively high values of COP and SCP coefficients.

Author Contributions: The co-authors contribution to the paper was: conceptualisation, P.B. and T.B.; formal analysis, Ł.M. and K.S.; writing—original draft preparation, P.B. and T.B.; writing—review and editing, Ł.M. and K.S.; visualisation, Ł.M., K.S., P.B. and T.B.; supervision, Ł.M. and K.S.; funding acquisition, Ł.M. and K.S. All authors have read and agreed to the published version of the manuscript.

Funding: The paper was funded by government money Department of Energy and Fuels number 16.16.210.476 (research subsidy).

Data Availability Statement: Not applicable.

Conflicts of Interest: The authors declare no conflict of interest.

Nomenclature

T_{eva}	evaporation temperature, °C
T_{chill}	cooling water temperature, °C
T_{heat}	heating water temperature, °C
T_{ads}	adsorption temperature, °C
T_{des}	desorption temperature, °C

Abbreviations

AA	anodic alumina
AC	activated carbon
ACF	activated carbon fiber
COP	coefficient of performance
EG	expanded graphite
GNP	graphene nanoplatelet
GWP	global warming potential
HDACF	high-density activated carbon fibre
HEC	hydroxyethylcellulose
IUPAC	International Union of Pure and Applied Chemistry
MOF	metal organic framework
ODP	ozone depletion potential
PTFE	polytetrafluoroethylene
PVA	polyvinyl alcohol
PVP	polyvinyl pyrrolidone
SCP	specific cooling power
SG	silica gel
SGP	silica gel powder

References

1. International Energy Agency World Energy Balances. *World Energy Balances Overview 2019*; International Energy Agency World Energy Balances: Paris, France, 2019.
2. Streck, C.; Von Unger, M.; Krämer, N. From Paris to Katowice: Cop-24 Tackles the Paris Rulebook. *J. Eur. Environ. Plan. Law* **2019**, *16*, 165–190. [CrossRef]
3. Hassan, H.; Mohamad, A.; Alyousef, Y.; Al-Ansary, H. A review on the equations of state for the working pairs used in adsorption cooling systems. *Renew. Sustain. Energy Rev.* **2015**, *45*, 600–609. [CrossRef]
4. Elsheniti, M.B.; Elsamni, O.A.; Al-Dadah, R.K.; Mahmoud, S.; Elsayed, E.; Saleh, K. Adsorption Refrigeration Technologies. *Sustain. Air Cond. Syst.* **2018**, 71–95. [CrossRef]
5. Demir, H.; Mobedi, M.; Ülkü, S. A review on adsorption heat pump: Problems and solutions. *Renew. Sustain. Energy Rev.* **2008**, *12*, 2381–2403. [CrossRef]
6. Council of European Energy Regulators (CEER). *Report on Power Losses 2017*; Council of European Energy Regulators (CEER): Brussels, Belgium, 2017.
7. Sztekler, K.; Kalawa, W.; Nowak, W.; Mika, L.; Gradziel, S.; Krzywanski, J.; Radomska, E. Experimental Study of Three-Bed Adsorption Chiller with Desalination Function. *Energies* **2020**, *13*, 5827. [CrossRef]
8. Sztekler, K.; Kalawa, W.; Mika, Ł.; Mlonka-Medrala, A.; Sowa, M.; Nowak, W. Effect of Additives on the Sorption Kinetics of a Silica Gel Bed in Adsorption Chiller. *Energies* **2021**, *14*, 1083. [CrossRef]

9. Thommes, M.; Kaneko, K.; Neimark, A.V.; Olivier, J.P.; Rodriguez-Reinoso, F.; Rouquerol, J.; Sing, K.S.W. Physisorption of gases, with special reference to the evaluation of surface area and pore size distribution (IUPAC Technical Report). *Pure Appl. Chem.* **2015**, *87*, 1051–1069. [CrossRef]
10. European Union Council Decision (EU). 2017/1541 of 17 July 2017—on the Conclusion, on Behalf of the European Union, of the Kigali Amendment to the Montreal Protocol on Substances That Deplete the Ozone Layer. *Off. J. Eur. Union* **2017**, *L 236*, 1–2.
11. Engineering ToolBox. 2001. Available online: https://www.engineeringtoolbox.com (accessed on 24 April 2021).
12. Engineering ToolBox. Argon—Density and Specific Weight. 2018. Available online: https://www.engineeringtoolbox.com/argon-density-specific-weight-temperature-pressure-d_2089.html (accessed on 24 April 2021).
13. Allouhi, A.; Kousksou, T.; Jamil, A.; El Rhafiki, T.; Mourad, Y.; Zeraouli, Y. Optimal working pairs for solar adsorption cooling applications. *Energy* **2015**, *79*, 235–247. [CrossRef]
14. Ojha, M.K.; Shukla, A.K.; Verma, P.; Kannojiya, R. Recent progress and outlook of solar adsorption refrigeration systems. In *Proceedings of the Materials Today*; Elsevier BV: Amsterdam, The Netherlands, 2020.
15. Wang, R.; Wang, L.; Wu, J. *Adsorption Refrigeration Technology*; Wiley: Hoboken, NJ, USA, 2014.
16. Sharma, R.; Kumar, E.A. Performance evaluation of simple and heat recovery adsorption cooling system using measured NH_3 sorption characteristics of halide salts. *Appl. Therm. Eng.* **2017**, *119*, 459–471. [CrossRef]
17. Jiang, L.; Lu, Y.; Tang, K.; Roskilly, A.P.; Wang, Y.; Yuan, Y.; Wang, R.; Wang, L. Investigation of a novel composite sorbent for improved sorption characteristic. *Energy Procedia* **2017**, *142*, 1455–1461. [CrossRef]
18. Boman, D.; Hoysall, D.C.; Pahinkar, D.G.; Ponkala, M.J.; Garimella, S. Screening of working pairs for adsorption heat pumps based on thermodynamic and transport characteristics. *Appl. Therm. Eng.* **2017**, *123*, 422–434. [CrossRef]
19. Chan, K.C.; Tso, C.Y.; Wu, C.; Chao, C.Y. Enhancing the performance of a zeolite 13X/$CaCl_2$–water adsorption cooling system by improving adsorber design and operation sequence. *Energy Build.* **2018**, *158*, 1368–1378. [CrossRef]
20. Grekova, A.; Girnik, I.; Nikulin, V.; Tokarev, M.; Gordeeva, L.; Aristov, Y. New composite sorbents of water and methanol "salt in anodic alumina": Evaluation for adsorption heat transformation. *Energy* **2016**, *106*, 231–239. [CrossRef]
21. Brancato, V.; Frazzica, A.; Sapienza, A.; Gordeeva, L.; Freni, A. Ethanol adsorption onto carbonaceous and composite adsorbents for adsorptive cooling system. *Energy* **2015**, *84*, 177–185. [CrossRef]
22. Dzigbor, A.; Chimphango, A. Evaluating the potential of using ethanol/water mixture as a refrigerant in adsorption cooling system by using activated carbon—sodium chloride composite adsorbent. *Int. J. Refrig.* **2019**, *97*, 132–142. [CrossRef]
23. Dias, J.M.; Costa, V.A. Adsorption heat pumps for heating applications: A review of current state, literature gaps and development challenges. *Renew. Sustain. Energy Rev.* **2018**, *98*, 317–327. [CrossRef]
24. Kumita, M.; Yamawaki, N.; Shinohara, K.; Higashi, H.; Kodama, A.; Kobayashi, N.; Seto, T.; Otani, Y. Methanol adsorption behaviors of compression-molded activated carbon fiber with PTFE. *Int. J. Refrig.* **2018**, *94*, 127–135. [CrossRef]
25. Mitra, S.; Muttakin, M.; Thu, K.; Saha, B.B. Study on the influence of adsorbent particle size and heat exchanger aspect ratio on dynamic adsorption characteristics. *Appl. Therm. Eng.* **2018**, *133*, 764–773. [CrossRef]
26. Rupam, T.; Islam, A.; Pal, A.; Saha, B.B. Adsorption thermodynamics and performance indicators of selective adsorbent/refrigerant pairs. *Appl. Therm. Eng.* **2020**, *175*, 115361. [CrossRef]
27. Palomba, V.; Vasta, S.; Giacoppo, G.; Calabrese, L.; Gulli, G.; La Rosa, D.; Freni, A. Design of an Innovative Graphite Exchanger for Adsorption Heat Pumps and Chillers. *Energy Procedia* **2015**, *81*, 1030–1040. [CrossRef]
28. Frazzica, A.; Palomba, V.; Dawoud, B.; Gullì, G.; Brancato, V.; Sapienza, A.; Vasta, S.; Freni, A.; Costa, F.; Restuccia, G. Design, realization and testing of an adsorption refrigerator based on activated carbon/ethanol working pair. *Appl. Energy* **2016**, *174*, 15–24. [CrossRef]
29. Brancato, V.; Gordeeva, L.; Sapienza, A.; Freni, A.; Frazzica, A. Dynamics study of ethanol adsorption on microporous activated carbon for adsorptive cooling applications. *Appl. Therm. Eng.* **2016**, *105*, 28–38. [CrossRef]
30. Singh, V.K.; Kumar, E.A.; Saha, B.B. Adsorption isotherms, kinetics and thermodynamic simulation of CO2-CSAC pair for cooling application. *Energy* **2018**, *160*, 1158–1173. [CrossRef]
31. Singh, V.K.; Kumar, E.A. Thermodynamic analysis of single-stage and single-effect two-stage adsorption cooling cycles using indigenous coconut shell based activated carbon-CO2 pair. *Int. J. Refrig.* **2017**, *84*, 238–252. [CrossRef]
32. Girnik, I.; Okunev, B.; Aristov, Y. Dynamics of pressure- and temperature-initiated adsorption cycles for transformation of low temperature heat: Flat bed of loose grains. *Appl. Therm. Eng.* **2020**, *165*, 114654. [CrossRef]
33. Hamrahi, S.E.; Goudarzi, K.; Yaghoubi, M. Experimental study of the performance of a continues solar adsorption chiller using Nano-activated carbon/methanol as working pair. *Sol. Energy* **2018**, *173*, 920–927. [CrossRef]
34. Yaïci, W.; Entchev, E. Coupled unsteady computational fluid dynamics with heat and mass transfer analysis of a solar/heat-powered adsorption cooling system for use in buildings. *Int. J. Heat Mass Transf.* **2019**, *144*, 118648. [CrossRef]
35. Vodianitskaia, P.J.; Soares, J.J.; Melo, H.; Gurgel, J.M. Experimental chiller with silica gel: Adsorption kinetics analysis and performance evaluation. *Energy Convers. Manag.* **2017**, *132*, 172–179. [CrossRef]
36. Mohammed, R.H.; Mesalhy, O.; Elsayed, M.L.; Hou, S.; Su, M.; Chow, L.C. Physical properties and adsorption kinetics of silica-gel/water for adsorption chillers. *Appl. Therm. Eng.* **2018**, *137*, 368–376. [CrossRef]
37. Mohammed, R.H.; Mesalhy, O.; Elsayed, M.L.; Chow, L.C. Performance evaluation of a new modular packed bed for adsorption cooling systems. *Appl. Therm. Eng.* **2018**, *136*, 293–300. [CrossRef]

38. Hong, S.W.; Ahn, S.H.; Chung, J.D.; Bae, K.J.; Cha, D.A.; Kwon, O.K. Characteristics of FAM-Z01 compared to silica gels in the performance of an adsorption bed. *Appl. Therm. Eng.* **2016**, *104*, 24–33. [CrossRef]
39. Li, M.; Zhao, Y.; Long, R.; Liu, Z.; Liu, W. Computational fluid dynamic study on adsorption-based desalination and cooling systems with stepwise porosity distribution. *Desalination* **2021**, *508*, 115048. [CrossRef]
40. Seol, S.-H.; Nagano, K.; Togawa, J. Modeling of adsorption heat pump system based on experimental estimation of heat and mass transfer coefficients. *Appl. Therm. Eng.* **2020**, *171*, 115089. [CrossRef]
41. Liu, Y.; Yuan, Z.; Wen, X.; Du, C. Evaluation on performance of solar adsorption cooling of silica gel and SAPO-34 zeolite. *Appl. Therm. Eng.* **2021**, *182*, 116019. [CrossRef]
42. Radu, A.; Defraeye, T.; Ruch, P.; Carmeliet, J.; Derome, D. Insights from modeling dynamics of water sorption in spherical particles for adsorption heat pumps. *Int. J. Heat Mass Transf.* **2017**, *105*, 326–337. [CrossRef]
43. Kayal, S.; Baichuan, S.; Saha, B. Adsorption characteristics of AQSOA zeolites and water for adsorption chillers. *Int. J. Heat Mass Transf.* **2016**, *92*, 1120–1127. [CrossRef]
44. Girnik, I.; Grekova, A.; Gordeeva, L.; Aristov, Y. Dynamic optimization of adsorptive chillers: Compact layer vs. bed of loose grains. *Appl. Therm. Eng.* **2017**, *125*, 823–829. [CrossRef]
45. Girnik, I.S.; Aristov, Y. Dynamics of water vapour adsorption by a monolayer of loose AQSOA™-FAM-Z02 grains: Indication of inseparably coupled heat and mass transfer. *Energy* **2016**, *114*, 767–773. [CrossRef]
46. Teo, H.W.B.; Chakraborty, A.; Han, B. Water adsorption on CHA and AFI types zeolites: Modelling and investigation of adsorption chiller under static and dynamic conditions. *Appl. Therm. Eng.* **2017**, *127*, 35–45. [CrossRef]
47. Kim, S.H.; Cho, K.; Kim, J.-N.; Beum, H.T.; Yoon, H.C.; Lee, C.H. Improvement of cooling performance of water adsorption chiller by using aluminophophate adsorbent. *Microporous Mesoporous Mater.* **2020**, *309*, 110572. [CrossRef]
48. Fan, W.; Chakraborty, A. Isosteric heat of adsorption at zero coverage for AQSOA-Z01/Z02/Z05 zeolites and water systems. *Microporous Mesoporous Mater.* **2018**, *260*, 201–207. [CrossRef]
49. Wang, S.; Xia, X.; Li, S. Cooling performance of metal organic framework-water pairs in cascaded adsorption chillers. *Appl. Therm. Eng.* **2021**, *189*, 116707. [CrossRef]
50. Shmroukh, A.N.; Ali, A.H.H.; Ookawara, S. Adsorption working pairs for adsorption cooling chillers: A review based on adsorption capacity and environmental impact. *Renew. Sustain. Energy Rev.* **2015**, *50*, 445–456. [CrossRef]
51. Han, B.; Chakraborty, A. Adsorption characteristics of methyl-functional ligand MOF-801 and water systems: Adsorption chiller modelling and performances. *Appl. Therm. Eng.* **2020**, *175*, 115393. [CrossRef]
52. Ma, L.; Rui, Z.; Wu, Q.; Yang, H.; Yin, Y.; Liu, Z.; Cui, Q.; Wang, H. Performance evaluation of shaped MIL-101–ethanol working pair for adsorption refrigeration. *Appl. Therm. Eng.* **2016**, *95*, 223–228. [CrossRef]
53. Solovyeva, M.V.; Aristov, Y.; Gordeeva, L.G. NH_2-MIL-125 as promising adsorbent for adsorptive cooling: Water adsorption dynamics. *Appl. Therm. Eng.* **2017**, *116*, 541–548. [CrossRef]
54. Gordeeva, L.G.; Tu, Y.D.; Pan, Q.; Palash, M.; Saha, B.B.; Aristov, Y.I.; Wang, R.Z. Metal-organic frameworks for energy conversion and water harvesting: A bridge between thermal engineering and material science. *Nano Energy* **2021**, *84*, 105946. [CrossRef]
55. Karmakar, A.; Prabakaran, V.; Zhao, D.; Chua, K.J. A review of metal-organic frameworks (MOFs) as energy-efficient desiccants for adsorption driven heat-transformation applications. *Appl. Energy* **2020**, *269*, 115070. [CrossRef]
56. Youssef, P.; Mahmoud, S.; Al-Dadah, R.; Elsayed, E.; Elsamni, O. Numerical Investigation of Aluminum Fumarate MOF adsorbent material for adsorption desalination/cooling application. *Energy Procedia* **2017**, *142*, 1693–1698. [CrossRef]
57. Solovyeva, M.; Gordeeva, L.; Krieger, T.; Aristov, Y. MOF-801 as a promising material for adsorption cooling: Equilibrium and dynamics of water adsorption. *Energy Convers. Manag.* **2018**, *174*, 356–363. [CrossRef]
58. Younes, M.M.; El-Sharkawy, I.I.; Kabeel, A.; Uddin, K.; Miyazaki, T.; Saha, B.B. Characterization of silica gel-based composites for adsorption cooling applications. *Int. J. Refrig.* **2020**, *118*, 345–353. [CrossRef]
59. Yagnamurthy, S.; Rakshit, D.; Jain, S.; Rocky, K.A.; Islam, A.; Saha, B.B. Adsorption of difluoromethane onto activated carbon based composites: Thermophysical properties and adsorption characterization. *Int. J. Heat Mass Transf.* **2021**, *171*, 121112. [CrossRef]
60. Pal, A.; Uddin, K.; Rocky, K.A.; Thu, K.; Saha, B.B. CO2 adsorption onto activated carbon–graphene composite for cooling applications. *Int. J. Refrig.* **2019**, *106*, 558–569. [CrossRef]
61. Chen, Q.; Du, S.; Yuan, Z.; Sun, T.; Li, Y. Experimental study on performance change with time of solar adsorption refrigeration system. *Appl. Therm. Eng.* **2018**, *138*, 386–393. [CrossRef]
62. Wang, Q.; Gao, X.; Xu, J.; Maiga, A.; Chen, G. Experimental investigation on a fluidized-bed adsorber/desorber for the adsorption refrigeration system. *Int. J. Refrig.* **2012**, *35*, 694–700. [CrossRef]
63. Younes, M.M.; El-Sharkawy, I.I.; Kabeel, A.E.; Uddin, K.; Pal, A.; Mitra, S.; Thu, K.; Saha, B.B. Synthesis and characterization of silica gel composite with polymer binders for adsorption cooling applications. *Int. J. Refrig.* **2019**, *98*, 161–170. [CrossRef]
64. Qadir, N.U.; Said, S.A.; Ben Mansour, R. Modeling the performance of a two-bed solar adsorption chiller using a multi-walled carbon nanotube/MIL-100(Fe) composite adsorbent. *Renew. Energy* **2017**, *109*, 602–612. [CrossRef]
65. Kim, D.-S.; Chang, Y.-S.; Lee, D.-Y. Modelling of an adsorption chiller with adsorbent-coated heat exchangers: Feasibility of a polymer-water adsorption chiller. *Energy* **2018**, *164*, 1044–1061. [CrossRef]
66. Ammann, J.; Michel, B.; Ruch, P.W. Characterization of transport limitations in SAPO-34 adsorbent coatings for adsorption heat pumps. *Int. J. Heat Mass Transf.* **2019**, *129*, 18–27. [CrossRef]

67. Grabowska, K.; Sosnowski, M.; Krzywanski, J.; Sztekler, K.; Kalawa, W.; Zylka, A.; Nowak, W. Analysis of heat transfer in a coated bed of an adsorption chiller. *MATEC Web Conf.* **2018**, *240*, 01010. [CrossRef]
68. Freni, A.; Bonaccorsi, L.; Calabrese, L.; Caprì, A.; Frazzica, A.; Sapienza, A. SAPO-34 coated adsorbent heat exchanger for adsorption chillers. *Appl. Therm. Eng.* **2015**, *82*, 1–7. [CrossRef]
69. Chang, K.-S.; Chen, M.-T.; Chung, T.-W. Effects of the thickness and particle size of silica gel on the heat and mass transfer performance of a silica gel-coated bed for air-conditioning adsorption systems. *Appl. Therm. Eng.* **2005**, *25*, 2330–2340. [CrossRef]
70. Restuccia, G.; Freni, A.; Russo, F.; Vasta, S. Experimental investigation of a solid adsorption chiller based on a heat exchanger coated with hydrophobic zeolite. *Appl. Therm. Eng.* **2005**, *25*, 1419–1428. [CrossRef]
71. Restuccia, G.; Freni, A.; Maggio, G. A zeolite-coated bed for air conditioning adsorption systems: Parametric study of heat and mass transfer by dynamic simulation. *Appl. Therm. Eng.* **2002**, *22*, 619–630. [CrossRef]
72. Seol, S.-H.; Nagano, K.; Togawa, J. A new experimental method to separate interfacial and internal mass transfer on coated adsorbent. *Appl. Therm. Eng.* **2019**, *159*, 113869. [CrossRef]
73. Kumita, M.; Oya, T.; Matsui, K.; Suwa, Y.; Kodama, A.; Higashi, H.; Seto, T.; Otani, Y. Deposition of thick, rigid and size-controlled silica particle layer on aluminum sheet for water vapor adsorption. *Appl. Therm. Eng.* **2017**, *124*, 815–819. [CrossRef]
74. Li, A.; Thu, K.; Bin Ismail, A.; Shahzad, M.W.; Ng, K.C. Performance of adsorbent-embedded heat exchangers using binder-coating method. *Int. J. Heat Mass Transf.* **2016**, *92*, 149–157. [CrossRef]
75. Wang, L.; Bu, X.; Ma, W. Experimental study of an Adsorption Refrigeration Test Unit. *Energy Procedia* **2018**, *152*, 895–903. [CrossRef]
76. Kummer, H.; Füldner, G.; Henninger, S.K. Versatile siloxane based adsorbent coatings for fast water adsorption processes in thermally driven chillers and heat pumps. *Appl. Therm. Eng.* **2015**, *85*, 1–8. [CrossRef]
77. Pinheiro, J.M.; Salústio, S.; Geraldes, V.; Valente, A.A.; Silva, C.M. Copper foam coated with CPO-27(Ni) metal–organic framework for adsorption heat pump: Simulation study using OpenFOAM. *Appl. Therm. Eng.* **2020**, *178*, 115498. [CrossRef]
78. Pahinkar, D.G.; Boman, D.B.; Garimella, S. High performance microchannel adsorption heat pumps. *Int. J. Refrig.* **2020**, *119*, 184–194. [CrossRef]
79. Albaik, I.; Al-Dadah, R.; Mahmoud, S.; Almesfer, M.K.; Ismail, M.A.; Elsayed, E.; Saleh, M. A comparison between the packed and coated finned tube for adsorption system using aluminium fumarate: Numerical study. *Therm. Sci. Eng. Prog.* **2021**, *22*, 100859. [CrossRef]
80. Duong, X.Q.; Cao, N.V.; Lee, W.S.; Park, M.Y.; Chung, J.D.; Bae, K.J.; Kwon, O.K. Effect of coating thickness, binder and cycle time in adsorption cooling applications. *Appl. Therm. Eng.* **2021**, *184*, 116265. [CrossRef]
81. Sztekler, K.; Kalawa, W.; Mlonka-Medrala, A.; Nowak, W.; Mika, Ł.; Krzywanski, J.; Grabowska, K.; Sosnowski, M.; Debniak, M. The Effect of Adhesive Additives on Silica Gel Water Sorption Properties. *Entropy* **2020**, *22*, 327. [CrossRef]
82. McCague, C.; Huttema, W.; Fradin, A.; Bahrami, M. Lab-scale sorption chiller comparison of FAM-Z02 coating and pellets. *Appl. Therm. Eng.* **2020**, *173*, 115219. [CrossRef]
83. Rezk, A.; Al-Dadah, R.; Mahmoud, S.; Elsayed, A. Effects of contact resistance and metal additives in finned-tube adsorbent beds on the performance of silica gel/water adsorption chiller. *Appl. Therm. Eng.* **2013**, *53*, 278–284. [CrossRef]
84. Wittstadt, U.; Füldner, G.; Laurenz, E.; Warlo, A.; Große, A.; Herrmann, R.; Schnabel, L.; Mittelbach, W. A novel adsorption module with fiber heat exchangers: Performance analysis based on driving temperature differences. *Renew. Energy* **2017**, *110*, 154–161. [CrossRef]
85. Sapienza, A.; Gullì, G.; Calabrese, L.; Palomba, V.; Frazzica, A.; Brancato, V.; La Rosa, D.; Vasta, S.; Freni, A.; Bonaccorsi, L.; et al. An innovative adsorptive chiller prototype based on 3 hybrid coated/granular adsorbers. *Appl. Energy* **2016**, *179*, 929–938. [CrossRef]
86. Ouchi, T.; Hamamoto, Y.; Mori, H.; Takata, S.; Etoh, A. Water vapor adsorption equilibrium and adsorption/desorption rate of porous alumina film adsorbent synthesized with anodization on heat transfer plate. *Appl. Therm. Eng.* **2014**, *72*, 219–228. [CrossRef]
87. Tatlier, M.; Kummer, H.; Henninger, S.K. Crystallization of zeolite X coatings on stainless steel by microwave heating. *J. Porous Mater.* **2015**, *22*, 347–352. [CrossRef]
88. Caprì, A.; Frazzica, A.; Calabrese, L. Recent Developments in Coating Technologies for Adsorption Heat Pumps: A Review. *Coatings* **2020**, *10*, 855. [CrossRef]
89. Dias, J.M.; Costa, V.A. Evaluating the performance of a coated tube adsorber for adsorption cooling. *Int. J. Refrig.* **2020**, *118*, 21–30. [CrossRef]
90. Frazzica, A.; Füldner, G.; Sapienza, A.; Freni, A.; Schnabel, L. Experimental and theoretical analysis of the kinetic performance of an adsorbent coating composition for use in adsorption chillers and heat pumps. *Appl. Therm. Eng.* **2014**, *73*, 1022–1031. [CrossRef]
91. Bahremand, H.; Ahmadi, M.; Bahrami, M. Analytical modeling of oscillatory heat transfer in coated sorption beds. *Int. J. Heat Mass Transf.* **2018**, *121*, 1–9. [CrossRef]
92. Li, M.; Zhao, Y.; Long, R.; Liu, Z.; Liu, W. Gradient porosity distribution of adsorbent bed for efficient adsorption cooling. *Int. J. Refrig.* **2021**, *128*, 153–162. [CrossRef]
93. Younes, M.M.; El-Sharkawy, I.I.; Kabeel, A.; Saha, B. A review on adsorbent-adsorbate pairs for cooling applications. *Appl. Therm. Eng.* **2017**, *114*, 394–414. [CrossRef]

94. Goyal, P.; Baredar, P.; Mittal, A.; Siddiqui, A.R. Adsorption refrigeration technology—An overview of theory and its solar energy applications. *Renew. Sustain. Energy Rev.* **2016**, *53*, 1389–1410. [CrossRef]
95. Shabir, F.; Sultan, M.; Miyazaki, T.; Saha, B.B.; Askalany, A.; Ali, D.I.; Zhou, Y.; Ahmad, R.; Shamshiri, R.R. Recent updates on the adsorption capacities of adsorbent-adsorbate pairs for heat transformation applications. *Renew. Sustain. Energy Rev.* **2020**, *119*, 109630. [CrossRef]
96. Papakokkinos, G.; Castro, J.; López, J.; Oliva, A. A generalized computational model for the simulation of adsorption packed bed reactors—Parametric study of five reactor geometries for cooling applications. *Appl. Energy* **2019**, *235*, 409–427. [CrossRef]
97. Palomba, V.; Nowak, S.; Dawoud, B.; Frazzica, A. Dynamic modelling of Adsorption systems: A comprehensive calibrateddataset for heat pump and storage applications. *J. Energy Storage* **2021**, *33*, 102148. [CrossRef]
98. Xu, S.; Wang, L.; Wang, R. Thermodynamic analysis of single-stage and multi-stage adsorption refrigeration cycles with activated carbon–ammonia working pair. *Energy Convers. Manag.* **2016**, *117*, 31–42. [CrossRef]
99. Dakkama, H.; Elsayed, A.; Al-Dadah, R.; Mahmoud, S.; Youssef, P. Investigation of Cascading Adsorption Refrigeration System with Integrated Evaporator-Condenser Heat Exchanger Using Different Working Pairs. *Energy Procedia* **2015**, *75*, 1496–1501. [CrossRef]
100. Ma, L.; Yang, H.; Wu, Q.; Yin, Y.; Liu, Z.; Cui, Q.; Wang, H. Study on adsorption refrigeration performance of MIL-101-isobutane working pair. *Energy* **2015**, *93*, 786–794. [CrossRef]
101. Modi, N.; Pandya, B. Integration of evacuated solar collectors with an adsorptive ice maker for hot climate region. *Energy Built Environ.* **2021**. [CrossRef]
102. Girnik, I.S.; Aristov, Y. Dynamic optimization of adsorptive chillers: The "AQSOA™-FAM-Z02—Water" working pair. *Energy* **2016**, *106*, 13–22. [CrossRef]
103. Sinha, S.; Milani, D.; Luu, M.T.; Abbas, A. Enhancing the performance of a solar-assisted adsorption chiller using advanced composite materials. *Comput. Chem. Eng.* **2018**, *119*, 406–424. [CrossRef]
104. Tso, C.; Chan, K.; Chao, C.Y.; Wu, C. Experimental performance analysis on an adsorption cooling system using zeolite 13X/$CaCl_2$ adsorbent with various operation sequences. *Int. J. Heat Mass Transf.* **2015**, *85*, 343–355. [CrossRef]
105. Elsheniti, M.B.; El-Hamid, A.A.; El-Samni, O.A.; Elsherbiny, S.; Elsayed, E. Experimental evaluation of a solar two-bed lab-scale adsorption cooling system. *Alex. Eng. J.* **2021**, *60*, 2747–2757. [CrossRef]
106. Papakokkinos, G.; Castro, J.; Capdevila, R.; Damle, R. A comprehensive simulation tool for adsorption-based solar-cooled buildings—Control strategy based on variable cycle duration. *Energy Build.* **2021**, *231*, 110591. [CrossRef]
107. Chan, K.; Chao, C.Y.; Sze-To, G.; Hui, K.S. Performance predictions for a new zeolite 13X/$CaCl_2$ composite adsorbent for adsorption cooling systems. *Int. J. Heat Mass Transf.* **2012**, *55*, 3214–3224. [CrossRef]
108. Qadir, N.U.; Said, S.; Mansour, R.; Imran, H.; Khan, M. Performance comparison of a two-bed solar-driven adsorption chiller with optimal fixed and adaptive cycle times using a silica gel/water working pair. *Renew. Energy* **2020**, *149*, 1000–1017. [CrossRef]
109. Velte, A.; Füldner, G.; Laurenz, E.; Schnabel, L. Advanced Measurement and Simulation Procedure for the Identification of Heat and Mass Transfer Parameters in Dynamic Adsorption Experiments. *Energies* **2017**, *10*, 1130. [CrossRef]
110. Sah, R.P.; Choudhury, B.; Das, R.K. A review on low grade heat powered adsorption cooling systems for ice production. *Renew. Sustain. Energy Rev.* **2016**, *62*, 109–120. [CrossRef]
111. Rogala, Z. Adsorption chiller using flat-tube adsorbers—Performance assessment and optimization. *Appl. Therm. Eng.* **2017**, *121*, 431–442. [CrossRef]
112. Jribi, S.; Miyazaki, T.; Saha, B.; Koyama, S.; Maeda, S.; Maruyama, T. CFD simulation and experimental validation of ethanol adsorption onto activated carbon packed heat exchanger. *Int. J. Refrig.* **2017**, *74*, 345–353. [CrossRef]

energies

Article

The Effects of Using Steam to Preheat the Beds of an Adsorption Chiller with Desalination Function

Karol Sztekler *, Wojciech Kalawa, Lukasz Mika, Lukasz Lis, Ewelina Radomska and Wojciech Nowak

Department of Thermal and Fluid Flow Machines, Faculty of Energy and Fuels, AGH University of Science and Technology, Mickiewicza 30 St., 30-059 Krakow, Poland; kalawa@agh.edu.pl (W.K.); lmika@agh.edu.pl (L.M.); llis@agh.edu.pl (L.L.); radomska@agh.edu.pl (E.R.); wnowak@agh.edu.pl (W.N.)
* Correspondence: sztekler@agh.edu.pl

Abstract: Adsorption chillers are a promising alternative to traditional compressor-based devices. Adsorption chillers can be supplied with low- or medium-temperature waste heat or heat from renewable energy sources. In addition, they can be used for water desalination purposes. Thus, the adsorption chillers are unique devices that meet essential problems of the modern world: a need to limit the negative impact of humankind on the natural environment and growing problems with access to drinking water. However, adsorption chillers also have disadvantages, including ineffective operation and large size. Therefore, in this paper, the influence of steam utilization on the operation of an adsorption chiller with water desalination function was investigated experimentally, which has not been done before. The research was carried out on the adsorption chiller, working on a silica gel–water pair, installed in the AGH UST Center of Energy. The chiller was modified to preheat the sorbent with the use of steam. The results show that the use of steam instead of water for preheating the bed leads to higher temperatures in the heat exchanger and the bed. As a result, heat transfer from the heating medium to the bed is more intense, and a significant shortening of the desorption process is observed. In the case of using steam for preheating, the desorption time was about 30 s, while for water, it was 300 s. Thanks to this solution, it is possible to reduce the size of the device and increase its efficiency. The proposed solution opens a new course of research on adsorption chillers and broadens the horizon of their applications, as steam is a by-product of many industrial processes.

Keywords: adsorption chiller; desalination; preheating; silica gel; steam

Citation: Sztekler, K.; Kalawa, W.; Mika, L.; Lis, L.; Radomska, E.; Nowak, W. The Effects of Using Steam to Preheat the Beds of an Adsorption Chiller with Desalination Function. *Energies* **2021**, *14*, 6454. https://doi.org/10.3390/en14206454

Academic Editor: Ron Zevenhoven

Received: 24 August 2021
Accepted: 29 September 2021
Published: 9 October 2021

1. Introduction

Cooling processes are used in many industries, including food, pharmaceutical, heavy and energy industries. The International Institute of Refrigeration (IIR) estimated that the total number of refrigeration, air-conditioning and heat pump systems worldwide in 2019 was around 5 billion [1], with electrically powered compressors being the most abundant. As a result, electricity consumption in the refrigeration industry accounts for around 20% of total electricity consumption in the world [1]. Moreover, the dynamically progressing economic development, population growth, observed climate changes and the increasing desire to increase thermal comfort among people mean that by 2050 the global demand for electricity in the refrigeration industry will double [1].

It should be noted that more than 65% of electricity in the world is generated by burning fossil fuels [2], which adversely affects the natural environment due, among others, to greenhouse gas emissions. Additionally, the operation of compressor refrigeration equipment is based on the appropriate thermodynamic transformations of refrigerants such as hydro-chlorofluorocarbons (HCFC) and chlorofluorocarbons (CFC). These refrigerants are characterized by a high ozone depletion potential (ODP) and a high global warming potential (GWP) [3]; therefore, various legal regulations are introduced regarding the possibility of using these refrigerants [4].

Another challenge faced by more and more people around the world is the growing problem of a lack of access to drinking water. It is estimated that currently around 3.6 billion people live in water-poor areas for at least one month a year, and this number could rise to 5.7 billion by 2050 [5]. Therefore, in order to meet the demand for drinking water, desalination of sea or ocean water seems inevitable. However, conventional methods of desalination of water, such as multi-effect distillation (MED) or reverse osmosis (RO), are energy-intensive processes. To obtain 1 m^3 of desalinated water using MED and RO technologies, 14.2–21.6 kWh and 1.5–6.0 kWh of energy is required, respectively [6].

As a result, alternative and environmentally friendly technologies are being sought in both the refrigeration and desalination sectors. One of them is the use of adsorption chillers with the function of water desalination. Table 1 presents a comparison of water desalination using adsorption chillers and other technologies. Comparisons are made for solar energy coupled desalination technologies. As can be seen, water desalination using adsorption chillers is competitive with conventional water desalination technologies listed in Table 1.

Table 1. Comparison of selected water desalination technologies coupled with solar energy. Based on [6].

Desalination Technology	Electrical Energy Consumption, kWh/m^3	Thermal Energy Consumption, kWh/m^3	Cost, USD/m^3	Heat Source Temperature, °C
Adsorption chiller	1.4	39.8	0.7	55–85
Multi-stage flash desalination	2.5–5.0	81.0–144.0	1–5	90–110
Multi-effect distillation	1.5–2.0	60.0–70.0	2–9	79–90
Passive solar still	0	1106 [1]	1.3–6.5	-
Reverse osmosis	41–45	0	3–27	-

[1] Calculated with the assumption that the latent heat of water vaporization is 2390 kJ/kg [7] and the thermal efficiency of the solar still is 60% [8].

Adsorption chillers, unlike compressor chillers, are supplied with low-temperature heat, the minimum required temperature of which is only 55–80 °C [9]. Therefore, to power adsorption chillers, low-temperature waste heat from industrial processes [10] or heat obtained from solar energy [11] can be used. Moreover, the adsorption devices do not use harmful HCFCs and HFCs, but are based on environmentally friendly sorbent-sorbate pairs, such as silica gel–water [12], zeolites–water [13], activated carbon–ethanol [14]. Additional advantages of adsorption chillers include low noise level, lack of system vibrations and no moving parts [15]. On the other hand, one of the main disadvantages of adsorption refrigeration equipment is discontinuity of operation and low coefficient of performance (COP), usually not exceeding a value of 0.6 [16], as well as a low specific cooling power (SCP) [17].

The solution to the first problem, i.e., the discontinuity in the operation of adsorption chillers, is the use of two- [11], three- [18], four- [19] or even six-bed chillers [20]. On the other hand, efforts to improve the COP and SCP include many different methods, some of which only consist in changing the operating conditions of the device, and some require interference with the construction of the chiller.

The simplest methods of improving the COP and SCP of adsorption chillers, which do not require modification of the chillers' structure, are the selection of the appropriate time of the adsorption/desorption cycles [21], or the use of a heat source with a higher temperature [18]. However, it should be kept in mind that the heat supplying adsorption chillers is usually waste heat with a given temperature, which usually cannot be changed. Therefore, another solution to improve COP may be, for example, the use of mass recovery, which requires the connection of sorbent beds [22]. As demonstrated by Ng et al. [22], the use of mass recovery can improve the COP of the chiller by up to 20%. Another possibility is to use heat recovery, which takes heat from the bed during the adsorption process and transfers it to another bed, which is in the desorption or preheating phase [23]. Such action can improve the COP of the chiller by about 40% [23].

The heat exchange between the sorbent and the bed heating/cooling medium has a significant impact on the course of the desorption/adsorption process and, therefore, on the effectiveness of adsorption chillers [24,25]. The intensification of heat transfer in beds can be implemented in several ways: Firstly, by using a heat exchanger of an appropriate design [26], as well as being made of an appropriate material, which should be characterized by a high thermal conductivity coefficient, low heat capacity and low density [27]. Secondly, the intensification of heat exchange can take place by reducing the thermal resistance between the sorbent layer and the heat exchanger, which can be achieved through the use of various types of glues [25]. The heat transfer can also be intensified in the bed itself by adding various materials with high thermal conductivity to it, such as graphite flakes [28], copper, aluminum or carbon nanotubes [29,30].

Therefore, based on the above paragraph, it can be concluded that much attention is paid to the intensification of heat transfer within the bed itself or between the bed and the heat exchanger. However, the rate of heat flow from the heat transfer medium (the medium heating the bed during desorption) to the bed is also dependent on the thermal resistance between the medium and the heat exchanger as well as on the temperature difference between the medium and the bed [24]. Therefore, it seems justified to look for the possibility of intensifying heat transfer on the side of the heat transfer medium.

The effectiveness of desorption depends on the inlet temperature of the heating medium, and the abovementioned effectiveness augments with increasing temperature of the heating medium [31]. This observation was also reported by Ghilen et al. [32], who concluded that desorption is much faster for high temperatures of the heating medium compared to low temperatures. As a result, when the heating water inlet temperature increased from 55 °C to 95 °C, the chiller COP increased by up to 80%. A similar result was obtained experimentally by Sekret and Turski [33], as well as Sztekler et al. [18]. Sekret and Turski [33] found that the chiller COP increased by 17% when the heating water inlet temperature increased from 80 °C to 95 °C, while Sztekler et al. [18] reported a COP improvement of 190% when the heating water inlet temperature increased from 57 °C to 85 °C.

It is worth noticing that in most cases of adsorption chillers, the medium that heats or preheats the bed is water and, in some cases, also thermal oil [34,35]. Ng et al. [36] suggested that adsorption chillers can be driven by low-temperature steam, but they did not provide any information about the effects of using steam for bed regeneration on the operating parameters, e.g., temperatures, pressures and COP, of the chiller. In turn, Shahzad et al. [37] investigated the performance of the adsorption chiller integrated with two ejectors driven by steam at a pressure of 2 bar. However, to the best of the authors' knowledge, the literature lacks research concerning the effect of using steam to directly preheat the bed of the adsorption chiller, without using additional devices such as ejectors. Steam can have a much higher temperature than water, which, as shown in the above paragraph, is expected to improve the performance of the adsorption chiller. Such an innovative solution has not been investigated before.

This article presents the results of experimental research on the use of steam for the preheating of the beds of a three-bed adsorption chiller with water desalination function. Such a solution has never been used in the technology of adsorption chillers and is undoubtedly a novelty that will contribute to the development of this direction of research. As a result of using high-temperature steam, the heat exchange between the medium and the bed takes place faster, thanks to which the desorption process can take place in a much shorter time. Additionally, steam is generated in many technological processes and, therefore, the use of steam for preheating adsorption chiller beds can significantly expand the scope of their applications, as well as develop a new direction of research on adsorption chillers.

2. Methodology

2.1. Test Stand

The test stand, the basis of which is an adsorption chiller with a water desalination function, is located at the AGH UST Center of Energy. The tested chiller is a three-bed chiller in which silica gel and water are used as a sorbent–sorbate pair. A detailed description of the test stand and the general principle of operation of the tested chiller can be found in the papers [18,21]. Nevertheless, for the needs of this experiment, the chiller worked in a two-bed configuration, and the test stand was modified so that it was possible to investigate the effect of bed preheating with steam on the parameters of the chiller operation. The diagram and photos of the test stand with applied modifications are presented in Figures 1 and 2, respectively.

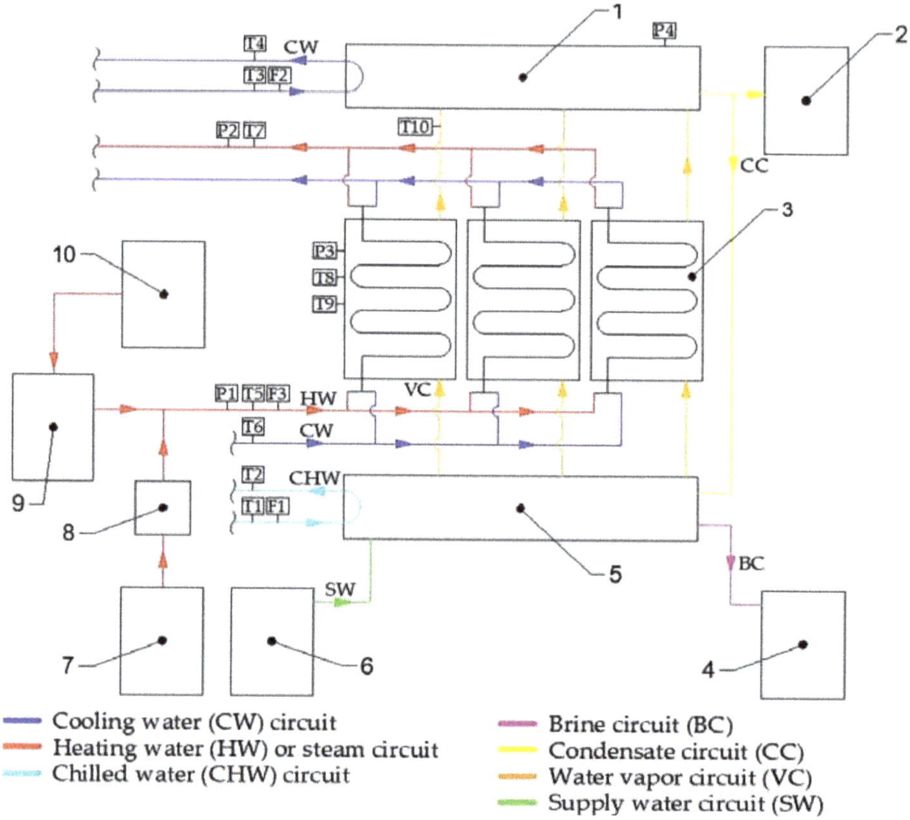

Figure 1. Simplified diagram of the test stand. 1—condenser; 2—distillate tank; 3—adsorption bed; 4—brine tank; 5—evaporator; 6—degasification tank; 7—steam generator; 8—manifold; 9—hot water tank; 10—electric heater; T1—chilled water inlet temperature; T2—chilled water outlet temperature; T3—cooling water inlet (to the condenser) temperature; T4—cooling water outlet (from the condenser) temperature; T5—heating water/steam inlet temperature; T6—cooling water inlet (to the bed) temperature; T7—heating water/steam outlet temperature; T8—temperature inside the heat exchanger; T9—temperature of the free space of the bed; T10—temperature of water vapor leaving the bed and flowing to the condenser; F1—chilled water flow rate; F2—cooling water flow rate; F3—heating water/steam flow rate; P1—steam inlet pressure; P2—steam outlet pressure; P3—pressure inside the bed; P4—pressure inside the condenser.

The main components of the chiller are three adsorption beds (3), an evaporator (5) and a condenser (1). Each bed is equipped with a finned Tichelmann lamella heat exchanger with a total length of 880 mm and a total width of 450 mm. The construction of the evaporator and condenser is cylinder-shaped with a length and diameter of 1500 and 350 mm (evaporator) and 1450 and 220 mm (condenser), respectively. In addition, the system is equipped, among other things, with a distillate tank (2), a brine tank (4), a degasification tank (6) and an expansion valve not shown in Figure 1. Additionally, a steam generator (7) and manifold (8) were installed at the test stand to enable investigating the use of steam for preheating the bed. In addition to the main components, the investigated unit is equipped with three circulating pumps (not shown in Figure 1), one for heating, one for cooling and one for chilled water circuits. However, these pumps are not an integral part of the adsorption chiller in contrast to, for example, a compressor in the case of compressor chillers. Theoretically, it is even possible to completely exclude the above-mentioned pumps thanks to the gravity flow of water. Fluids in various forms and under different conditions flow through the system in seven main circuits, as marked in Figure 1.

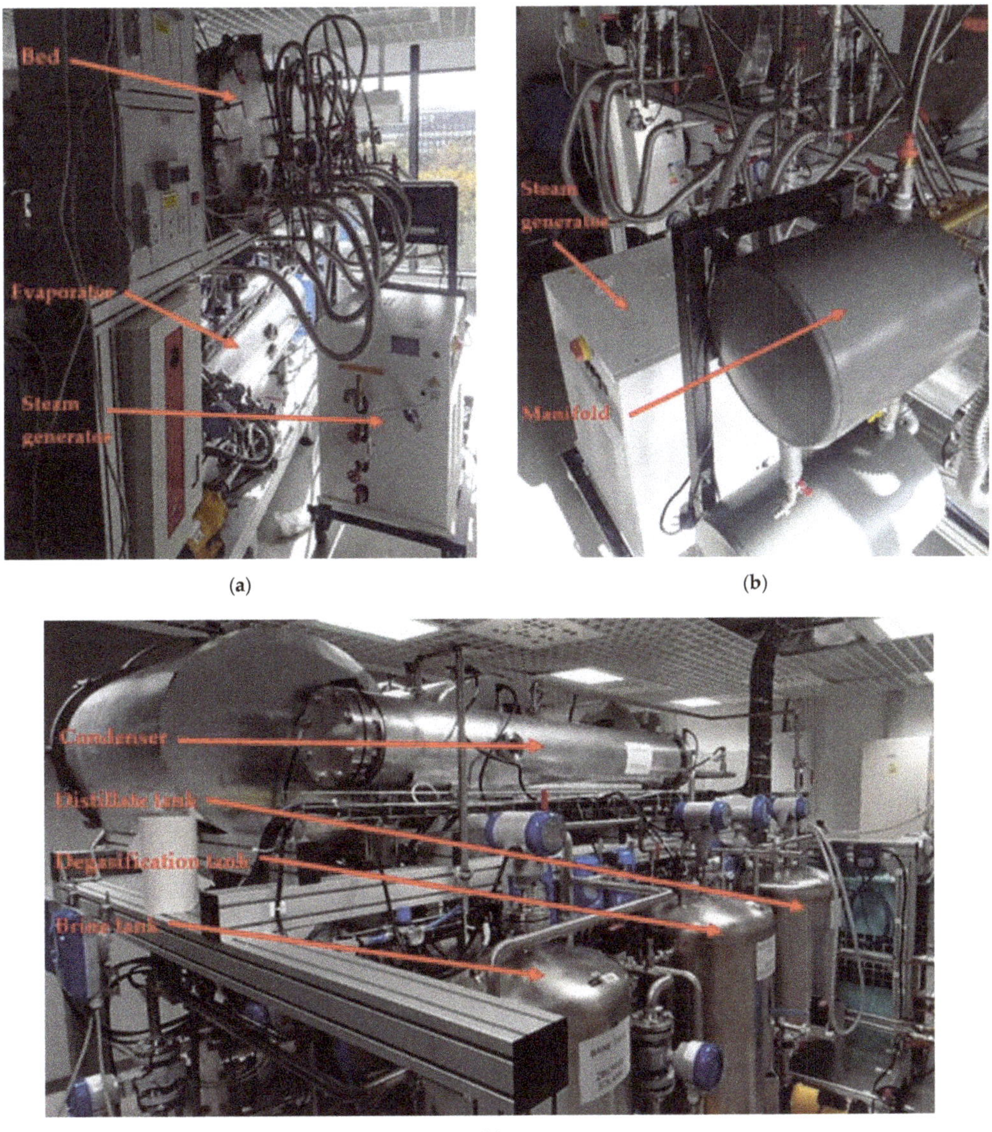

Figure 2. Pictures of the test stand: (**a**) front view on the beds, evaporator and steam generator; (**b**) front view on the steam generator and manifold; (**c**) back view on the condenser, distillate tank, degasification tank and brine tank.

In order to test the use of steam for the preheating of the chiller beds, the electric steam generator MAXI 24 (marked with the number 7 in Figure 1) was installed on the test stand. Detailed technical parameters of the steam generator are summarized in Table 2. The device has modulated power, thanks to which it is possible to control the parameters of the steam supplied to the chiller. The steam generator was connected to the distribution manifold (8), which in turn was connected to the heating water circuit. Thanks to this, two variants of bed preheating are possible:

1. With the use of hot water: the water is heated in the electric heater (10) and directed to the hot water tank (9) and, then, to the bed (3), while the steam generator (7) is not used and is cut off from the system by using special valves (not shown in Figure 1).
2. With the use of steam: steam is generated in the steam generator (7) and directed to the distribution manifold (8) and, then, to the bed (3). Steam flows through the bed for about 50 s (until its inlet and outlet temperatures are equal). Then, the supply of steam to the bed is cut off, and the supply of hot water, used for the desorption process, is opened. After leaving the bed, the steam is returned to the hot water tank.

Table 2. Technical parameters of the steam generator MAXI 24.

Parameter	Value
Maximum heating power	14 kW
Steam output	18.5 kg/h
Supply voltage	230 V
Boiler capacity	24.9 L
Maximum steam pressure	7 bar
Water pump motor power	0.55 kW

2.2. Measuring Devices and Uncertainty Analysis

The following parameters were measured during the experiments:

- The inlet and outlet temperatures of heating water or steam (depending on the operation variant of the chiller);
- The temperature inside the heat exchanger;
- The temperature of the free space in the bed;
- The temperature of the water vapor (sorbate) at the outlet of the bed;
- The pressure in the bed;
- The pressure in the condenser;
- The flow rates of heating water, cooling water, chilled water and steam.

The measuring sensors used are listed in Table 3.

Table 3. Technical parameters of measuring sensors.

Parameter	Sensor	Measurement Range	Measurement Uncertainty
Temperature	Pt-100 [1]	−80 °C to 250 °C	±0.1 °C
	Pt-1000	−80 °C to 150 °C	±0.1 °C
Pressure	Pressure transducer	0–99 kPa	±0.5%
Flow rate	Electromagnetic flow meter	1–100 L/min	±0.5%

[1] The Pt-100 sensor was used to measure the temperature of steam for the preheating of bed. The remaining temperatures were measured with Pt-1000 sensors.

All values were measured in the steady state of the chiller operation, achieved after about one hour from the moment of its start-up, and then, 300 adsorption/desorption cycles were performed. The values were saved every 5 s on a personal computer thanks to the use of special software. The test results were exported to csv files, which enabled their further processing. In addition, the test stand is equipped with an appropriate control system, thanks to which it is possible to fully control the parameters of the chiller.

The total measurement uncertainties were calculated as a quadratic sum of the random and systematic uncertainties [38]:

$$u(k) = \sqrt{\delta k_{ran} + \delta k_{sys}}, \tag{1}$$

where k, $u(k)$, δk_{ran} and δk_{sys} are the temperature or the pressure or the mass flow rate, the total measurement uncertainty, the random uncertainty and the systematic uncertainty, respectively. The systematic uncertainties are equal to the uncertainties of the measuring devices (Table 3), while the random uncertainties were calculated as follows:

$$\delta k_{ran} = \sqrt{\frac{1}{N \cdot (N-1)} \cdot \sum (k_n - k_{avg})^2}, \tag{2}$$

where N, k_n and k_{avg} are the number of measurements (N = 300), the value of n-th measurement and average from 300 measurements, respectively.

The coefficient of performance of the investigated chiller was calculated as [39]:

$$COP = \frac{\dot{m}_c \cdot c_{p,c} \cdot (T_{i,c} - T_{o,c})}{\dot{m}_h \cdot c_{p,h} \cdot (T_{i,h} - T_{o,h})}, \tag{3}$$

where m, c_p and T are the mass flow rate, (kg/s), the specific heat capacity, (J/kg/K) and the temperature, (°C), respectively. The subscripts c, h, i and o correspond to the chilled water, heating medium, inlet and outlet, respectively. The uncertainty of the COP was calculated from the general formula for error propagation [38]:

$$u(COP) = \frac{\sqrt{\sum \left(\frac{\partial COP}{\partial k_i} \cdot u(k_i) \right)^2}}{COP} \cdot 100\%, \tag{4}$$

where k_i stands for quantities from Equation (3).

The calculated measurement uncertainty of temperatures, pressures, mass flow rates and COP is 0.2 °C, 0.1 kPa, 0.01 kg/s and 9.7%, respectively.

2.3. Experimental Procedure

During the experiments, the influence of the use of steam for preheating the adsorption chiller beds on its operating parameters was investigated. For this purpose, two variants of the chiller operation were investigated, during which the beds were preheated with hot water (variant 1) or steam (variant 2). The working parameters of the chiller in variants 1 and 2 are presented in Table 4.

As mentioned before, the investigated adsorption chiller has three beds. Nevertheless, for this study, the chiller operated in a two-bed mode. Thus, the adsorption and desorption processes occur in the beds periodically. Primarily, the refrigerant from the evaporator flows to the bed, and the adsorption, which is an exothermic process, begins. The cooling water flows through the bed and receives heat from the bed through the heat exchanger. When the adsorption is completed, the bed is firstly preheated with water (variant 1) or steam (variant 2) for 50 s (until its inlet and outlet temperatures are equalized). Next, the desorption starts. As desorption is an endothermic phenomenon, heat in the form of hot water is supplied to the bed through the heat exchanger. After finishing the desorption, the refrigerant (desorbed water vapor) flows to the condenser, while the heat recovery process starts in the bed. Then, the entire process (adsorption, preheating, desorption, heat recovery) repeats. During all processes (adsorption, preheating, desorption and heat recovery), supply water is continuously supplied to the evaporator, and to ensure the continuous cooling capacity of the adsorption chiller, two beds are being used and working alternately. That is, desorption takes place in one bed, while adsorption takes place in the second bed. The preheating/desorption (adsorption)/heat recovery cycle time was 50/300/30 s in both variant 1 and 2.

Table 4. Chiller operating parameters in two variants of operation.

	Parameter	Variant 1 (Water Preheating)	Variant 2 (Steam Preheating)
Evaporator	Cooling capacity	1.10 kW	1.10 kW
	The chilled water inlet temperature T1	20 °C	20 °C
	The chilled water outlet temperature T2	18 °C	18 °C
	The chilled water mass flow rate F1	0.125 kg/s	0.125 kg/s
Condenser	Capacity	2.00 kW	2.00 kW
	The cooling water inlet temperature T3	25 °C	25 °C
	The cooling water outlet temperature T4	27 °C	27 °C
	The cooling water mass flow rate F2	0.250 kg/s	0.250 kg/s
Beds	The heating water inlet temperature T5	80 °C	80 °C
	The heating water mass flow rate F3	0.250 kg/s	0.250 kg/s
	Steam inlet temperature T5	-	120 °C
	Steam inlet pressure P1	-	1.864 bar
	Steam outlet pressure P2	-	1.386 bar
	Steam mass flow rate F3	-	0.0021 kg/s
	The cooling water inlet temperature T6	25 °C	25 °C
	Mass of the sorbent	12 kg	12 kg
Cycle time	Preheating	50	50
	Adsorption/desorption	300	300
	Heat recovery	30	30

3. Results and Discussion

3.1. Preheating with Steam and Water

Figure 3 shows the temperatures of heating water (variant 1) and steam (variant 2) at the inlet to the bed, while the outlet temperatures of the media (heating water and steam) are shown in Figure 4. As the preheating process starts, the inlet and outlet temperatures of the media increase in the first 50 s due to the cooling water circuit closure and the heating water/steam circuit opening. When analyzing the preheating with steam, it can be noticed that after opening the valve supplying water to the bed (transition from high to low temperature and vice versa), there is a certain time shift related to the change in the medium. This shift time is approximately 5 s.

In the case of variant 1, the heating water temperature at the bed inlet reaches about 80 °C after 50 s and remains at this level throughout the desorption process (300 s). When steam is used to preheat the bed, it can be observed that its maximum temperature at the inlet to the bed is about 120 °C after about 50 s from the beginning of the process. Thereafter, a decrease in the temperature of the medium to about 80 °C is visible, caused by the closure of the steam flow and the opening of the flow of heating water.

After 300 s, the desorption process is completed and the cooling water flow is opened, as evidenced by the drop in its inlet and outlet temperatures. In about 320 s, a sharp rise in temperature is visible in the steam temperature curve. A possible reason for such an undesirable system behavior is improperly selected solenoid valves shutting off the steam circuit from the bed.

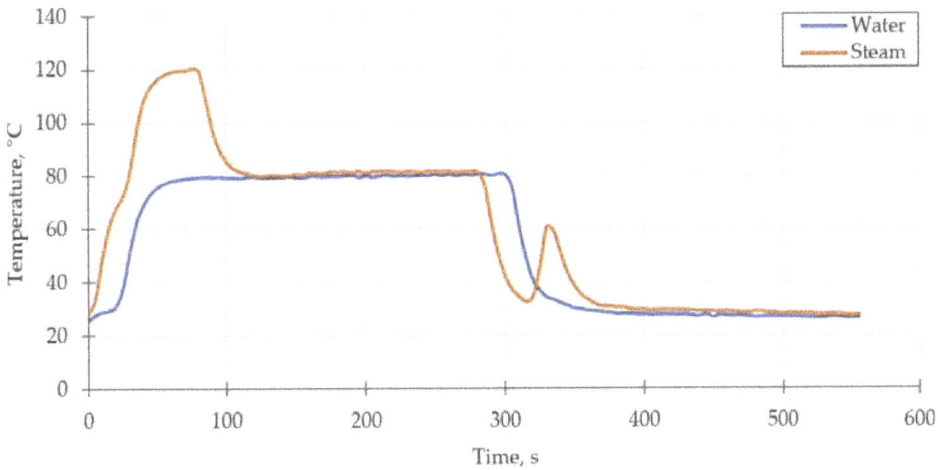

Figure 3. Temperature of the medium inlet (T5) to the adsorption chiller bed during desorption.

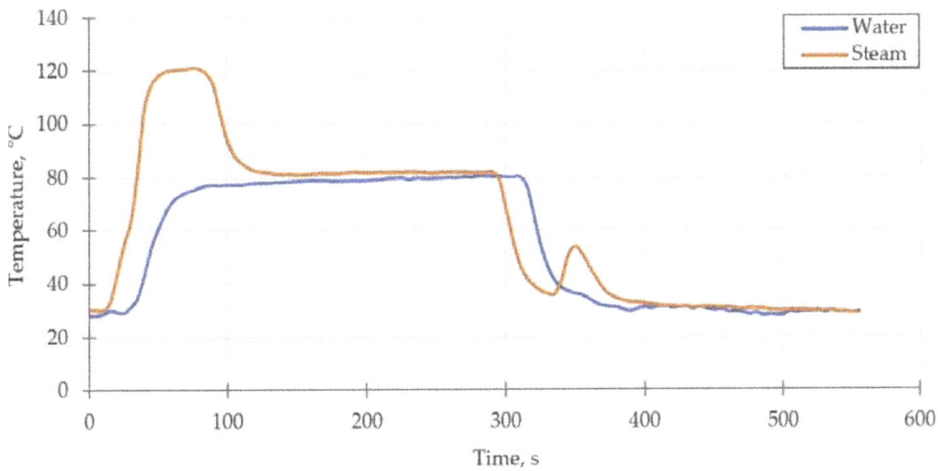

Figure 4. Temperature of the medium outlet (T7) from the adsorption chiller bed during desorption.

Figure 5 shows the change in pressure inside the bed as a function of time for preheating with water and steam. During steam preheating, the pressure in the bed increases much faster than in the case of water preheating. In addition, preheating the bed with steam leads to a greater maximum pressure inside the bed (about 12 kPa) than preheating with water (maximum pressure of about 8 kPa), which indicates that the desorption process is faster when the bed is preheated with steam.

Faster heat transfer between steam and silica gel than between water and silica gel is also confirmed by the temperature changes inside the heat exchanger, which is presented in Figure 6. The temperature rises and stabilizes faster for steam preheating. The use of steam to preheat the bed leads to higher temperatures inside the heat exchanger than in the case of the use of heating water. This is because the steam at the inlet to the bed has a higher temperature than heating water. Furthermore, the steam condenses in the heat exchanger, which additionally intensifies heat transfer. The high temperature of the steam enables faster heat transfer to the silica gel, which in turn leads to an intensification of the desorption process. It is also worth paying attention to the time after which the heat exchanger temperature reaches its maximum. In the case of steam preheating, the maximum temperature of the heat exchanger is about 70 °C, reached after about 100 s from the beginning of the process, and remains constant until the end of the desorption cycle. On the other hand, for water preheating, the temperature in the heat exchanger increases throughout the desorption phase until it reaches its maximum (approximately 65 °C) after about 300 s.

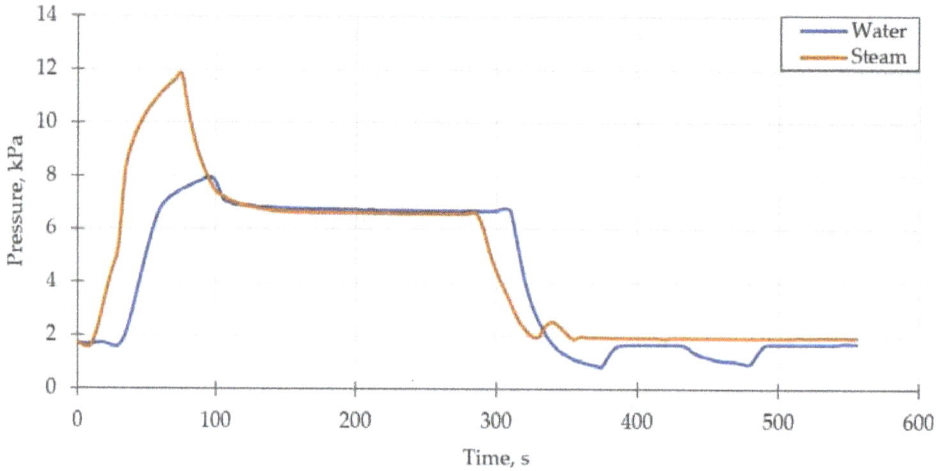

Figure 5. Pressure inside the bed (P3) during desorption.

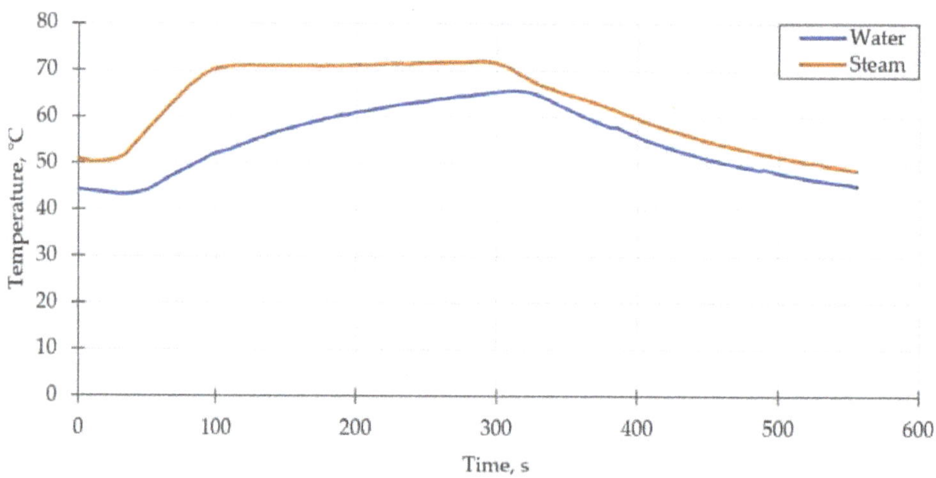

Figure 6. Temperature inside the heat exchanger (T8).

As already mentioned, the use of steam to preheat the bed leads to the intensification of heat transfer and acceleration of the desorption process. Therefore, in the case of preheating with steam, the desorption ends practically a few seconds after the preheating process. As a result, the desorption lasts for about 30 s for steam preheating, while in the classic cycle (water preheating), it lasts for about 300 s.

The temperature of the free space of the bed is presented in Figure 7. It can be noticed that the temperature of the free space in the bed preheated with steam increases and stabilizes faster and is higher than in the case of preheating with water. With the use of steam, the temperature of the free space in the bed reaches a constant value of 50 °C already after 80 s from the beginning of the cycle. On the other hand, the temperature of the free space in the bed preheated with water increases gradually, reaching its maximum (about 48 °C) after 300 s.

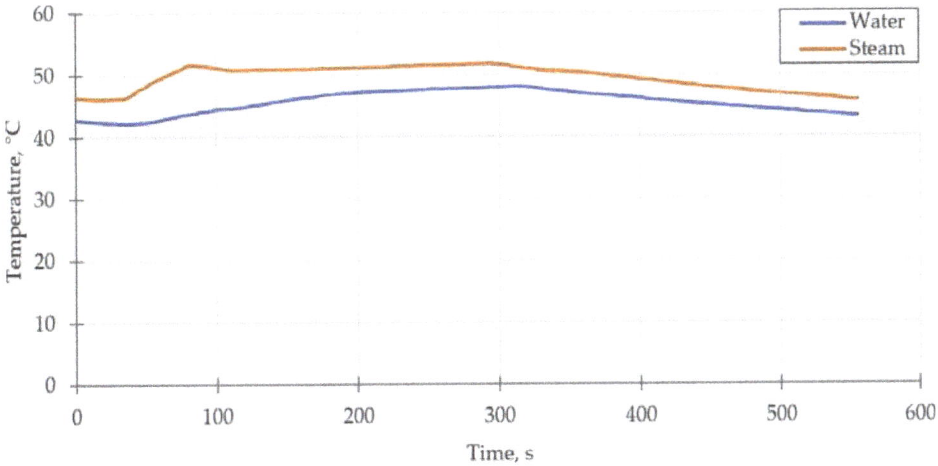

Figure 7. Temperature of the free space of the bed (T9) during desorption.

Figure 8 shows the temperature of water vapor, i.e., the sorbate after desorption, leaving the bed and flowing to the condenser. It can be noticed that the sorbate temperature increases faster in the case of steam preheating than water preheating. Such a course of the curves proves that the desorption of water vapor from the silica gel surface is more intense and occurs earlier in the case of steam preheating compared to the preheating with water. Thus, the transfer of sorbate from the bed to the condenser takes place more quickly, resulting in an increased COP (Equation (3)). The COP for water and steam preheating is 0.49 and 0.66, respectively, which means an increase of 35.2%.

As shown in the Introduction, numerous researchers reported that the COP of adsorption chillers increases with an increasing temperature of the heating medium. Among the works cited in the Introduction, the greatest improvement (by 190%) was obtained by Sztekler et al. [18]. Thus, the improvement of the COP achieved in this work (by about 35%) seems to be modest. However, it should be noted that the COP of adsorption chillers depends not only on the temperature of the heat source but on many other factors, including the number of beds, sorbent–sorbate pair, the temperature of chilled water, the pressure in the condenser and evaporator, the time of adsorption/desorption cycles, the construction of the chiller and others. Therefore, the improvements of the COP reported by different researchers cannot be compared directly, since, as explained above, the COP depends on several factors. Thus, the reported values of the COP enhancement could be compared only if the tests were carried out on the same chillers and with the same operational parameters. Additionally, the reference temperature of the heating source is meaningful, e.g., Sztekler et al. [18] compared the COP for the heat source temperatures of 57 °C and 85 °C, while in this paper, the temperatures of the heat source are 80 °C and 120 °C. As reported by Ng et al. [40], the rate of water desorption from silica gel is temperature-dependent but this relationship is not linear, i.e., the enhancement of desorption rate is much higher when the heat source temperature is increased from 40 °C to 80 °C than from 80 °C to 120 °C.

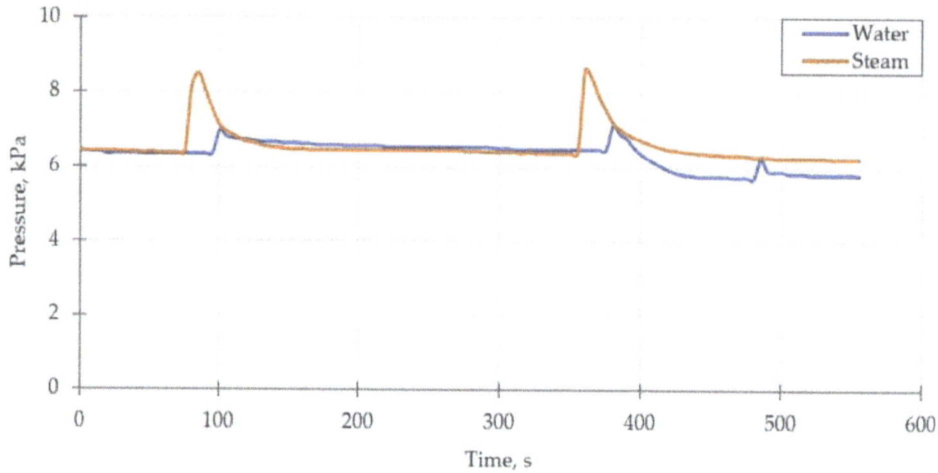

Figure 8. Temperature of the water vapor leaving the bed and flowing to the condenser (T10).

Such observations are confirmed by the pressure distribution in the condenser, which is shown in Figure 9. For bed preheating with steam, the pressure change in the condenser is more dynamic than in the case of water. Regarding steam preheating, after the initial pressure increase (facilitating the formation of condensate in the condenser), the pressure drops quite quickly, which leads to the flow of larger amounts of water vapor from the bed and, therefore, enhances the effectiveness of water desalination.

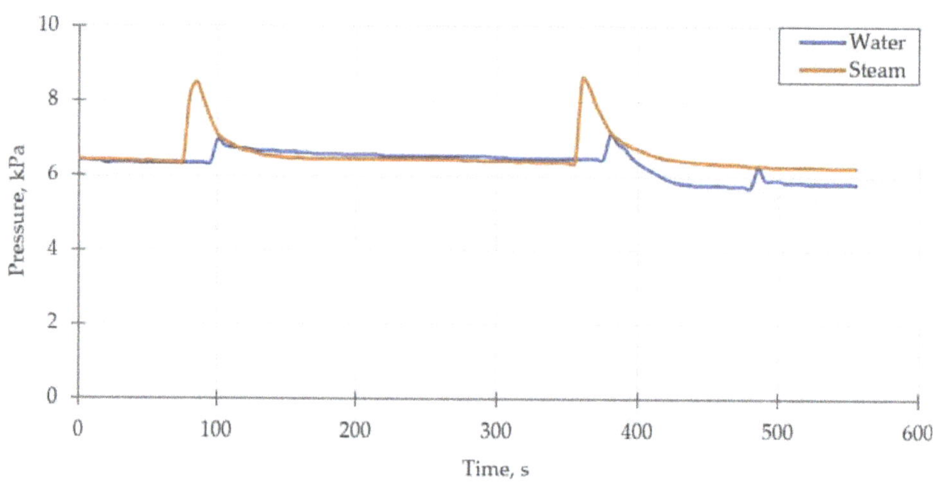

Figure 9. Pressure inside the condenser (P4).

3.2. Cost Calculation

As presented in Section 3.1., using steam for bed preheating can be beneficial in the terms of improving heat transfer and reducing the desorption time. Therefore, if free waste steam is easily available, the proposed solution seems to be cost-effective (ignoring the costs of chiller rebuilding). However, if there is no access to waste steam, the steam must be generated. Thus, a simple cost analysis is done to estimate the cost-effectiveness of this solution.

Assuming the initial water temperature T_0, heating water temperature T_w and steam temperature T_s as 20 °C, 80 °C and 120 °C, respectively, their specific enthalpies are i_0 = 83.9 kJ/kg, i_w = 335 kJ/kg and i_s = 2705.9 kJ/kg. Therefore, the energy necessary to increase the temperature of 1 kg of water from 20 °C (initial condition) to 80 °C and to generate 1 kg of steam (120 °C) is E_w = 251.1 kJ/kg and E_s = 2622 kJ/kg, respectively. The cost of generating steam and preparing heating water can be calculated from Equation (5):

$$C = c_{fuel} \ldots E/Q, \tag{5}$$

where C, Q and c_{fuel} are the cost of generating the steam (or preparing the heating water), the calorific value of fuel and the unit price of fuel, respectively. The assumed values of fuel unit price, calorific value and the results of calculation are summarized in Table 5.

Table 5. Cost calculation of generating steam and heating water from different fuels. Based on data from [41].

Fuel	Calorific Value, (kJ/kg)	Unit Price, (PLN/kg)	Cost of Generating the Steam, (PLN/kg)	Cost of Preparing the Heating Water, (PLN/kg)
Black coal	21,240	0.89	0.11	0.01
Lignite	9470	0.30	0.08	0.01
Oil	42,300	3.02	0.19	0.02
Natural gas	48,000	1805.56	98.63	9.45
Waste steam from industrial processes	-	0	0	-

Based on Table 5, it can be concluded that the cost of generating 1 kg of steam is approximately 10 times higher than the cost of preparing 1 kg of heating water with predetermined parameters. Nevertheless, if steam being a by-product of, e.g., the industrial process is available, then the cost of steam is zero.

3.3. Challenges and Design Guidelines

As explained in the Introduction and shown in the Results and Discussion, higher temperatures of the medium used for bed preheating, achieved by using steam instead of water, lead to faster heat transfer and shorten the desorption time from 300 s to 30 s. As a result, the COP of the adsorption chiller is improved by 35.2%, and more adsorption/desorption cycles can be completed in a given time. Therefore, the size of the chiller can be reduced maintaining the same cooling capacity. Nevertheless, using steam instead of water for bed preheating proposed in this paper involves a completely different approach to the desorption process and the material requirements of the device. Under given operating conditions of the chiller (Table 4), the average velocity of the steam is about 7.3 times higher than the average velocity of the water during the preheating. As a result, during the supply of steam to the bed, a very strong mechanical reaction of the system, especially of the pipelines supplying the steam to the bed, takes place, manifested by strong vibrations of the system.

The heat transfer rate between the heating medium and bed is directly proportional to the overall heat transfer coefficient, heat transfer surface area and temperature difference between the medium and the bed. Therefore, based on the results presented in this paper, the following general guidelines for designing steam-preheated adsorption chillers can be outlined:

- The steam supply system must be taken into account when designing new devices or, in the case of the existing ones, the adsorption chillers should be redesigned and rebuilt;
- An appropriate heat exchanger that enables the condensation of the steam should be designed or selected;

- Due to the condensation of steam, a heat exchanger should allow free drainage of the condensate and deaeration, which can be achieved by using straight pipes arranged parallelly and inclining the heat exchanger by about 3–5%;
- The steam inlet and water inlet should be on two sides of the heat exchanger to allow deaeration and condensate drainage;
- An appropriate heat exchanger and steam injection method that ensures uniform steam flow and reduces the vibrations should be designed or selected;
- Flow resistance during steam preheating is about 50 times greater than the flow resistance during water preheating. Therefore, it is advisable to increase the cross-sectional area of the pipes in the heat exchanger by at least 50%, which will result in a two-fold reduction in the flow resistance;
- The temperature difference between the steam and bed is about two times greater than the temperature difference between the heating water and bed. Additionally, the main barrier of heat transfer from the heating medium to the bed is the thermal contact resistance between the surface of the heat exchanger and silica gel grains [24]. Thus, changing the material of which the heat exchanger is made can be considered (e.g., from copper with high thermal conductivity to aluminum with lower thermal conductivity);
- Size reduction in the adsorption chiller, maintaining the same cooling capacity, should be kept in mind;
- The cost of preparing steam is about 10 times higher than the cost of preparing heating water using fossil fuels (Table 5). Thus, if there is a lack of waste steam from industry processes (or renewable energy sources), a thorough economic analysis should be conducted to ensure that using steam for bed preheating is cost-effective.

In addition to the design guidelines listed above, the results presented in this work may also be a basis for creating and validating a mathematical model of the desorption in the adsorption chillers, which may be a topic of the next paper. Furthermore, although the research was carried out for one adsorption chiller on a laboratory scale, it is expected that the presented results translate into other adsorption chillers, including the industrial-scale ones.

4. Conclusions

This article analyzes the possibilities of bed preheating with steam at a pressure of 3 bar and the maximum temperature of 120 °C.

During the experimental tests, a significant effect of the use of steam for preheating on the reduction in the desorption time was observed. The maximum temperature of the inlet steam to the bed was 120 °C, and a dynamic equalization of the outlet temperature to the inlet temperature was observed. The temperatures equalization lasted up to 30 s, which proves a very intense heat exchange process between the bed and the heating medium.

The heat exchanger temperature reaches its maximum after about 80 s from the start of the cycle (a few seconds after the end of steam feeding to the heat exchanger) and remains constant until the end of the desorption cycle, while the temperature in the case of water preheating slowly increases to reach its maximum at the end of desorption.

In the case of steam preheating, the pressure in the bed reaches a maximum value of 12 kPa, where in the case of heating water, it is 8 kPa. A higher pressure facilitates the desorption process, which is much more dynamic in the case of steam preheating.

When analyzing the use of steam, it should be noted that when using steam for preheating, the desorption process is shortened to about 30 s compared to 300 s for heating water. Reducing the desorption time to 30 s causes the device, using the same amount of sorbent, to increase its cooling capacity, because at the same time the sorption/desorption cycle can run several times in relation to the preheating the bed using hot water. Thus, the shortening of the desorption time can also lead to the reduction in the dimensions of the device. This is because the same cooling capacity can be obtained from a smaller amount of sorbent. Additionally, the COP of the chiller increased by 35.2% when steam is used instead of water for bed preheating.

If steam is used as a heat source for bed preheating, special preparation of the entire steam supply system to the device is required, and structural changes in the adsorption chiller are necessary. Nevertheless, using steam to preheat the beds of the adsorption chiller shows the potential of increasing its efficiency by reducing the desorption time.

Author Contributions: Conceptualization, K.S. and W.K.; methodology, K.S., L.M., L.L. and W.K.; formal analysis, K.S.; investigation, W.K. and L.L.; data curation, K.S., W.K. and E.R.; writing—original draft preparation, K.S., L.M., E.R. and L.L.; writing—review and editing, K.S. and E.R.; supervision L.M. and W.N.; project administration, K.S. and W.N.; funding acquisition, L.M. All authors have read and agreed to the published version of the manuscript.

Funding: The article was funded by the Faculty of Energy and Fuels AGH UST number 16.16.210.476 (research subsidy).

Data Availability Statement: The data presented in this paper are available on request from the corresponding author.

Conflicts of Interest: The authors declare no conflict of interest.

References

1. Dupont, J.L.; Domanski, P.; Lebrun, P.; Ziegler, F. The Role of Refrigeration in the Global Economy. In Proceedings of the 38th Note on Refrigeration Technologies, International Institute of Refrigeration, Paris, France, 1 June 2019.
2. International Energy Agency. Electricity Information: Overview. Available online: https://www.iea.org/reports/electricity-information-overview (accessed on 6 June 2021).
3. El-Sayed, A.R.; El Morsi, M.; Mahmoud, N.A. A Review of the Potential Replacements of HCFC/HFCs Using Environment-Friendly Refrigerants. *Int. J. Air-Cond. Refrig.* **2018**, *26*, 1830002. [CrossRef]
4. Regulation (EU) No 517/2014 of the European Parliament and of the Council of 16 April 2014 on Fluorinated Greenhouse Gases and Repealing Regulation (EC) No 842/2006. Available online: https://eur-lex.europa.eu/legal-content/EN/TXT/?uri=uriserv%3AOJ.L_.2014.150.01.0195.01.POL (accessed on 10 January 2020).
5. UNESCO. *The United Nations World Water Development Report 2018: Nature Based Solutions for Water*; UNESCO: Paris, France, 2018.
6. Nassrullah, H.; Anis, S.F.; Hashaikeh, R.; Hilal, N. Energy for desalination: A state-of-the-art review. *Desalination* **2020**, *491*. [CrossRef]
7. Incropera, F.P.; DeWitt, D.P.; Bergman, T.L.; Lavine, A.S. *Fundamentals of Heat and Mass Transfer*; John Wiley & Sons: Hoboken, NJ, USA, 2007.
8. Sharshir, S.W.; Elsheikh, A.H.; Peng, G.; Yang, N.; El-Samadony, M.O.A.; Kabeel, A.E. Thermal performance and exergy analysis of solar stills–A review. *Renew. Sustain. Energy Rev.* **2017**, *73*, 521–544. [CrossRef]
9. Lu, Z.S.; Wang, R.Z. Performance improvement by mass-heat recovery of an innovative adsorption air-conditioner driven by 50–80 °C hot water. *Appl. Therm. Eng.* **2013**, *55*, 113–120. [CrossRef]
10. Brückner, S.; Liu, S.; Miró, L.; Radspieler, M.; Cabeza, L.F.; Lävemann, E. Industrial waste heat recovery technologies: An economic analysis of heat transformation technologies. *Appl. Energy* **2015**, *151*, 157–167. [CrossRef]
11. Alahmer, A.; Wang, X.; Alam, K.C.A. Dynamic and Economic Investigation of a Solar Thermal-Driven Two-Bed Adsorption Chiller under Perth Climatic Conditions. *Energies* **2020**, *13*, 1005. [CrossRef]
12. Manila, M.R.; Mitra, S.; Dutta, P. Studies on dynamics of two-stage air cooled water/silica gel adsorption system. *Appl. Therm. Eng.* **2020**, *178*, 115552. [CrossRef]
13. Du, S.W.; Li, X.H.; Yuan, Z.X.; Du, C.X.; Wang, W.C.; Liu, Z.B. Performance of solar adsorption refrigeration in system of SAPO-34 and ZSM-5 zeolite. *Sol. Energy* **2016**, *138*, 98–104. [CrossRef]
14. Sha, A.A.; Baiju, V. Thermodynamic analysis and performance evaluation of activated carbon-ethanol two-bed solar adsorption cooling system. *Int. J. Refrig.* **2021**, *123*, 81–90. [CrossRef]
15. Sztekler, K.; Kalawa, W.; Stefanski, S.; Krzywanski, J.; Grabowska, K.; Sosnowski, M.; Nowak, W.; Makowski, M. Using adsorption chillers for utilising waste heat from power plants. *Therm. Sci.* **2019**, *23*, S1143–S1151. [CrossRef]
16. Sah, R.P.; Choudhury, B.; Das, R.K. A review on adsorption cooling systems with silica gel and carbon as adsorbents. *Renew. Sustain. Energy Rev.* **2015**, *45*, 123–134. [CrossRef]
17. Boruta, P.; Bujok, T.; Mika, Ł. Adsorbents, Working Pairs and Coated Beds for Natural Refrigerants in Adsorption Chillers—State of the Art. *Energies* **2021**, *14*, 4707. [CrossRef]
18. Sztekler, K.; Kalawa, W.; Nowak, W.; Mika, L.; Gradziel, S.; Krzywanski, J.; Radomska, E. Experimental study of three-bed adsorption chiller with desalination function. *Energies* **2020**, *13*, 5827. [CrossRef]
19. Jribi, S.; Saha, B.B.; Koyama, S.; Bentaher, H. Modeling and simulation of an activated carbon–CO2 four bed based adsorption cooling system. *Energy Convers. Manag.* **2014**, *78*, 985–991. [CrossRef]
20. Saha, B.B.; Koyama, S.; Kashiwagi, T.; Akisawa, A.; Ng, K.C.; Chua, H.T. Waste heat driven dual-mode, multi-stage, multi-bed regenerative adsorption system. *Int. J. Refrig.* **2003**, *26*, 749–757. [CrossRef]
21. Sztekler, K. Optimisation of Operation of Adsorption Chiller with Desalination Function. *Energies* **2021**, *14*, 2668. [CrossRef]

22. Ng, K.C.; Wang, X.; Lim, Y.S.; Saha, B.B.; Chakarborty, A.; Koyama, S.; Akisawa, A.; Kashiwagi, T. Experimental study on performance improvement of a four-bed adsorption chiller by using heat and mass recovery. *Int. J. Heat Mass Transf.* **2006**, *49*, 3343–3348. [CrossRef]
23. Chekirou, W.; Boukheit, N.; Karaali, A. Heat recovery process in an adsorption refrigeration machine. *Int. J. Hydrogen Energy* **2016**, *41*, 7146–7157. [CrossRef]
24. Rezk, A.; Al-Dadah, R.K.; Mahmoud, S.; Elsayed, A. Effects of contact resistance and metal additives in finned-tube adsorbent beds on the performance of silica gel/water adsorption chiller. *Appl. Therm. Eng.* **2013**, *53*, 278–284. [CrossRef]
25. Grabowska, K.; Krzywanski, J.; Nowak, W.; Wesolowska, M. Construction of an innovative adsorbent bed configuration in the adsorption chiller-Selection criteria for effective sorbent-glue pair. *Energy* **2018**, *151*, 317–323. [CrossRef]
26. Sharafian, A.; Bahrami, M. Assessment of adsorber bed designs in waste-heat driven adsorption cooling systems for vehicle air conditioning and refrigeration. *Renew. Sustain. Energy Rev.* **2014**, *30*, 440–451. [CrossRef]
27. Cao, N.V.; Duong, X.Q.; Lee, W.S.; Park, M.Y.; Chung, J.D.; Hong, H. Effect of heat exchanger materials on the performance of adsorption chiller. *J. Mech. Sci. Technol.* **2020**, *34*, 2217–2223. [CrossRef]
28. Bahrehmand, H.; Khajehpour, M.; Bahrami, M. Finding optimal conductive additive content to enhance the performance of coated sorption beds: An experimental study. *Appl. Therm. Eng.* **2018**, *143*, 308–315. [CrossRef]
29. Kulakowska, A.; Pajdak, A.; Krzywanski, J.; Grabowska, K.; Zylka, A.; Sosnowski, M.; Wesolowska, M.; Sztekler, K.; Nowak, W. Effect of Metal and Carbon Nanotube Additives on the Thermal Diffusivity of a Silica Gel-Based Adsorption Bed. *Energies* **2020**, *13*, 1391. [CrossRef]
30. Sztekler, K.; Kalawa, W.; Mika, Ł.; Mlonka-Medrala, A.; Sowa, M.; Nowak, W. Effect of Additives on the Sorption Kinetics of a Silica Gel Bed in Adsorption Chiller. *Energies* **2021**, *14*, 1083. [CrossRef]
31. Szyc, M.; Nowak, W. Operation of an adsorption chiller in different cycle time conditions. *Chem. Process Eng.* **2014**, *35*, 109–119. [CrossRef]
32. Ghilen, N.; Gabsi, S.; Benelmir, R.; Ganaoui, M. El Performance Simulation of Two-Bed Adsorption Refrigeration Chiller with Mass Recovery. *J. Fundam. Renew. Energy Appl.* **2017**, *7*. [CrossRef]
33. Sekret, R.; Turski, M. Research on an adsorption cooling system supplied by solar energy. *Energy Build.* **2012**, *51*, 15–20. [CrossRef]
34. Grzebielec, A.; Rusowicz, A.; Laskowski, R. Experimental Study On Thermal Wave Type Adsorption Refrigeration System Working On A Pair Of Activated Carbon And Methanol. *Chem. Process Eng.* **2015**, *36*, 395–404. [CrossRef]
35. Schwamberger, V.; Desai, A.; Schmidt, F.P. Novel Adsorption Cycle for High-Efficiency Adsorption Heat Pumps and Chillers: Modeling and Simulation Results. *Energies* **2019**, *13*, 19. [CrossRef]
36. Ng, K.C.; Thu, K.; Kim, Y.; Chakraborty, A.; Amy, G. Adsorption desalination: An emerging low-cost thermal desalination method. *Desalination* **2013**, *308*, 161–179. [CrossRef]
37. Shahzad, M.W.; Ybyraiymkul, D.; Burhan, M.; Oh, S.J.; Ng, K.C. An innovative pressure swing adsorption cycle. *AIP Conf. Proc.* **2019**, *2062*. [CrossRef]
38. Taylor, J.R. *An Introduction to Error Analysis. The Study of Uncertainties in Physical Mesurements*, 2nd ed.; University Science Books: Sausalito, CA, USA, 1997.
39. Sztekler, K.; Kalawa, W.; Nowak, W.; Mika, L.; Grabowska, K.; Krzywanski, J.; Sosnowski, M.; Al-Harbi, A.A. Performance evaluation of a single-stage two-bed adsorption chiller with desalination function. *J. Energy Resour. Technol.* **2020**, *143*, 082101. [CrossRef]
40. Ng, K.C.; Chua, H.T.; Chung, C.Y.; Loke, C.H.; Kashiwagi, T.; Akisawa, A.; Saha, B.B. Experimental investigation of the silica gel-water adsorption isotherm characteristics. *Appl. Therm. Eng.* **2001**, *21*, 1631–1642. [CrossRef]
41. The National Centre for Emissions Management. *Wartości opałowe (WO) i Wskaźniki Emisji CO2 (WE) w roku 2018 do Raportowania w Ramach Systemu Handlu Uprawnieniami do Emisji za rok 2021*; The National Centre for Emissions Management: Warszawa, Polska, 2020.

energies MDPI

Article

Effect of Metal Additives in the Bed on the Performance Parameters of an Adsorption Chiller with Desalination Function

Karol Sztekler *, Wojciech Kalawa, Łukasz Mika and Marcin Sowa

Faculty of Energy and Fuels, AGH University of Science and Technology, Mickiewicza 30, 30-059 Krakow, Poland; kalawa@agh.edu.pl (W.K.); lmika@agh.edu.pl (Ł.M.); masowa@agh.edu.pl (M.S.)
* Correspondence: sztekler@agh.edu.pl

Abstract: Adsorption chillers with desalination functionality, being devices characterised by very low electricity consumption, provide an alternative to conventional sources of cooling and water. The option of desalinating water means that the use of a single device enables obtaining two useful products. Adsorption chillers are not widely used at present. due to their low performance characteristics; these are, however, constantly being improved. This paper presents a verification of the possibility of increasing the cooling coefficient of performance (COP) and specific cooling power (SCP) of a laboratory adsorption chiller by optimising the length of cycle times and using a copper additive to silica gel with a mass fraction of 15% to increase heat transport in the bed. The choice of copper among other considered additives was determined by the conclusions from the research on the sorption kinetics of various mixtures, price and availability, and a high thermal conductivity. The device was operated in a two-bed mode aimed at producing cooling. The adsorbate was distilled water. The results were compared with those obtained under similar conditions when the beds were only filled with silica gel. As a result of the testing, it was found that the use of the copper additive with the sorbent increased both the COP and SCP. The tests were performed for different cycle times, of 100, 200, 300 and 600 s. With increasing cycle time COP also increased. In contrast, the specific cooling power increased only up to a certain point, whereafter its value decreased.

Keywords: adsorption; adsorption chiller; desalination; coefficient of performance; specific cooling power; low-temperature heat; silica gel; copper; additives; cycle time

Citation: Sztekler, K.; Kalawa, W.; Mika, Ł.; Sowa, M. Effect of Metal Additives in the Bed on the Performance Parameters of an Adsorption Chiller with Desalination Function. *Energies* **2021**, *14*, 7226. https://doi.org/10.3390/en14217226

Academic Editor: Christopher Micallef

Received: 5 October 2021
Accepted: 28 October 2021
Published: 2 November 2021

1. Introduction

The intensive development of the world economy and the expected population growth mean that the demand for cooling and water will continue to rise. The use of conventional technologies to meet this demand is associated with a huge expenditure of electricity, which still comes mainly from non-renewable sources [1]. The energy demand for cooling buildings has more than tripled since 1990 [2]. The IIR (International Institute of Refrigeration) estimates that, currently, about 20% of total electricity generation is used for refrigeration and air conditioning processes, of which about 8.5% is used to power air conditioners during space cooling [2,3]. It is also important to take measures to reduce peak energy demand, which can increase significantly during the summer period [4]. The IIR also predicts that power consumption of cooling equipment may double by 2050 [3]. In 2017, the number of new air conditioning devices was 110 million units, an increase of 8.1% over 2016 [5].

Climate change caused, among other things, by greenhouse gas emissions from anthropogenic sources has led to an increase in the global average temperature. Its increase, in turn, intensifies the demand for cooling, which is most often generated in a way contributing indirectly to negative climate change [6]. In most cases, the chillers used today are compressor systems driven solely by electricity and using synthetic refrigerants [7].

Alternatives to this type of refrigeration equipment include ejector systems as well as adsorption and absorption chillers [8]. The development of non-conventional chillers has accelerated with the emergence of more stringent requirements imposed by the Vienna Convention and the Montreal Protocol relating to limiting the production and use of CFC (chlorofluorocarbon) and HCFC (hydrochlorofluorocarbon) refrigerants as well as energy efficiency requirements for equipment [9,10].

The operation of adsorption chillers is based on thermal compression using the phenomena of adsorption and desorption. Thanks to the replacement of classical electrical compression by thermal compression, they do not consume as much electrical energy as compressor systems that are only driven by electricity. Adsorption chillers represent an environmentally friendly option for generating cooling and desalinated water while using low-quality heat, usually with a temperature in the range of 50–90 °C, e.g., waste heat from industrial processes, heat obtained as a result of solar energy conversion in solar collectors or heat from geothermal energy [11–13]. Adsorption chillers are characterised by many advantages that distinguish them from other cooling technologies. A very important feature is the possibility of using environmentally neutral, non-ozone-depleting refrigerants, such as methanol, ethanol or water [4]. An important advantage of adsorption chillers is also the possibility of their operation in tri-generation systems capable of producing electricity, heating and cooling simultaneously, thus contributing to more efficient fuel consumption [14,15]. In addition, the low number of moving mechanical parts means that the risk of failure of this type of installation is significantly lower [16]. Adsorption chillers are also quiet, easy to operate and vibration-free, which makes them even more attractive [4,14]. Importantly, temperature fluctuations of the heating water supplying adsorption chillers do not significantly affect their operation [17]. This makes them ideal for powering with solar energy [18].

A unique feature of adsorption chillers is the water desalination option. Increasing water consumption and decreasing water availability are causing water scarcity in more and more places in the world. It is estimated that more than 700 million people experience it, and another 500 million are close to it. By 2025, almost 1.8 billion people might be affected by water scarcity, and as many as 5.7 billion by 2050 [19,20]. The WHO (World Health Organization) predicts that, by 2025, as much as half the population may live in areas with poor access to drinking water [21]. Meeting this growing demand necessitates the use of desalination technologies. Desalination is considered particularly promising, also, because about 39% of the population lives within 100 km of saline waters, which makes this method of water treatment even more advantageous and cost-effective [22]. Conventional desalination methods require very high energy inputs, so adsorption desalination technology, being of minimal electricity consumption, is particularly attractive and promising for improving water security worldwide [23].

However, adsorption chillers are not currently used on a large scale. This is due to their main disadvantage, which is their set of relatively low performance parameters. The COP (coefficient of performance) of adsorption chillers is usually in the range of 0.5–0.6 while the SCP (specific cooling power) usually reaches up to 200 W/kg [24,25]. In this respect, therefore, this technology cannot compete with more efficient electrically driven compressor chillers, with COPs estimated at 3–5 [24]. Adsorption chillers' low performance coefficients are due to such factors as poor mass flow and heat transfer within the adsorbent bed and low adsorption capacity of the sorbent [26,27]. The consequence of the low cooling capacity of adsorption chillers is their large size and weight [18,26]. Reducing the size and weight of the unit is very important due to the commercialisation of the adsorption refrigeration technology.

The performance of an adsorption chiller depends on many factors. These include the design of the device including the bed's heat exchanger and the choice of the adsorbent/adsorbate pair. The most common working pairs used in adsorption chillers are silica gel/water, zeolite/water, activated carbon with methanol, ethanol or ammonia, and carbon fibre with menthol or ammonia [28]. Of key importance for increasing the efficiency of adsorption chillers are also the temperatures and flow rates of chilled water, heating water and cooling water [8,9,29,30]. The basic methods of increasing the operating efficiencies of adsorption chillers are briefly described further in this paper.

One way to improve the operating efficiency is by modifying the adsorption refrigeration cycle [28]. The basic adsorption refrigeration cycle is the simplest and also the most reliable [31]. However, it is characterised by low efficiency related to the large losses of thermal energy. For this reason, more complex cycles began to be developed; these include cycles with heat or mass recovery [28]. By using them simultaneously, maximum COP and SCP values can be achieved [32,33]. Even more complex cycles include the thermal wave cycle, the forced convection thermal wave cycle and the cascade cycle [34]. When an adsorption chiller is operating in desalination mode only, it is also possible to use advanced heat recovery where the condensation heat recovered is transferred to the working medium in the evaporator [35].

A method often used in practice to improve performance involves designing multi-component systems. They ensure stable production of cooling and make it possible to reduce fluctuations in the temperature of the chilled water obtained, and to increase cooling capacity [28]. Additionally, they offer the possibility of implementing various operational strategies in order to maximise efficiency [36]. A number of studies have been carried out to this end, the results of which indicate an improvement in heat recovery efficiency of 70% when increasing the number of beds from two to four, and of 40% for six beds compared with a chiller with four beds [37]. Further, findings presented by Saha et al. found a 35% improvement in the heat recovery efficiency of a three-bed chiller compared with a two-bed chiller [9].

An effective method of optimising the operation of an adsorption chiller and maximising its performance parameters is the proper selection of adsorption/desorption cycle times, switching times associated with precooling or preheating, and times allocated for heat and mass recovery [38]. First of all, it must be stressed that the length of the adsorption and desorption times directly affects the amount of adsorbate flowing from the evaporator to the bed and from the bed to the condenser. If it is too short, not enough adsorbate is evaporated, resulting in incomplete and inefficient adsorption and desorption, leading to the wrong chilled water temperature or insufficient production of distillate. Incomplete desorption, in turn, reduces the affinity of the adsorbent for adsorption of vapour in the next cycle. It should also be noted that this time must not be too long. In a situation where the adsorbate has been completely desorbed from the sorbent surface, the heat energy supplied in the heating water circuit will be unnecessarily consumed for further bed heating [39].

It has been repeatedly proven that the cooling coefficient of performance COP increases monotonically with increasing cycle times [4,13]. This is due to the fact that when the cycle time increases, the heat consumption of the chiller during the desorption process decreases, and the time allocated for switching the bed also decreases [9,40,41]. On the other hand, increasing cycle time is associated with a decrease in cooling power. Therefore, there are operating parameters of a chiller for which it is possible to achieve its maximum performance [15]. In the case of the experiment conducted by Saha et al., the maximum SCP value was achieved for adsorption/desorption times in the range from 180 to 300 s [9]. In a paper by El-Sharkawy et al., the adsorption and desorption times for which the highest value of cooling capacity were obtained were between 350 and 450 s [42]. Studies in search of the optimum sorption cycle time have also been carried out for a three-bed adsorption chiller located at the Energy Centre of the AGH University of Science and Technology in Krakow, Poland. The experiment was performed for cycles of 100, 200, 300, 500, 600 and 900 s. The results found that the COP increased over time, while the SCP initially increased, reaching a maximum value for a time equal to 500 s before decreasing [15].

Many research groups have also focused on the notion of allocating time to each phase of chiller operation. The results of their research have clearly indicated that the performance of the device will be higher if the time allocated to regeneration is shorter [43].

The results of studies on the rate of adsorption and desorption processes confirmed that bed regeneration can be as short as 2.2–3.5 times shorter than the adsorption process [43]. Aristov et al. carried out experimental studies from which their results showed that reducing the duration of desorption relative to adsorption resulted in higher device efficiency. For adsorption and desorption times of 255 and 100 s, respectively, more than double the SCP and COP values were obtained compared with a test run with adsorption and desorption times of 178 and 177 s [44]. The experimental study also proved that the operational stability of the chiller is higher with a longer time for each cycle [15].

A particularly important method for achieving more favourable performance parameters is to improve the heat and mass transfer within the bed. Heat transfer within the bed is hampered by the high thermal resistance of the porous adsorbent [45]. Its improvement can be achieved in several ways, which mainly include increasing the heat transfer surface in the bed, increasing the heat transfer at the heat exchanger-adsorbent interface and increasing the heat transfer inside the bed by increasing the thermal conductivity of the material filling the bed [5,24,28,31]. The first method is related to increasing the surface area of the heat exchanger. For this purpose, different types and designs of heat exchangers are considered for use in adsorption chiller beds [25]. The second method involves decreasing the thermal resistance by using coatings deposited on the surface of the heat exchanger or using binders to coat the heat exchanger with sorption material [24,46].

However, the last method is the most important from the point of view of this paper. Heat conduction through the adsorbent has a crucial impact on the overall thermal resistance of the bed [47]. An improvement in this property without a significant deterioration of the sorption capacity will result in a much better heat exchange in the bed and, consequently, in an increase in the rate of adsorption and desorption processes as well as in an increase in the specific cooling power (SCP) [18,31]. This is possible e.g., by replacing the pure sorption material with a mixture of the sorbent and materials characterised by higher thermal conductivity. This additive can be in the form of powder, chips, shavings or chunks [45,48,49]. A number of studies have already been carried out on this method of improving heat transfer [48]. They checked e.g., the effect of copper powders on the thermal conductivity of a mixture with activated carbon, and an increase in thermal conductivity in the range of 2 to 25% was achieved [31]. Improved heat transfer is also achieved by mixing a sorbent with a metal foam [47]. In their paper, Demir et al. presented the results of a study on improving heat transfer in a bed by adding pieces of metal to silica gel. The authors decided to investigate four metals, aluminium, copper, brass and stainless steel, in two chip size ranges of 1.0–2.8 mm and 2.80–4.75 mm and at different percentages of shares of the additive, i.e., 5%, 10% or 15%. The findings indicated that of all the additives used, it was aluminium, at a percentage share of 15% in the mixture, that provided the greatest benefit in terms of heat transfer. The improvement in thermal diffusivity and heat transfer coefficient over pure silica gel was 157% and 242%, respectively. However, increasing the additive percentage in the mixture contributes to a significant reduction in the adsorption capacity of the bed. Thus, the search for the most suitable percentage should take into account the gain associated with improved heat transfer, but also the loss of sorption capacity associated with using less sorption material [48]. In their paper, Askalany et al. focused on investigating the thermal conductivity of activated carbon with additives in the form of aluminium, copper and iron filings with a mass fraction of 10 to 30%. The highest increase in thermal conductivity in comparison with pure activated carbon was observed for 30% aluminium addition. For an aluminium addition of 20% by weight, the heat transfer coefficient increased from 1 to 20 W/(mK). As part of the study, a simulation of operation of a two-bed chiller was also carried out to check the effect of metal additives on the cycle time, SCP and COP of the device. An increase in the mass fraction of any of the additives contributes to a reduction in cycle time and an increase in the SCP [49].

With the development of materials science, more and more modern materials are being developed. These include hybrid MOF (metal-organic frameworks) and SWS (selective water sorbents) materials [24,40]. The ideal sorbent should combine good thermal conductivity and high sorption capacity.

The group of innovative methods influencing the improvement of performance parameters of chillers also includes the possibility of fluidisation of the adsorbent bed and optimisation of chiller operation through the use of neural networks [24,50]. It is also possible to combine individual stages of thermal vapour compression to form multi-stage adsorption chillers, but this method only works under certain operating conditions [51].

Thus, as shown, the simultaneous generation of cold and purification of water in adsorption chillers depends on many factors which are already described individually in the literature. However, so far there has been a lack of experiments in which two or more methods are used simultaneously to improve the performance of such devices. Therefore, this paper presents the results of tests carried out for a laboratory adsorption chiller operating in the two-bed mode with beds filled with silica gel with added copper at a mass fraction of 15%. The tests were carried out for four cycle times of 100, 200, 300 and 600 s in one operating configuration of the chiller. The results were compared with those obtained when the beds were only filled with silica gel, and the experiment was carried out under similar conditions.

2. Materials and Methods

In several previous publications, the authors presented the results of studies on the properties of mixtures of sorption material with additives characterised by high thermal conductivity. The experiments examined e.g., the sorption kinetics and thermal diffusivity of silica gel with copper and aluminium additives and carbon nanotube additives. Sztekler et al. published results that make it possible to determine the magnitude of the loss of sorption properties after using additives in the individual mixtures. Figure 1 presents a graph of sorption kinetics for mixtures of silica gel (SG) with additives of aluminium (Al), copper (Cu) and carbon nanotubes (CN) in mass fractions equal to 5% and 15%, tested at 60 °C. Details of the experiment and a detailed analysis of the results can be found in the paper [45].

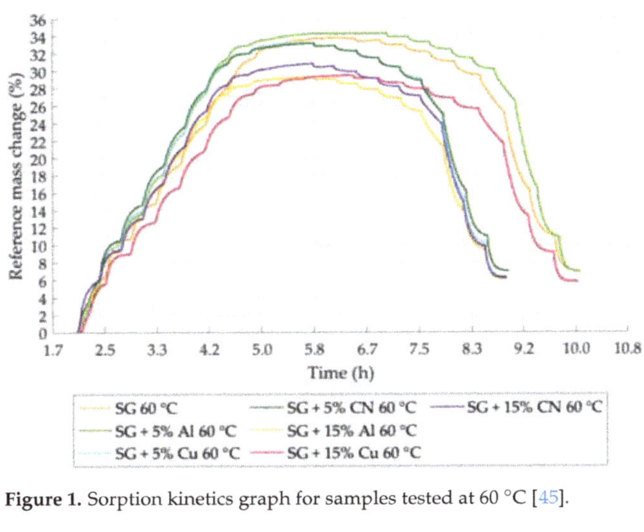

Figure 1. Sorption kinetics graph for samples tested at 60 °C [45].

The analysis of sorption kinetics curves clearly shows the loss of sorption properties of the samples with additives in comparison with the reference sample containing only silica gel. Naturally, for the mass fraction of the additive being 5%, this loss will be smaller, but in this case the focus was on its higher value. The sorption capacity of the material with added aluminium, copper and carbon nanotubes at a mass fraction of 15% decreases by 13.1%, 12.7% and 8.8%, respectively, compared with the mass change of the sample containing pure silica gel. Based on the sorption kinetics graph, it can also be seen that the sample with copper added at a mass fraction of 15% has the highest amount of adsorbate desorbed from its surface.

The next stage of the study was to assess the effect of the use of additives in terms of the increase of the cooling coefficient of performance and the specific cooling power of the desalinating adsorption chiller. This makes possible checking the actual effect of the additives on the operation of the adsorption chiller. The additive used was copper, at a mass fraction of 15%. The choice of this particular admixture was guided by, among other things, the conclusions from the study on additive sorption kinetics and the price and availability of the material. Particularly important for the selection of copper was its higher thermal conductivity compared to aluminium. For copper, it is 401 W/(mK), and for aluminium 237 W/(mK) [48]. The value of the heat conductivity coefficient plays a key role in the heat transfer mechanism. Based on Fourier's law, it can be concluded that its increase contributes to an increase in the density of the heat flux.

Of the two admixture mass fractions considered, equal to 5% and 15%, the latter was selected due to the, then, greater amount of material characterised by a high thermal conductivity coefficient in the mixture. At the cost of a greater loss of sorption capacity, it was decided to improve heat transport within the bed.

2.1. Description of the Adsorption Chiller with Desalination Functionality in the Test

The experiment was carried out for a laboratory adsorption chiller with desalination functionality with the option of operating with two or three adsorbent beds in cooling and water-desalination modes. The device is located at the Energy Centre of the AGH University of Science and Technology in Krakow, Poland. Its simplified diagram is shown in Figure 2. The main components of the adsorption chiller are the adsorbent beds (3), the evaporator (5), the condenser (1) and the expansion valve, which is responsible for maintaining the correct pressure difference between the condenser and the evaporator during the condensate flow from the condenser to the evaporator. The system under consideration is also equipped with a distillate tank (2), a brine tank (4) and a deaerator (6).

The description of the principle of operation of an adsorption chiller is presented for a laboratory three-bed adsorption chiller operating in a two-bed mode. First, saline water is supplied to the deaerator, where it is deaerated. Then it flows inside the supply water system to the evaporator where there is a system of exchanger pipes responsible for the heat exchange between the chilled water flowing in the chilled water circuit and the saline water in the evaporator. As a result of absorbing heat from the chilled water and transferring it to the saline water, this changes its state of the matter, which evaporates. In this system, the adsorbate circulates in a closed circuit and is sprayed onto the heat exchanger pipes through nozzles. Due to the low absolute pressure in the evaporator, evaporation can take place at a low temperature, thus allowing the desired cooling effect to be achieved. After reaching the appropriate level of salinity, the water that accumulates at the bottom of the evaporator flows from the evaporator to the brine tank. The next stage begins with the opening of the electromagnetic valves that separate the bed from the evaporator. The resulting water vapour then flows from the evaporator to the adsorber. At this point, the exothermic process of adsorption begins. In order for the adsorption to take place in an uninterrupted and maximally efficient way, the adsorption heat being released must be transferred into the cooling water circuit. Achieving the right temperature inside the adsorbent bed, both during adsorption and desorption, is possible due to the presence of the heat exchanger. Each bed is equipped with one finned coil heat exchanger unit configured in Tichelmann arrangement. The design of the heat exchanger consists of two main collectors, which are connected to each other by smaller diameter tubes. In order to increase the heat transfer surface area, the tubes are maximally elongated taking a meandering shape. A photograph of the heat exchanger is shown in Figure 3.

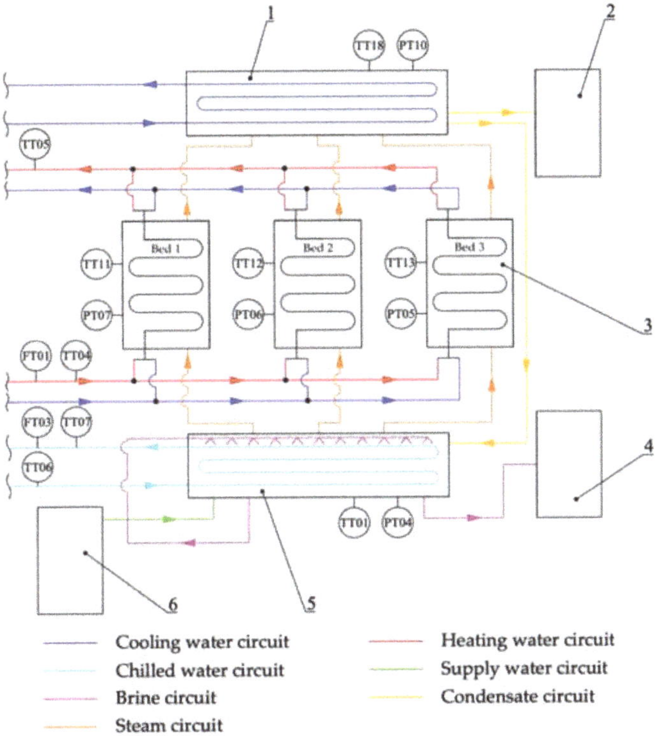

Figure 2. Simplified diagram of the three-bed adsorption chiller with desalination function. 1—condenser; 2—distillate tank; 3—adsorbent bed; 4—brine tank; 5—evaporator; 6—deaerator; TT01—temperature in the evaporator; TT04—hot water inlet temperature; TT05—hot water outlet temperature; TT06—chilled water inlet temperature; TT07—chilled water outlet temperature; TT11—temperature in bed 1; TT12—temperature in bed 2; TT13—temperature in bed 3; TT18—temperature in the condenser; PT04—pressure in the evaporator; PT07—pressure in bed 1; PT06—pressure in bed 2; PT05—pressure in bed 3; PT10—pressure in the condenser; FT01—hot water flow; FT03—chilled water flow.

Figure 3. Finned coil heat exchanger unit configured in a Tichelmann arrangement.

After the adsorption is complete, the valve between the evaporator and the bed is closed, thus preventing the further flow of evaporated water. This initiates the pre-heating of the bed before the desorption process. Appropriate adjustment of the three-way valves on the inlet and outlet of the adsorber enables the flow of water from the heating circuit through the exchanger inside the bed. The correct temperature of the heating water is obtained in a cyclically operating electric boiler with a capacity of 18 kW. This device is responsible for simulating a source of low-temperature heat, which, in actual conditions man be, for example, waste heat. The high-temperature water first flows into a buffer tank and then feeds the beds, which are in the regeneration phase. Desorption starts as soon as the temperature in the bed is sufficient to evaporate water from the surface of the adsorbent. During desorption, the bed continues to be heated, such that the right desorption temperature is maintained at all times. The heat energy supplied causes an increase in pressure of the adsorbate previously adsorbed and consequently desorption from the silica gel surface. In the next stage, an electromechanical valve between the bed and the condenser is opened allowing the flow of the desorbed water vapour. The design of the condenser is similar to that of the evaporator. Inside the condenser there is a heat exchanger in which water from the cooling circuit flows. The gaseous adsorbate condenses and gives off heat on its surface. As a result, condensate accumulates in the lower part of the condenser. Depending on whether the adsorption chiller operates only in cooling mode only or whether desalinated water is to be an additional product, the distillate flows through the expansion valve back to the evaporator or is collected in a distillate tank as a product of the desalination process. Once desorption is complete, the bed is cooled before the next cycle begins. The three-way valve is set in such manner that the water supply from the heating circuit is closed and the supply from the cooling circuit is opened. With this particular device, basic recovery of heat from water is possible as the water passes through the heat exchanger in the bed and heats the bed during desorption. After opening the cooling water supply to the exchanger, heat recovery is carried out for a certain time, and the water, still stored in the bed, is fed into the heating water circuit, thus minimising energy losses. Water flow is forced into the heating, cooling and chilled water circuits by circulation pumps with a capacity of up to 500 L/h. The system provides the option to manually control the flow in selected circuits by means of ball valves with 1/2 - or 3/4 -inch thread diameters. Figure 4 shows photographs of the adsorption chiller with desalination function used to carry out the tests.

The adsorption chiller under investigation is classified in the group of low-pressure devices as the pressure ranges in the evaporator, condenser and beds are 0.5–3 kPa, 3–7 kPa and 0.5–7 kPa, respectively. Table 1 presents the rated parameters of the tested adsorption chiller investigated.

Table 1. Basic nominal parameters of the investigated adsorption chiller with desalination functionality.

	Parameter	Value	Unit
Evaporator	Chilled water outlet temperature	7	°C
	Chilled water inlet temperature	12	°C
	Cooling capacity	1.10	kW
	Chilled water mass flow	0.052	kg/s
Condenser	Cooling water outlet temperature	22	°C
	Cooling water inlet temperature	20	°C
	Capacity	2.00	kW
	Cooling water mass flow	0.250	kg/s
	Daily distillate production	40	kg

The sorption material used in the experimental study was silica gel with a granulation of 700–800 μm manufactured in the Republic of Korea by KD Corporation. Its basic properties are presented in Table 2. The additive responsible for improving heat transport inside the bed were pieces of copper classified as grade Cu-ETP. The length and width of the metal pieces were adapted to the size of the gap between the exchanger fins, and were approx. 5 mm and approx. 0.2 mm, respectively. The adsorbate was distilled water.

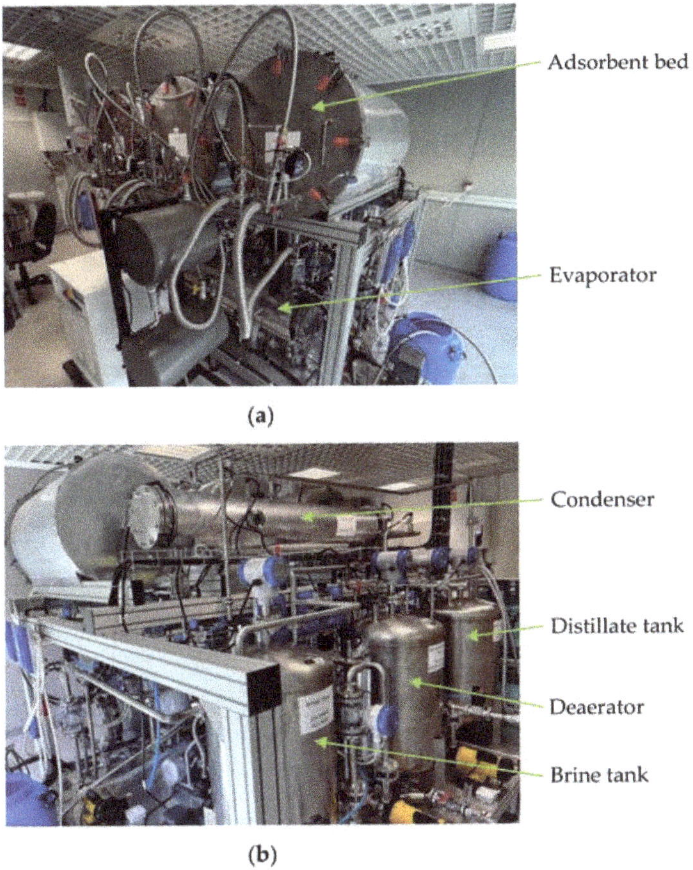

Figure 4. Three-bed adsorption chiller with desalination function: (**a**) front view, (**b**) back view working in a two-bed configuration.

Table 2. Properties of silica gel.

Manufacturer	Granulation (μm)	Thermal Conductivity (W/(mK))	Specific Surface Area (m^2/g)	Bulk Density (g/mL)	Pore Volume (cm^3/g)	Pore Diameter (Avg. Value) (nm)
KD Corporation	700–800	0.177	750–800	0.73	0.44	1.0–3.0

2.2. Measuring Instruments

During the experiment, the parameters of the adsorption chiller were recorded under stabilised operating conditions at an interval of 5 s. The SCADA control system was responsible for recording the measurement data on a personal computer. The data recorded were exported to CSV format in the form of a table. Analysis of the results was possible by editing them in Microsoft Excel.

The adsorption chiller control system is responsible for controlling all the parameters of the device and controlling some of the components, e.g., the pneumatic valves between the evaporator and the beds, and between the beds and the condenser. In addition, the chiller control system allows the modification of cycle times, preheating and cooling times and energy recovery times. The software can also be used to control the operation of pumps and valve settings.

Parameters such as temperature, pressure and flow rate were measured using appropriate measuring instruments. The location of the measuring instruments is shown schematically in Figure 2. All the measuring instruments of the adsorption chiller were supplied from a specially allocated switchboard. Temperature measurements were taken using PT-1000 temperature sensors. Water and vapour flow rates were measured by an electromagnetic flow meter. Pressure values were recorded by pressure transducers. The most important parameters of the above-mentioned measuring devices are presented in Table 3.

Table 3. Characteristics of the measuring apparatus.

Parameter Measured	Instrument	Operating Range	Accuracy
Temperature	Temperature sensor PT-1000	−80 °C to 150 °C	0.1 °C
Pressure	Pressure transducer	0–99 kPa	0.50%
Flow rate	Electromagnetic flow meter	1–100 L/min	0.50%

2.3. Conditions for the Experiment

The experiment was performed using two of the three beds of the adsorption chiller (beds 2 and 3). Each bed alternately and cyclically carried out adsorption and desorption processes. The performance coefficients of the chiller were determined for fixed parameters, such as heating water temperature, cooling water temperature and the temperature of the chilled water obtained. The length of the cycle time was variable. Table 4 summarises the most important operating parameters of the chiller under investigation for the experiment carried out.

Table 4. Basic operating parameters of the adsorption chiller for the tests.

Parameter	Value
Temperature—heating water inlet	80 °C
Temperature—cooling water inlet	19–21 °C
Temperature—chilled water outlet	15 °C
Heating water flow rate	0.22 kg/s
Chilled water flow rate	0.13 kg/s
Duration of adsorption and desorption	100, 200, 300, 600 s
Duration of pre-heating and cooling	50 s
Duration of heat recovery	30 s

One chiller cycle configuration was used for the testing. The full cycle consisted of pre-heating, desorption, heat recovery and adsorption. Figure 5 shows the chiller operation configuration used during the tests.

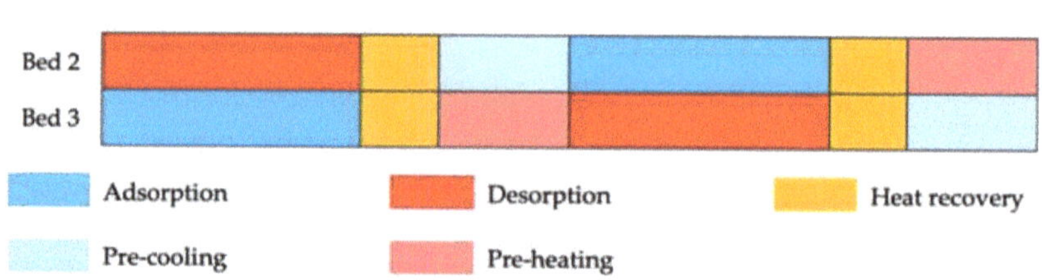

Figure 5. Operation cycle configuration for the adsorption under testing.

The tests were carried out for different cycle durations, i.e., 100, 200, 300 and 600 s. The heating water temperature was approx. 80 °C. The total mass of the sorbent, together with the copper additive at a mass fraction of 15% in the two beds, was 6 kg.

2.4. Course of the Experiment

The experiment was preceded by the proper organisation of the test station. This mainly involved placing the adsorbent with added copper in a heat exchanger and proper preparation of the adsorption chiller. The heat exchanger was filled in such a manner as to create seven layers of alternating adsorbent and metal additive. The experiment was started by starting the adsorption chiller in the automatic mode, bringing the parameters to a stable state required according to the test conditions adopted and starting the measurements. The tests for a fixed temperature of heating water and all the cycle times took about 48 h including preparation of the entire system.

2.5. Performance Parameters of the Adsorption Chiller

The operation of an adsorption chiller is cyclical, which implies that its parameters are unstable. This causes difficulties in the experimental determination of the COP and the SCP. Therefore, in order to calculate the efficiency and power of the device, it is necessary to average the experimental data for the selected complete cycles. Taking into account the values obtained for the chiller under testing, the experimental cooling coefficient of performance (COP) and specific cooling power (SCP) were calculated based on the Equations (1) and (2), presented below:

$$\mathrm{COP} = \frac{\mathrm{CC}}{\mathrm{HP}} \tag{1}$$

$$\mathrm{SCP} = \frac{\mathrm{CC}}{m_{ads}} \tag{2}$$

where:

CC—cooling capacity;
HP—heating power supplied to the system;
m_{ads}—the mass of sorbent.

The cooling capacity CC and the total energy supply required to remove the water vapour adsorbed on the adsorbent surface (heating power, HP) required to determine the COP and SCP of the chiller were calculated according to Equations (3) and (4):

$$\mathrm{CC} = \dot{m}_c \cdot c_{p,c} \cdot \Delta T_c = \dot{m}_c \cdot c_{p,c} \cdot (T_{i,c} - T_{o,c}) = \mathrm{FT03} \cdot c_{p,c} \cdot (\mathrm{TT06} - \mathrm{TT07}) \tag{3}$$

$$\mathrm{HP} = \dot{m}_h \cdot c_{p,h} \cdot \Delta T_c = \dot{m}_h \cdot c_{p,h} \cdot (T_{i,h} - T_{o,h}) = \mathrm{FT01} \cdot c_{p,h} \cdot (\mathrm{TT04} - \mathrm{TT05}) \tag{4}$$

where:

$\dot{m}$—mass flow rate;
c_p—specific heat of water;
T—temperature;
FT03—chilled water flow rate;
TT06—chilled water inlet temperature at evaporator;
TT07—chilled water outlet temperature the evaporator;
FT01—hot water flow rate;
TT04—hot water inlet temperature at the beds;
TT05—hot water outlet temperature at the beds.

The following indices were used in the equations: c—chilled water; h—heating water; i—inlet; o—outlet. For all the calculations, the specific heat of the chilled and heating water was equal to $c_{p,h} = c_{p,c} = 4200\ \frac{J}{kg \cdot K}$.

3. Results

The results of the tests are presented in the form of graphs showing the relation between the pressures, temperatures and flow rates of the selected operating parameters of the adsorption chiller for each cycle time.

Figures 6 and 7 present the pressure variations in the evaporator, in the beds and in the condenser for several stable cycles of operation of the chiller with beds filled with silica gel with added copper at a mass fraction of 15%. The curves are shown for cycle times of 100 and 600 s. Graphs for cycle times of 200 and 300 s are provided in Supplementary Materials as Figures S1 and S2.

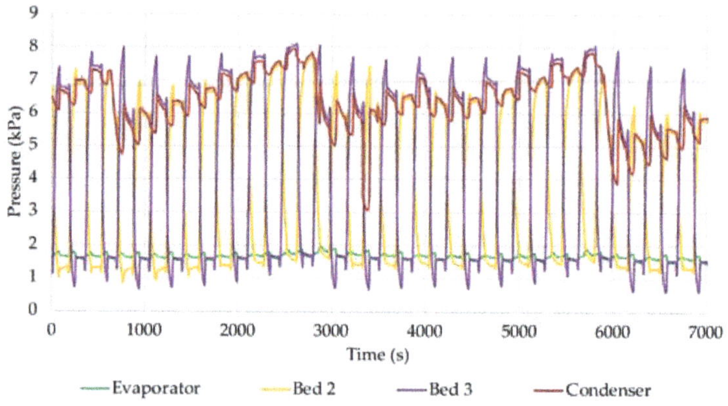

Figure 6. Pressure in the beds, evaporator and condenser for a time cycle of 100 s.

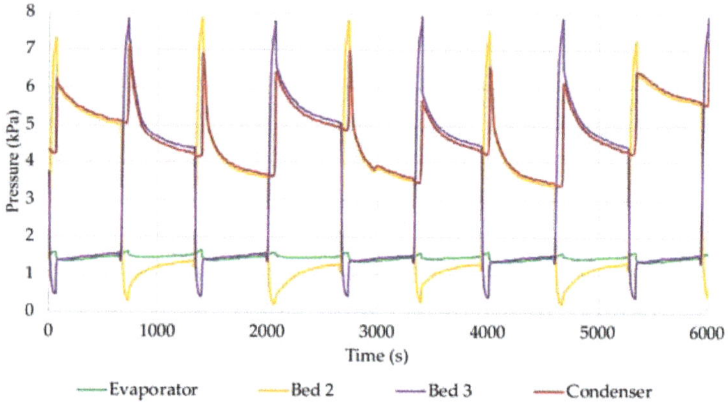

Figure 7. Pressure in the beds, evaporator and condenser for a time cycle of 600 s.

Figures 8–11 show the graphs of the temperature change inside the evaporator, the beds, the condenser, as well as the temperature of the heating water and cooling water at the inlet to the beds for stable cycles of the adsorption chiller with beds filled with silica gel with added copper in a proportion of 15%. The runs were determined for cycle times of 100, 200, 300 and 600 s.

Figures 12–15 present graphs showing the temperature of the heating water at the inlet and outlet of the beds, the temperature at the inlet and outlet of the evaporator, and the chilled water's flow rate. The results are shown for stable cycles of the adsorption chiller operation with beds filled with silica gel with 15% added copper. The individual graphs were made for different cycle times of 100, 200, 300 and 600 s.

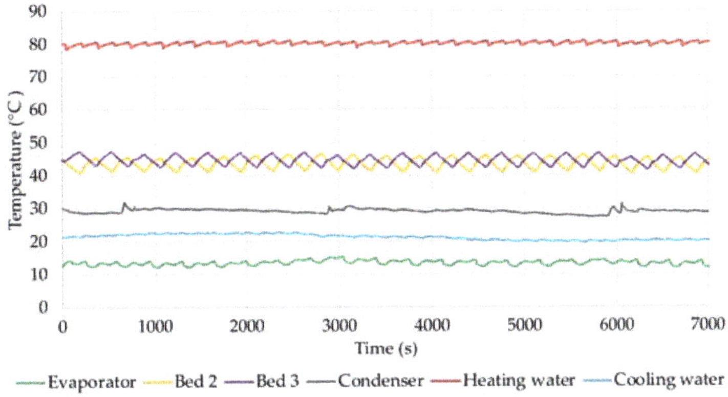

Figure 8. Temperature in the beds, evaporator and condenser and temperature of the heating and cooling water for a cycle of 100 s.

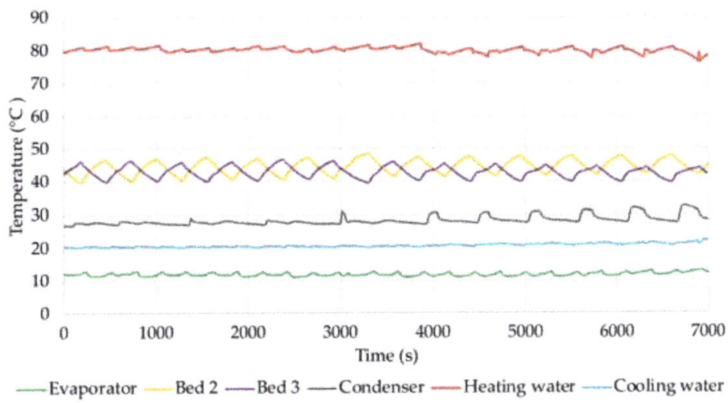

Figure 9. Temperature in the beds, evaporator and condenser and temperature of the heating and cooling water for a cycle of 200 s.

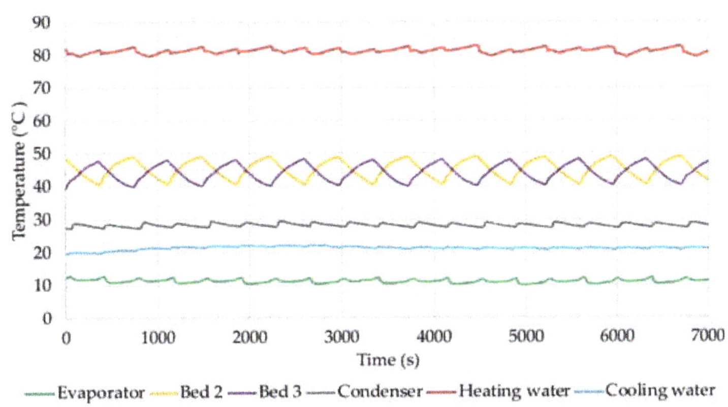

Figure 10. Temperature in the beds, evaporator and condenser and temperature of the heating and cooling water for a cycle of 300 s.

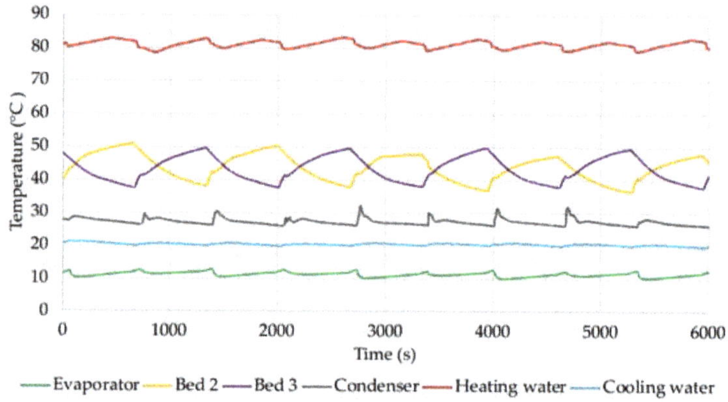

Figure 11. Temperature in the beds, evaporator and condenser and temperature of the heating and cooling water for a cycle of 600 s.

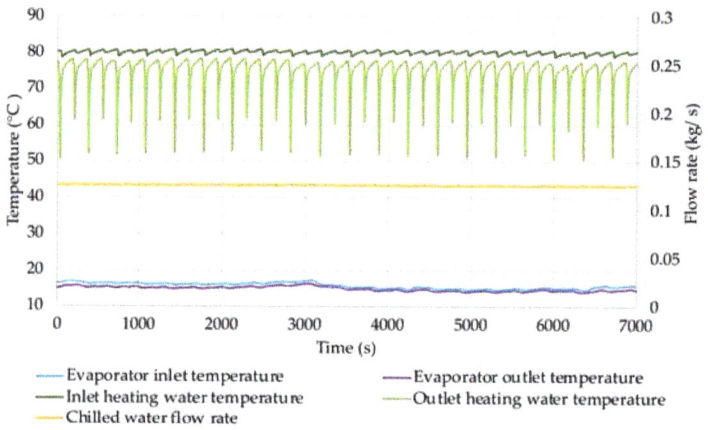

Figure 12. Water temperature at the inlet and outlet of the evaporator, heating water temperature at the inlet and outlet of the bed and the chilled water flow rate for a cycle of 100 s.

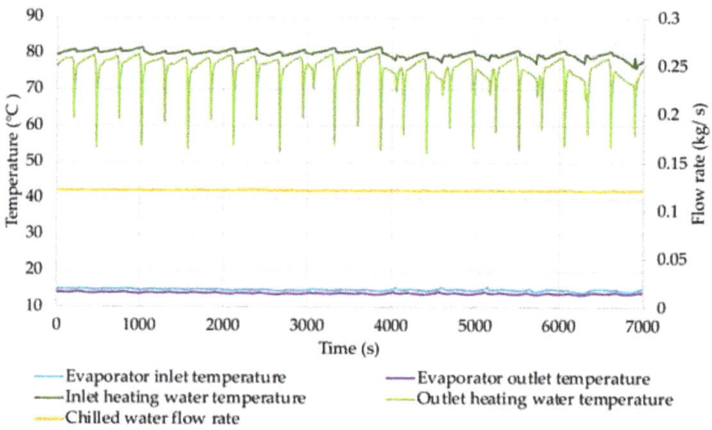

Figure 13. Water temperature at the inlet and outlet of the evaporator, heating water temperature at the inlet and outlet of the bed and the chilled water flow rate for a cycle of 200 s.

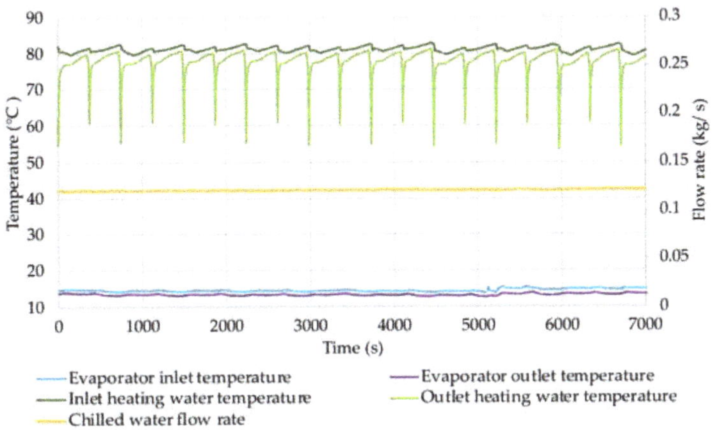

Figure 14. Water temperature at the inlet and outlet of the evaporator, heating water temperature at the inlet and outlet of the bed and the chilled water flow rate for a cycle of 300 s.

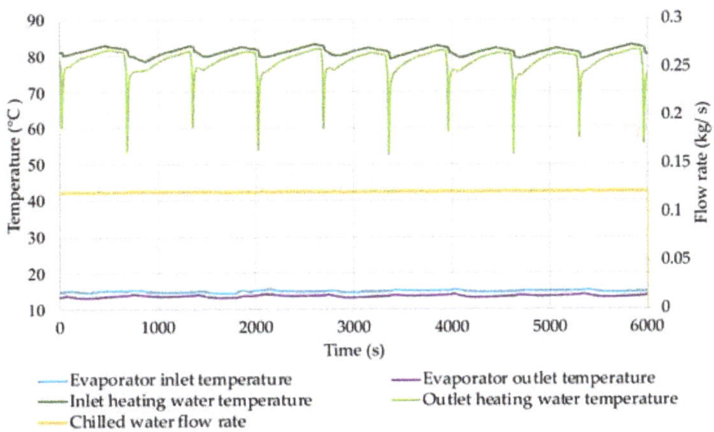

Figure 15. Water temperature at the inlet and outlet of the evaporator, heating water temperature at the inlet and outlet of the bed and the chilled water flow rate for a cycle of 600 s.

The results of water temperature measurements at the inlet and outlet of the beds were also compared for exchangers filled with a mixture of silica gel and copper, and for a bed filled only with silica gel. This makes it possible to assess the effect of the use of a metal additive on the pre-heating and cooling time of the bed during cycle-mode switching. This time will be reduced if the heat transfer inside the bed is more efficient. The graphs in Figures 16–23 show the change in heating- and cooling-water tempera-tures at the inlet and outlet of the beds for several stable cycles of the adsorption chiller. The course of the temperature variation is presented for the cycle times of 100, 200, 300 and 600 seconds. Figures marked with even numbers show the temperature course for beds filled with silica gel with copper addition, while figures marked with odd numbers refer to beds filled with silica gel without the addition.

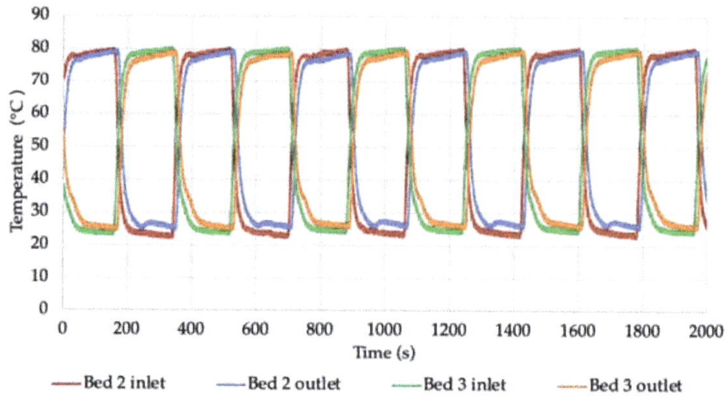

Figure 16. Water temperature at the inlet and outlet of the beds filled with silica gel with added copper at a mass fraction of 15% for a cycle of 100 s.

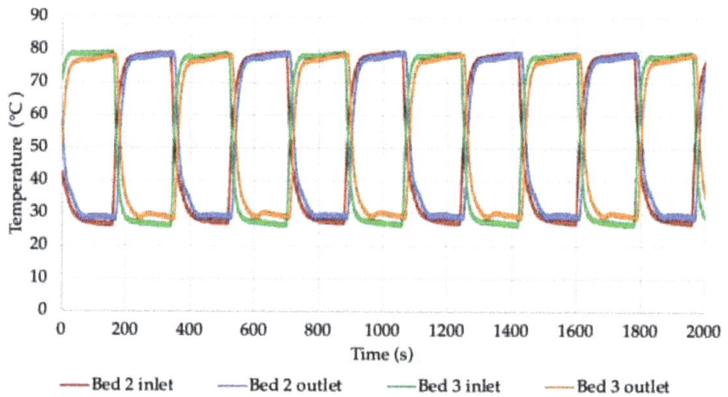

Figure 17. Water temperature at the inlet and outlet of the beds filled with silica gel for a cycle of 100 s.

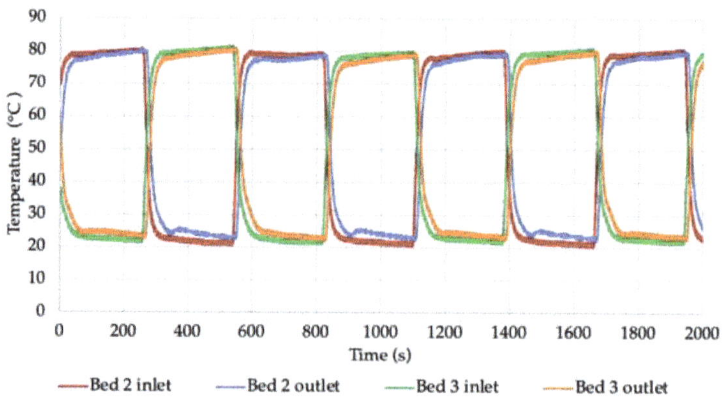

Figure 18. Water temperature at the inlet and outlet of the beds filled with silica gel with added copper at a mass fraction of 15% for a cycle of 200 s.

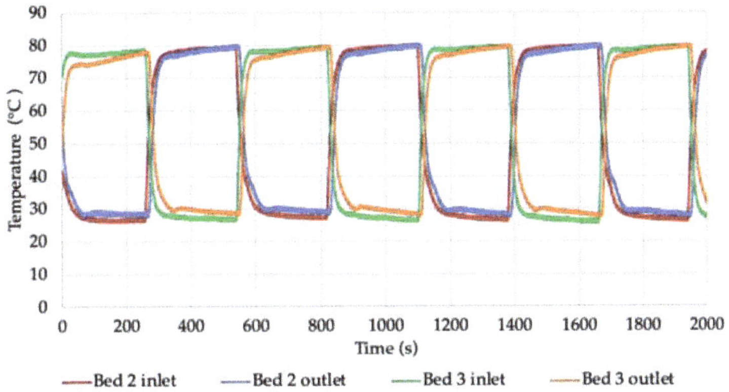

Figure 19. Water temperature at the inlet and outlet of the beds filled with silica gel for a cycle of 200 s.

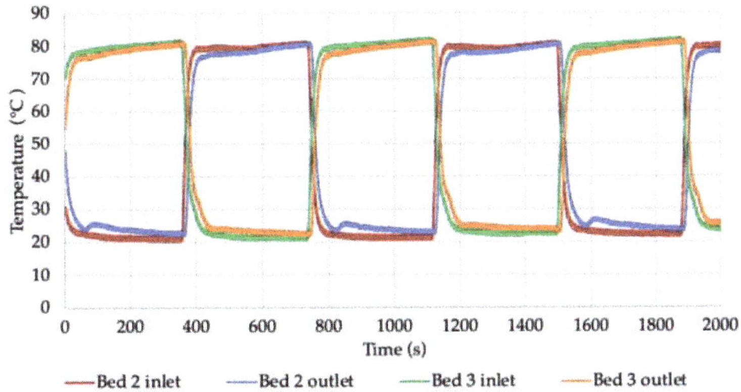

Figure 20. Water temperature at the inlet and outlet of the beds filled with silica gel with added copper at a mass fraction of 15% for a cycle of 300 s.

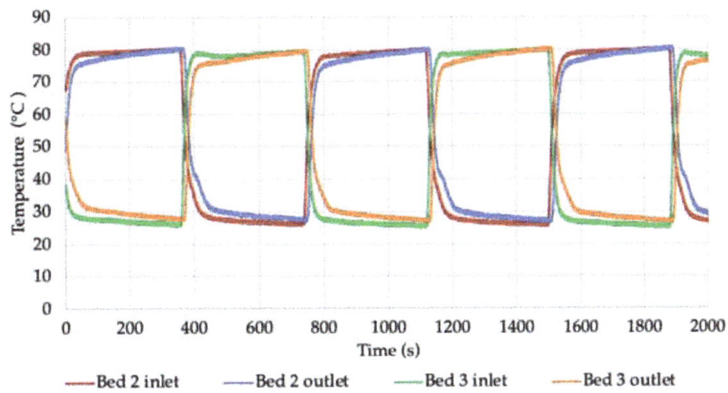

Figure 21. Water temperature at the inlet and outlet of the beds filled with silica gel for a cycle of 300 s.

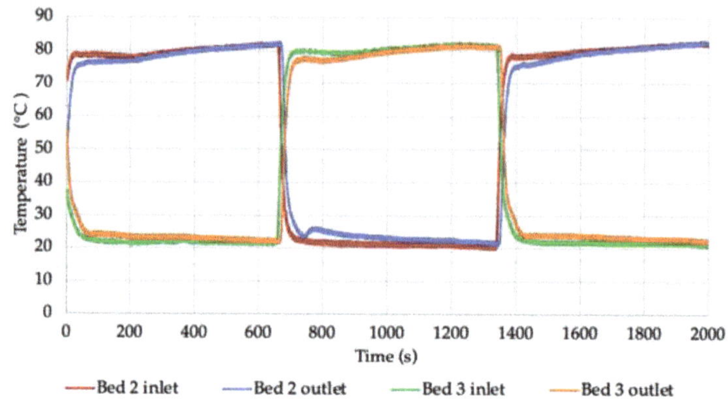

Figure 22. Water temperature at the inlet and outlet of the beds filled with silica gel with added copper at a mass fraction of 15% for a cycle of 600 s.

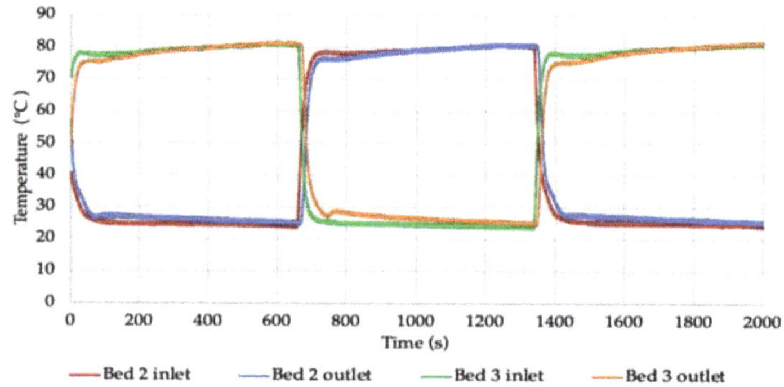

Figure 23. Water temperature at the inlet and outlet of the beds filled with silica gel for a cycle of 600 s.

Table 5 compares the average pre-heating and cooling times for pure silica gel (SG) and a mixture of silica gel and copper at a mass fraction of 15% (SG + 15% Cu) for a cycle time of 600 s. The comparison of pre-heating and cooling times was made for the longest cycle duration that was tested. The reason is that it is certain that stabilised temperature values will be observed then. This makes the results more reliable by eliminating the possibility of the bed not reaching the target temperature, which might occur for the shorter cycle times. The start and end times of pre-heating and cooling were estimated by recording the exact time of the beginning of the rise or fall of the temperature at the bed inlet, as well as the time during which the temperatures at the bed inlet and outlet were stable, close to each other and did neither rising nor falling. The bed was then assumed to be fully pre-heated or cooled and the temperature was considered to be the same throughout the entire volume. Then the difference between the two was determined, representing the sought values of the pre-heating and cooling times.

Table 5. Average pre-heating and cooling times of the beds.

Bed Filling	Pre-Heating Time (s)		Cooling Time (s)	
	Bed No. 2	Bed No. 3	Bed No. 2	Bed No. 3
SG	61	65	86	84
SG + 15% Cu	53	57	71	74

4. Discussion

By following the pressure graphs shown in Figures 6 and 7, a clear boundary between the successive cycles can be observed. At the point when the adsorption process starts in one of the beds, the pressure inside it increases and its value approaches the pressure in the evaporator. However, these pressures did not reach the same value for any of the times studied. It was found that, as the cycle time increased, the difference between the pressure in the evaporator and the pressure in the bed during adsorption decreased. This means that the time was too short for these values to reach the same level. On the other hand, during the desorption process, the pressure in the bed increases due to the formation of steam. For each cycle time, the pressure in the condenser was almost equal after opening the valves between the respective bed and the condenser. Its value decreased as the pressure in the bed decreased due to steam flow and condensation.

The change in temperature values in the beds shown in Figures 8–11 indicates the points at which the adsorption and desorption processes begin and end. With the start of adsorption, the temperature decreases because the bed is cooled by water from the cooling water circuit. The pre-heating of the bed is confirmed by the temperature increase, which is associated with the start of desorption. Evaporation of the water adsorbed on the surface of the sorption material then takes place. These processes are very dynamic, as indicated by the intense and frequent fluctuations and changes in temperature. The graphs also include curves showing the temperature of the heating and cooling water. Their value was relatively constant throughout the experiment at approx. 80 °C and 20 °C, respectively.

Furthermore, the changes in temperatures inside the beds presented in Figures 8–11 indicate that as the cycle time increases, the temperature of the bed increases during the desorption process and decreases during adsorption. For alonger cycle time, the bed pre-heating and cooling times are longer, which directly affects the temperature inside the adsorbent bed. For the shortest cycle time tested, the temperature inside the bed was approx. 41 °C and 46 °C, respectively, during adsorption and desorption. For the longest cycle time, these temperatures were approx. 36 °C for adsorption and 50 °C for desorption. The graphs also show the temperature in the evaporator, which corresponds to the evaporation temperature of the adsorbate. Its value depends on the adsorption/desorption cycle. It was observed that at the start of adsorption (which is equivalent to opening the valve between the evaporator and the bed) this temperature slightly increases. Furthermore, it was found that its value decreases with increasing cycle time. The analysis of the temperature curve in the condenser also shows that it depends on the course of the cycle. This is particularly evident for longer cycles.

From the graphs shown in Figures 12–15 it is possible e.g., to determine the chilled water flow rate. It was approx. 0.12 kg/s during all the tests carried out. The temperature of the chilled water was kept at a relatively stable level. The transfer of thermal energy to the refrigerant in the evaporator causes the chilled water temperature to decrease. The difference between the evaporator inlet and outlet temperatures is approx. 1.5 °C. The small changes in its value are caused by changes in pressure in the evaporator. The chilled water temperature increases at the mode switching in the cycle because then the cooling effect is smaller.

Additionally, to show the difference between pressures and temperatures in individual systems, their average values and standard deviation were determined. Table 6 shows the mean values and standard deviation for the measured pressures, while Table 7 shows the mean values and standard deviation for the measured temperatures.

Based on the results obtained according to Equations (1) and (2), instantaneous values of the cooling coefficient of performance COP and specific cooling power SCP were calculated. Figure 24 shows the curves of the calculated cooling coefficients of performance COP for the chiller with beds filled with silica gel with the addition of copper at a mass fraction of 15% for the individual cycle times.

Table 6. Arithmetic mean and standard deviation of the measured pressure values.

Cycle Time (s)	Parameter	Mean (kPa)	Standard Deviation (kPa)
100	Evaporator pressure	1.68	0.08
	Bed 2 pressure	4.01	2.32
	Bed 3 pressure	4.15	2.65
	Condenser pressure	6.35	0.89
200	Evaporator pressure	1.53	0.07
	Bed 2 pressure	3.25	2.23
	Bed 3 pressure	3.60	2.24
	Condenser pressure	5.39	0.65
300	Evaporator pressure	1.45	0.07
	Bed 2 pressure	3.23	2.38
	Bed 3 pressure	3.37	2.07
	Condenser pressure	5.17	0.65
600	Evaporator pressure	1.55	0.06
	Bed 2 pressure	3.17	2.04
	Bed 3 pressure	3.34	1.98
	Condenser pressure	4.93	0.83

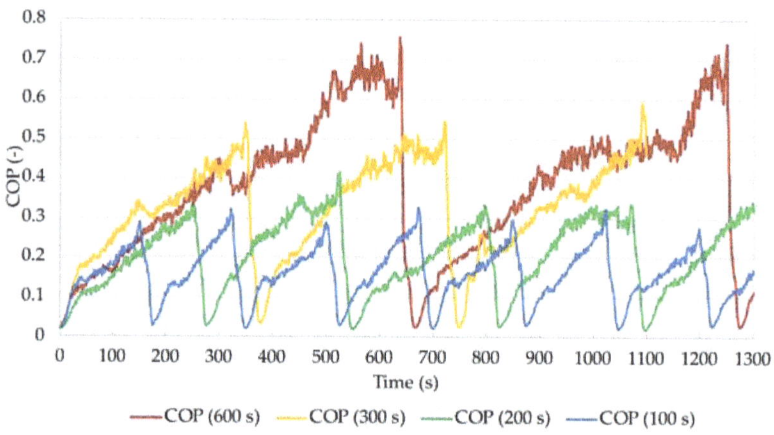

Figure 24. Comparison of changes in COP values depending on cycle time.

When analysing the above graph, special attention should be paid to the extremes. The minimum instantaneous COP value occurs at the point when the system is about to be switched to the regeneration mode and the bed is being pre-heated. As desorption begins, the cooling coefficient of performance gradually increases. This is when the most intensive heat exchange takes place between the heat exchanger and the material the bed is filled with. The temperature difference between the heating water at the bed inlet and outlet is then the greatest and decreases over the pre-heating time. The COP reaches its maximum at the end of the desorption process. The temperature difference between the bed inlet and outlet is then the smallest, which, according to Equation (4), is associated with the minimum value of the heating power HP, and consequently, according to Equation (1), with the highest value of the cooling coefficient of performance. It can also be seen from Figure 24 that the maximum instantaneous COP value increases with increasing cycle time. When analysing the curve for the shortest cycle time, it was noted that this time was too short for the stabilisation of the adsorption chiller parameters to occur. A cycle time of 100 s is insufficient to maintain the parameters at an appropriate level and to carry out the adsorption and desorption processes effectively and completely after the pre-heating and cooling of the bed. With longer cycle times, the COP value is clearly more stable. The cooling coefficient of performance reaches higher values and maintains them longer at the right level. This paper does not present the change in the specific cooling power achieved, but it was noted in the course of the experiments that the lowest instantaneous values are obtained during mode-switching in the cycle. The reason why these drops occur is the momentary reduction of the heat absorption in the evaporator and, consequently, of the difference between the chilled water temperature at the inlet and outlet of the evaporator when there is no steam flow to any of the beds.

Table 7. Arithmetic mean and standard deviation of the measured temperature values.

Cycle Time (s)	Parameter	Mean (°C)	Standard Deviation (°C)
100	Evaporator temperature	13.41	0.67
	Bed 2 temperature	43.53	1.56
	Bed 3 temperature	44.57	1.35
	Condenser temperature	29.04	0.72
	Heating water temperature	80.19	0.45
	Cooling water temperature	21.17	1.04
	Evaporator inlet temperature	15.68	0.71
	Evaporator outlet temperature	14.86	0.53
	Inlet heating water temperature	80.19	0.45
	Outlet heating water temperature	74.87	4.78
200	Evaporator temperature	11.82	0.48
	Bed 2 temperature	44.29	2.06
	Bed 3 temperature	42.75	1.75
	Condenser temperature	28.19	1.20
	Heating water temperature	79.91	0.88
	Cooling water temperature	20.66	0.36
	Evaporator inlet temperature	14.76	0.19
	Evaporator outlet temperature	13.80	0.16
	Inlet heating water temperature	79.91	0.88
	Outlet heating water temperature	75.65	4.10
300	Evaporator temperature	11.06	0.58
	Bed 2 temperature	45.02	2.66
	Bed 3 temperature	44.09	2.47
	Condenser temperature	28.18	0.50
	Heating water temperature	81.15	0.72
	Cooling water temperature	21.20	0.51
	Evaporator inlet temperature	14.58	0.30
	Evaporator outlet temperature	13.39	0.26
	Inlet heating water temperature	81.15	0.72
	Outlet heating water temperature	77.77	3.65
600	Evaporator temperature	11.28	0.60
	Bed 2 temperature	43.86	3.89
	Bed 3 temperature	43.33	3.76
	Condenser temperature	27.23	0.94
	Heating water temperature	81.18	1.05
	Cooling water temperature	20.41	0.28
	Evaporator inlet temperature	14.99	0.23
	Evaporator outlet temperature	13.65	0.25
	Inlet heating water temperature	81.18	1.05
	Outlet heating water temperature	78.35	3.60

Figure 25 shows the change in the average COP of the adsorption chiller for different cycle times. The graph is a comparison of the change in the cooling coefficient of performance when the beds are filled with silica gel and silica gel with copper additive. Based on the results obtained it was found that for the times studied, the average COP increases with increasing cycle time. The lowest value was achieved for a cycle time of 100 s. In the case of beds filled with silica gel, the average COP was 0.12, while for beds filled with silica gel with copper additive it was 0.21. The highest average cooling coefficient of performance was obtained for a cycle time of 600 s. For the bed without and with the additive it was

0.36 and 0.60, respectively. Furthermore, the difference between the COP values obtained for the different cycle times increases with increasing cycle time. For 100 s the average COP value was approx. 0.09, for 200 s approx. 0.15, for 300 s 0.16, and for 600 s it is the highest at 0.24.

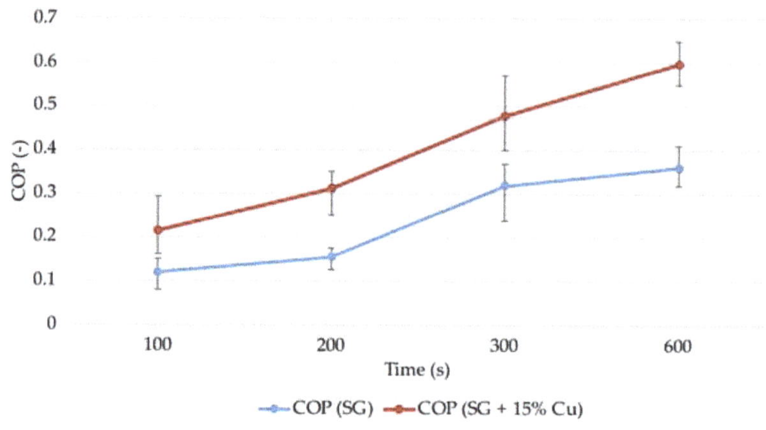

Figure 25. Change in the average COP value of the adsorption chiller for both bed fillings tested, depending on cycle time.

Figure 26 shows a comparison of the change in average specific cooling power SCP for different cycle times depending on whether the metal additive was used or the beds were filled with silica gel only. The lowest SCP value for both cases of bed filling was achieved for a cycle time of 100 s. It was approx. 43 W/kg for the beds filled with silica gel and 106 W/kg for the beds filled with silica gel with copper additive. The maximum specific cooling power was achieved for a cycle time of 300 s. The SCP values were then 56 W/kg and 128 W/kg for the beds filled with silica gel only and the beds with copper additive, respectively. As the SCP increases, the difference between the values obtained for silica gel and silica gel with copper increases and reaches 63 W/kg, 72 W/kg and 87 W/kg for cycle times of 100, 200 and 300 s, respectively. At the point of inflection of the specific cooling power curve for the longest cycle time, this difference is smaller and reaches the value of 71 W/kg.

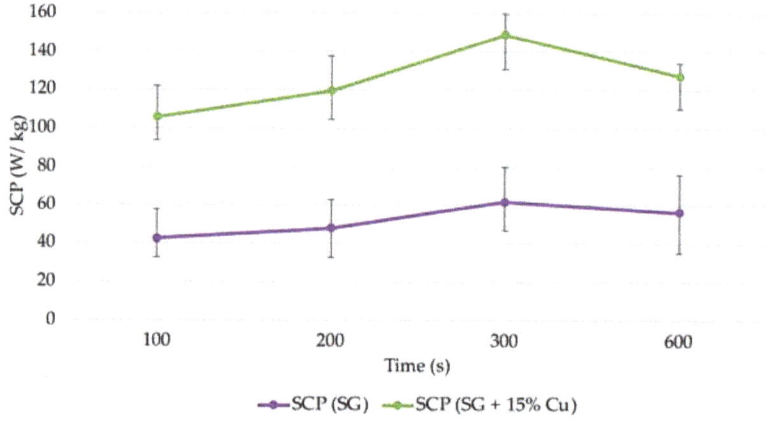

Figure 26. Change in the SCP value of the adsorption chiller for both bed fillings tested depending on cycle time.

Figure 27 shows the effect of cycle time on the average COP and SCP values obtained for beds filled with silica gel with copper added at a mass fraction of 15%.

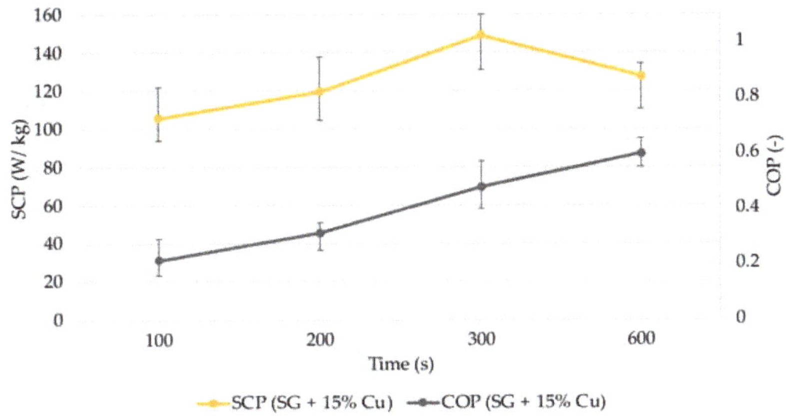

Figure 27. Change in mean SCP and COP values of the adsorption chiller with beds filled with silica with copper depending on cycle time.

Regardless of whether the additive was used with silica gel or only pure sorption material was used to fill the beds, longer adsorption and desorption times mean that more adsorbate is adsorbed and desorbed. This translates into an increase in the COP which can be observed in Figure 25. For the longest time tested, the efficiency is the highest. Its further extension to a certain level would probably contribute to an even higher COP. This hypothesis should be verified experimentally because according to research findings published in the literature, too long a cycle time is associated with no increase or even a decrease in the value of the cooling coefficient of performance [29]. Specific cooling power, SCP, increases up to a maximum value for a cycle time of 300 s, and then decreases with longer cycle times.

Considering the results obtained, it can be concluded that the use of long cycle times favours the achievement of relatively high COP values. On the other hand, increasing the cycle time is not advantageous due to the decreasing specific cooling power. In this situation, a workable compromise must be sought. Based on Figures 25–27, it was found that the most optimal cycle duration for the operating configuration of the adsorption chiller that was tested was 300 s. The highest SCP value and a satisfactory COP value were then obtained.

Tables 8 and 9 provide a summary of this part of the analysis, with all the average, minimum and maximum COP and SCP values obtained, depending on whether the beds were filled with silica gel (SG) or whether copper was added at a mass fraction of 15% (SG + 15% Cu).

Table 8. Average, minimum and maximum COP values, depending on cycle time and adsorbent bed filling.

Bed Filling	Cycle Time (s)	COP (-)	Minimum COP	Maximum COP
SG	100	0.12	0.08	0.15
	200	0.16	0.13	0.18
	300	0.32	0.24	0.37
	600	0.36	0.32	0.41
SG + 15% Cu	100	0.21	0.16	0.29
	200	0.31	0.25	0.35
	300	0.48	0.40	0.57
	600	0.60	0.55	0.65

Table 9. Average, minimum and maximum SCP values, depending on cycle time and adsorbent bed filling.

Bed Filling	Cycle Time (s)	SCP (W/kg)	Minimum SCP	Maximum SCP
SG	100	42.69	33.06	58.37
	200	47.95	31.88	62.21
	300	61.81	46.86	80.61
	600	56.44	35.01	76.04
SG + 15% Cu	100	106.00	93.91	122.28
	200	119.63	105.12	138.36
	300	148.87	131.09	160.20
	600	127.51	110.50	134.24

An analysis of Figures 16–23, showing changes in temperatures at the bed inlet and outlet, clearly indicates faster stabilisation of temperature values during pre-heating and cooling of the beds in the case of using a copper additive. With increasing cycle time, an increasingly longer duration of stabilised temperatures can also be seen. Based on a comparison of the pre-heating and cooling times shown in Table 5 for a cycle time of 600 s, it can be concluded that the addition of copper to silica gel at a mass fraction of 15% contributed to a reduction in the pre-heating and cooling times of the beds. In the case of the pre-heating and cooling of the beds, the reduction in time relative to the bed filled with silica gel only was approx. 13.1% and 17.4%, respectively.

5. Conclusions

The paper described options for improving the performance of an adsorption chiller. The main focus was on optimising the cycle time and increasing heat transfer through the bed. The aim of this study was to assess the effect of the use of a copper additive to silica gel, at a mass fraction of 15% and paired with water, on the performance of an adsorption chiller operating in a two-bed configuration, and to investigate the impact of the cycle time on the COP and SCP of the device. The tests were carried out for a laboratory adsorption chiller with desalination functionality located at the Energy Centre of the AGH University of Science and Technology in Krakow, Poland. The findings were compared with the results of tests carried out under similar conditions for a chiller with beds filled with pure silica gel.

Based on the results, it was found that the use of a copper additive, at a mass fraction of 15%, increased the COP and SCP values of the device as compared with the values obtained under similar conditions for a chiller with beds filled with silica gel. The cooling coefficients of performance calculated the chiller with beds filled with silica gel with the additive were up to twice as high as for the chiller with beds filled with silica gel. The SCP value determined for the chiller with beds filled with silica gel with added copper was maximally almost 2.5 times higher compared with the results obtained for beds filled with silica gel.

The effect of the cycle time on the performance parameters was also checked during the experiment. The cooling coefficient of performance COP increases with increasing cycle time for beds filled with silica gel only, as well as beds with added copper. The increase in the average COP for the range of cycle times from 100 to 600 s was from 0.12 to 0.36 and from 0.21 to 0.60 for the first and the second case, respectively. Specific cooling power increased, for both types of beds, with increasing cycle time, up to a maximum for a cycle time of 300 s, and then decreased with increasing cycle time.

During the research, the influence of the parallel use of two methods of improving the performance parameters of the adsorption chiller was verified. The simultaneous optimisation of the cycle time length and the improvement of heat transfer resulted in a significant increase in the COP and SCP of the device. In the case of the tests carried out for the longest cycle time of 600 s, a five-fold increase of COP was obtained for a refrigerator with beds filled with silica gel and copper addition compared to a refrigerator with beds filled only with silica gel and the shortest cycle duration. At that time, the specific cooling power increased threefold for a refrigerator with beds filled with a sorbent with the addition of copper. For the experiment carried out for the beds filled with sorbent with the additive and the cycle duration of 300 s, the COP increased 4 times and SCP 3.5 times, compared with the test, during which the cycle time was 100 s and the refrigerator beds were filled only with silica gel.

The average pre-heating and cooling times of the beds for both filling variants were also measured. The addition of copper to silica gel reduces the pre-heating and cooling time of the beds. The reductions in pre-heating and cooling time for beds 2 and 3 were 13.1% and 17.4%, respectively, compared with beds filled only with silica gel. This means that the use of the additive resulted in a faster temperature equilibration inside the bed, which could result in a shorter cycle time.

The conducted experiment, due to the characteristics of the research equipment used, had certain limitations. The scope of the measurements was limited by the temperature control system in the cooling water system, which is responsible for cooling the condenser and the beds. In this case, the difficulty was the value of the maximum achievable temperature of the cooling water, 35 °C. Moreover, the temperature in the adsorbent beds may be up to 95 °C. Another limitation is the need to keep the temperature in the evaporator higher than 4 °C. The electric boiler used as a heat source, also imposes a limitation in the form of the minimum heating water temperature, which is 45°C. The thermal compression process requires a sorption and desorption for each set parameter, which is related to a long stabilization time. This means that the entire experiment took a very long time. In addition, the system was very sensitive to changes in pressure and temperature, so similar experiments should be carefully conducted.

At present, the commercialisation of adsorption refrigeration systems is hampered by, inter alia, their low cooling coefficients of performance and low specific cooling powers, which translates into the considerable size and bulkiness of the chillers. The test results presented in this paper constitute a further step towards increasing their attractiveness and competitiveness on the market. The tests provide a basis for further experiments using other additives to sorption materials, such as aluminium, graphite or MOF materials.

Supplementary Materials: The following are available online at https://www.mdpi.com/article/10.3390/en14217226/s1, Figure S1: Pressure in the beds, evaporator and condenser for time cycle of 200 s, Figure S2: Pressure in the beds, evaporator and condenser for time cycle of 300 s.

Author Contributions: Conceptualization, K.S.; methodology, K.S.; validation, K.S. and W.K.; formal analysis, K.S., M.S. and Ł.M.; investigation, K.S. and W.K.; resources, Ł.M.; data curation, W.K., M.S. and K.S.; writing—original draft preparation, K.S.; writing—review and editing, K.S., M.S. and Ł.M.; visualization, W.K. and M.S.; supervision, Ł.M. and K.S.; project administration, Ł.M. and K.S.; funding acquisition, Ł.M. All authors have read and agreed to the published version of the manuscript.

Funding: This research was funded by Ministry of Science and Higher Education, Poland, grant AGH no 16.16.210.476 and partly supported by program "Excellence initiative—research university" for the AGH University of Science and Technology.

Institutional Review Board Statement: Not applicable.

Informed Consent Statement: Not applicable.

Data Availability Statement: The data that support the results reported in this study are available from the corresponding author upon request.

Conflicts of Interest: The authors declare no conflict of interest.

References

1. *Key World Energy Statistics*; Report of the International Energy Agency; International Energy Agency: Paris, France, 2020. Available online: https://www.iea.org/reports/key-world-energy-statistics-2020 (accessed on 20 September 2021).
2. *Cooling*; Report of the International Energy Agency. International Energy Agency: Paris, France, 2020. Available online: https://www.iea.org/reports/cooling (accessed on 20 September 2021).
3. Dupont, J.L.; Domanski, P.; Lebrun, P.; Ziegler, F. *The Role of Refrigeration in the Global Economy (2019), 38th Note on Refrigeration Technologies*; Informatory Note of the International Institute of Refrigeration; The International Institute of Refrigeration: Paris, France, 2019.
4. Pan, Q.W.; Wang, R.Z. Study on operation strategy of a silica gel-water adsorption chiller in solar cooling application. *Sol. Energy* **2018**, *172*, 24–31. [CrossRef]
5. Sztekler, K.; Kalawa, W.; Mlonka-Medrala, A.; Nowak, W.; Mika, Ł.; Krzywanski, J.; Grabowska, K.; Sosnowski, M.; Debniak, M. The Effect of Adhesive Additives on Silica Gel Water Sorption Properties. *Entropy* **2020**, *22*, 327. [CrossRef]
6. Stefański, S.; Mika, Ł.; Sztekler, K.; Kalawa, W.; Lis, Ł.; Nowak, W. Adsorption bed configurations for adsorption cooling application. *E3S Web Conf.* **2019**, *108*, 01010. [CrossRef]
7. Wang, L.; Bu, X.; Ma, W. Experimental study of an Adsorption Refrigeration Test Unit. *Energy Procedia* **2018**, *152*, 895–903. [CrossRef]
8. Gado, M.; Elgendy, E.; Elsayed, K.; Fatouh, M. Performance enhancement of an adsorption chiller by optimum cycle time allocation at different operating conditions. *Adv. Mech. Eng.* **2019**, *11*, 1–12. [CrossRef]
9. Saha, B.B.; Koyama, S.; Lee, J.B.; Kuwahara, K.; Alam, K.C.A.; Hamamoto, Y.; Akisawa, A.; Kashiwagi, T. Performance evaluation of a low-temperature waste heat driven multi-bed adsorption chiller. *Int. J. Multiph. Flow* **2003**, *29*, 1249–1263. [CrossRef]
10. Pan, Q.; Peng, J.; Wang, H.; Sun, H.; Wang, R. Experimental investigation of an adsorption air-conditioner using silica gel-water working pair. *Sol. Energy* **2019**, *185*, 64–71. [CrossRef]
11. Pan, Q.; Peng, J.; Wang, R. Experimental study of an adsorption chiller for extra low temperature waste heat utilization. *Appl. Therm. Eng.* **2019**, *163*, 114341. [CrossRef]
12. Ng, K.C.; Thu, K.; Kim, Y.; Chakraborty, A.; Amy, G. Adsorption desalination: An emerging low-cost thermal desalination method. *Desalination* **2013**, *308*, 161–179. [CrossRef]
13. Zajaczkowski, B. Optimizing performance of a three-bed adsorption chiller using new cycle time allocation and mass recovery. *Appl. Therm. Eng.* **2016**, *100*, 744–752. [CrossRef]
14. Sztekler, K.; Kalawa, W.; Mika, L.; Krzywanski, J.; Grabowska, K.; Sosnowski, M.; Nowak, W.; Siwek, T.; Bieniek, A. Modeling of a Combined Cycle Gas Turbine Integrated with an Adsorption Chiller. *Energies* **2020**, *13*, 515. [CrossRef]
15. Sztekler, K. Optimisation of Operation of Adsorption Chiller with Desalination Function. *Energies* **2021**, *14*, 2668. [CrossRef]
16. Ng, K.C.; Thu, K.; Chakraborty, A.; Saha, B.B.; Chun, W.G. Solar-assisted dual-effect adsorption cycle for the production of cooling effect and potable water. *Int. J. Low-Carbon Technol.* **2009**, *4*, 61–67. [CrossRef]
17. Younes, M.M.; El-Sharkawy, I.I.; Kabeel, A.E.; Saha, B.B. A review on adsorbent-adsorbate pairs for cooling applications. *Appl. Therm. Eng.* **2017**, *114*, 394–414. [CrossRef]
18. Choudhury, B.; Saha, B.B.; Chatterjee, P.K.; Sarkar, J.P. An overview of developments in adsorption refrigeration systems towards a sustainable way of cooling. *Appl. Energy* **2013**, *104*, 554–567. [CrossRef]
19. Ahmed, F.E.; Hashaikeh, R.; Hilal, N. Hybrid technologies: The future of energy efficient desalination—A review. *Desalination* **2020**, *495*, 114659. [CrossRef]
20. Elsayed, E.; AL-Dadah, R.; Mahmoud, S.; Anderson, P.; Elsayed, A. Experimental testing of aluminium fumarate MOF for adsorption desalination. *Desalination* **2020**, *475*, 114170. [CrossRef]
21. Drinking-Water. Available online: https://www.who.int/news-room/fact-sheets/detail/drinking-water (accessed on 20 September 2021).
22. Ghaffour, N.; Missimer, T.M.; Amy, G.L. Technical review and evaluation of the economics of water desalination: Current and future challenges for better water supply sustainability. *Desalination* **2013**, *309*, 197–207. [CrossRef]
23. Eke, J.; Yusuf, A.; Giwa, A.; Sodiq, A. The global status of desalination: An assessment of current desalination technologies, plants and capacity. *Desalination* **2020**, *495*, 114633. [CrossRef]
24. Grabowska, K.; Krzywanski, J.; Nowak, W.; Wesolowska, M. Construction of an innovative adsorbent bed configuration in the adsorption chiller - Selection criteria for effective sorbent-glue pair. *Energy* **2018**, *151*, 317–323. [CrossRef]
25. Rogala, Z. Adsorption chiller using flat-tube adsorbers—Performance assessment and optimization. *Appl. Therm. Eng.* **2017**, *121*, 431–442. [CrossRef]
26. Fernandes, M.S.; Brites, G.J.V.N.; Costa, J.J.; Gaspar, A.R.; Costa, V.A.F. Review and future trends of solar adsorption refrigeration systems. *Renew. Sustain. Energy Rev.* **2014**, *39*, 102–123. [CrossRef]
27. Thu, K.; Saha, B.B.; Chua, K.J.; Ng, K.C. Performance investigation of a waste heat-driven 3-bed 2-evaporator adsorption cycle for cooling and desalination. *Int. J. Heat Mass Transf.* **2016**, *101*, 1111–1122. [CrossRef]
28. Alahmer, A.; Ajib, S.; Wang, X. Comprehensive strategies for performance improvement of adsorption air conditioning systems: A review. *Renew. Sustain. Energy Rev.* **2019**, *99*, 138–158. [CrossRef]
29. Chang, W.S.; Wang, C.C.; Shieh, C.C. Experimental study of a solid adsorption cooling system using flat-tube heat exchangers as adsorption bed. *Appl. Therm. Eng.* **2007**, *27*, 2195–2199. [CrossRef]

30. Thu, K.; Saha, B.B.; Chakraborty, A.; Chun, W.G.; Ng, K.C. Study on an advanced adsorption desalination cycle with evaporator–condenser heat recovery circuit. *Int. J. Heat Mass Transf.* **2011**, *54*, 43–51. [CrossRef]
31. Wang, D.C.; Li, Y.H.; Li, D.; Xia, Y.Z.; Zhang, J.P. A review on adsorption refrigeration technology and adsorption deterioration in physical adsorption systems. *Renew. Sustain. Energy Rev.* **2010**, *14*, 344–353. [CrossRef]
32. Qu, T.F.; Wang, R.Z.; Wang, W. Study on heat and mass recovery in adsorption refrigeration cycles. *Appl. Therm. Eng.* **2001**, *21*, 439–452. [CrossRef]
33. Chan, K.C.; Tso, C.Y.; Chao, C.Y.H. Study of heat and mass recovery process on the performance of adsorption cooling systems. In Proceedings of the 10th International Conference on Heat Transfer, Fluid Mechanics and Thermodynamics, Orlando, FL, USA, 14–16 July 2014. Available online: http://hdl.handle.net/2263/44751 (accessed on 20 September 2021).
34. Wang, R.Z. Performance improvement of adsorption cooling by heat and mass recovery operation. *Int. J. Refrig.* **2001**, *24*, 602–611. [CrossRef]
35. Thu, K.; Yanagi, H.; Saha, B.B.; Ng, K.C. Performance analysis of a low-temperature waste heat-driven adsorption desalination prototype. *Int. J. Heat Mass Transf.* **2013**, *65*, 662–669. [CrossRef]
36. Sztekler, K.; Kalawa, W.; Nowak, W.; Mika, L.; Gradziel, S.; Krzywanski, J.; Radomska, E. Experimental Study of Three-Bed Adsorption Chiller with Desalination Function. *Energies* **2020**, *13*, 5827. [CrossRef]
37. Chua, H.T.; Ng, K.C.; Malek, A.; Kashiwagi, T.; Akisawa, A.; Saha, B.B. Multi-bed regenerative adsorption chiller—Improving the utilization of waste heat and reducing the chilled water outlet temperature fluctuation. *Int. J. Refrig.* **2001**, *24*, 124–136. [CrossRef]
38. Basdanis, T.; Tsimpoukis, A.; Valougeorgis, D. Performance optimization of a solar adsorption chiller by dynamically adjusting the half-cycle time. *Renew. Energy* **2021**, *164*, 362–374. [CrossRef]
39. Thu, K.; Ng, K.C.; Saha, B.B.; Chakraborty, A.; Koyama, S. Operational strategy of adsorption desalination systems. *Int. J. Heat Mass Transf.* **2009**, *52*, 1811–1816. [CrossRef]
40. Saha, B.B.; Chakraborty, A.; Koyama, S.; Aristov, Y.I. A new generation cooling device employing $CaCl_2$-in-silica gel–water system. *Int. J. Heat Mass Transf.* **2009**, *52*, 516–524. [CrossRef]
41. Qadir, N.U.; Said, S.A.M.; Mansour, R.B.; Imran, H.; Khan, M. Performance comparison of a two-bed solar-driven adsorption chiller with optimal fixed and adaptive cycle times using a silica gel/water working pair. *Renew. Energy* **2020**, *149*, 1000–1017. [CrossRef]
42. El-Sharkawy, I.I.; AbdelMeguid, H.; Saha, B.B. Towards an optimal performance of adsorption chillers: Reallocation of adsorption/desorption cycle times. *Int. J. Heat Mass Transf.* **2013**, *63*, 171–182. [CrossRef]
43. Glaznev, I.S.; Aristov, Y.I. The effect of cycle boundary conditions and adsorbent grain size on the water sorption dynamics in adsorption chillers. *Int. J. Heat Mass Transf.* **2010**, *53*, 1893–1898. [CrossRef]
44. Aristov, Y.I.; Sapienza, A.; Ovoshchnikov, D.S.; Freni, A.; Restuccia, G. Reallocation of adsorption and desorption times for optimisation of cooling cycles. *Int. J. Refrig.* **2012**, *35*, 525–531. [CrossRef]
45. Sztekler, K.; Kalawa, W.; Mika, Ł.; Mlonka-Medrala, A.; Sowa, M.; Nowak, W. Effect of Additives on the Sorption Kinetics of a Silica Gel Bed in Adsorption Chiller. *Energies* **2021**, *14*, 1083. [CrossRef]
46. Tatlier, M. Performances of MOF vs. zeolite coatings in adsorption cooling applications. *Appl. Therm. Eng.* **2017**, *113*, 290–297. [CrossRef]
47. Verde, M.; Harby, K.; Corberán, J.M. Optimization of thermal design and geometrical parameters of a flat tube-fin adsorbent bed for automobile air-conditioning. *Appl. Therm. Eng.* **2017**, *111*, 489–502. [CrossRef]
48. Demir, H.; Mobedi, M.; Ülkü, S. The use of metal piece additives to enhance heat transfer rate through an unconsolidated adsorbent bed. *Int. J. Refrig.* **2010**, *33*, 714–720. [CrossRef]
49. Askalany, A.A.; Henninger, S.K.; Ghazy, M.; Saha, B.B. Effect of improving thermal conductivity of the adsorbent on performance of adsorption cooling system. *Appl. Therm. Eng.* **2017**, *110*, 695–702. [CrossRef]
50. Krzywanski, J.; Grabowska, K.; Sosnowski, M.; Zylka, A.; Kulakowska, A.; Czakiert, T.; Sztekler, K.; Wesolowska, M.; Nowak, W. Heat Transfer in Adsorption Chillers with Fluidized Beds of Silica Gel, Zeolite, and Carbon Nanotubes. *Heat Transf. Eng.* **2021**, *43*, 1–15. [CrossRef]
51. Mitra, S.; Kumar, P.; Srinivasan, K.; Dutta, P. Simulation study of a two-stage adsorber system. *Appl. Therm. Eng.* **2014**, *72*, 283–288. [CrossRef]

energies

MDPI

Article

CFD Analysis of Elements of an Adsorption Chiller with Desalination Function

Karol Sztekler *, Tomasz Siwek, Wojciech Kalawa, Lukasz Lis, Lukasz Mika, Ewelina Radomska and Wojciech Nowak

Department of Thermal and Fluid Flow Machines, Faculty of Energy and Fuels, AGH University of Science and Technology, al. Mickiewicza 30, 30-059 Krakow, Poland; siwek@agh.edu.pl (T.S.); kalawa@agh.edu.pl (W.K.); llis@agh.edu.pl (L.L.); lmika@agh.edu.pl (L.M.); radomska@agh.edu.pl (E.R.); wnowak@agh.edu.pl (W.N.)

* Correspondence: sztekler@agh.edu.pl

Abstract: This paper presents the results of numerical tests on the elements of an adsorption chiller that comprises a sorption chamber with a bed, a condenser, and an evaporator. The simulation is based on the data and geometry of a prototype refrigeration appliance. The simulation of this problem is unique and has not yet been performed, and so far, no simulation of the phenomena occurring in the systems on a real scale has been carried out. The presented results are part of the research covering the entire spectrum of designing an adsorption chiller. The full process of numerical modeling of thermal and flow phenomena taking place in the abovementioned components is presented. The computational mesh sensitivity analysis combined in the k-ε turbulence model was performed. To verify and validate the numerical results obtained, they were compared with the results of tests carried out on a laboratory stand at the AGH Center of Energy. The results of numerical calculations are in good agreement with the results of the experimental tests. The maximum deviation between the pressure obtained experimentally and by simulations is 1.8%, while for temperatures this deviation is no more than 0.5%. The results allow the identification of problems and their sources, which allows for future structural modifications to optimize the operation of the device.

Keywords: CFD; adsorption; chiller; desalination

Citation: Sztekler, K.; Siwek, T.; Kalawa, W.; Lis, L.; Mika, L.; Radomska, E.; Nowak, W. CFD Analysis of Elements of an Adsorption Chiller with Desalination Function. *Energies* **2021**, *14*, 7804. https://doi.org/10.3390/en14227804

Academic Editor: Christopher Micallef

Received: 5 October 2021
Accepted: 18 November 2021
Published: 22 November 2021

1. Introduction

According to the report prepared by the Lawrence Livermore National Laboratory institute, humanity uses almost as much energy as it loses in all sectors of the economy [1]. Thus, many legislators and manufacturers strive to constantly increase the energy efficiency of devices and systems. One way to reduce energy losses, and thus increase the efficiency of systems, may be to use waste heat as an energy source. An example of a device that uses waste heat as an energy source is an adsorption chiller.

Adsorption chillers usually consist of one evaporator, a condenser, appropriate control valves, and a bed filled with a sorbent. The sorbent in the bed adsorbs and desorbs cyclically a working medium, called sorbate. The sorbent and sorbate together are known as a working pair, and their molecular structure and intermolecular interactions between them determine how specific adsorbent–adsorbate pairs will behave during adsorption and desorption [2]. Due to the low desorption temperature, low cost, and a lack of negative impact on the environment, the silica gel–water vapor working pair is commonly used in adsorption chillers [3]. Silica gel has a fairly large specific surface area (wide pores, 250–350 m^2/g; narrow porosity, 600–850 m^2/g), thanks to which it is possible to adsorb a large amount of water vapor [2,4,5].

Adsorption chillers use waste heat as an energy source. Additionally, they possess many other advantages, including lack of moving parts, low level of noise, no vibrations, and the possibility of water desalination [6]. On the other hand, intermittent operation, low coefficient of performance, big size and weight, and unequal time of desorption and adsorption processes can be accounted as the main drawbacks of adsorption chillers [7]. Hence, numerous scientific works are conducted to eliminate the abovementioned drawbacks.

The problem of intermittent operation can be eliminated by using multi-bed refrigerators with three [8] or more beds [9]. When using multiple beds, cycle lengths and their spacing can be adjusted to maximize cooling efficiency and working time of the evaporator in the cycle [8]. The other parameters influencing the performance, size, and weight of the adsorption chillers include, but are not limited to, the design of the heat exchanger in the bed [10] and the materials of which the bed is made [11], the size of sorbent particles [12], the velocity of heat transfer fluid that heats or cools the bed [13], and the type of evaporator [14,15]. As can be seen, the performance of adsorption chillers depends on numerous different factors. Additionally, the physical phenomena, i.e., evaporation, condensation, adsorption, and desorption, occurring in the main components of the adsorption chiller are intrinsically complex, as they are related to heat and mass transfer at various pressures and temperatures. Thus, fully understanding the abovementioned processes is essential to improve the performance of adsorption chillers.

As commonly known, the abovementioned processes (evaporation, condensation, adsorption, and desorption) are difficult to analyze using only experimental measurements. The measurements of temperatures, pressures, flow rates, etc., are crucial since they allow determining, e.g., the instantaneous and average efficiency of the chiller. However, these measurements are not sufficient for a comprehensive, and detailed scientific analysis of the chillers' operation is sometimes even impossible to perform. Furthermore, conducting experimental studies is very time and cost consuming. As a result, computational fluid dynamics (CFD) analysis becomes a necessity, and numerous scientific works use CFD to investigate either the evaporation, condensation, and adsorption/desorption processes themselves or the influence of defined parameters on the operation of the entire adsorption chiller.

Modeling the condensation and evaporation phenomena is a difficult issue since the characteristics of the fluid, both in the liquid and gaseous state and the relationships between them, should be taken into account. It is necessary to apply a two-phase model with appropriate consideration of the possible turbulence and the accompanying surface forces phenomenon [16]. It is also necessary to track the liquid–gas interface, which can be achieved by using the volume of fluid method [17–19]. Adsorption and desorption processes are also difficult to model, and a variety of different mathematical models of the sorption phenomenon are available [20]. Nevertheless, the linear driving force (LDF) model [21–23] is one of the most widely used.

Mohammed et al. [24] designed and investigated numerically, using FORTRAN software, a new modular bed for adsorption chillers. The bed consisted of an array of modules filled with adsorbent and placed in a metal case with vapor channels. After parametric analysis, it was concluded that the cooling power per unit volume was higher for the proposed bed compared to the commercially available beds. Hong et al. [22] used CFD to compare the performance of the adsorption chiller equipped with plate and fin-tube heat exchangers in the adsorption bed. It was found that the chiller with plate heat exchanger had a lower, by 19.9%, coefficient of performance (COP) but a higher, by 15.7%, specific cooling power, compared to the fin-tube heat exchanger. CFD analysis was also conducted by Mohammed et al. [25], who investigated heat and mass transfer in an open-cell aluminum foam packed with silica gel, which can be used as a bed in adsorption cooling applications. It was reported that the foam with 20 pores per inch (PPI) has a larger surface area and smaller cell size than the 10 PPI foam and thus is more advisable to be used in adsorption chillers. It was also found that the average bed temperature and adsorption rate was higher for 0.35 mm silica gel than for 0.70 mm silica gel. The effect of sorbent particle size was also investigated, using CFD, by Mitra et al. [21], who concluded that using a smaller sorbent does not always lead to faster adsorption, and the optimum geometry of a heat exchanger exists for a particular sorbent size.

Apart from the construction of the heat exchanger in the bed or the size of sorbent particles, the thermal conductivity of the bed strongly influences the overall performance of the adsorption chiller [23]. Therefore, different metal- or carbon-based additives can be added to the bed to increase its thermal conductivity [4]. Furthermore, the rate of adsorption depends on the convective heat transfer coefficient between the bed and the heat transfer fluid [26]. Therefore, the desorption and adsorption time can be reduced by increasing the velocity of the heat transfer fluid [27]. Elsheniti et al. [28], who used COMSOL Multiphysics software for investigations, reported that the COP and specific cooling capacity could be improved by 68% and 42%, respectively, when the flow character of the heat transfer fluid was changed from laminar to turbulent.

The literature review presented in the above paragraphs justifies the possibility of using CFD simulations for investigating adsorption chillers.

This paper presents a CFD analysis, carried out in the Ansys Academic Research CFD 19- Ignis Research Group ANSYS, Inc. 2600 ANSYS Drive Canonsburg, PA 15317 USA of the phenomena occurring in the main elements of the adsorption chiller, i.e., evaporator, condenser, and bed. The results of the CFD analysis are validated using data from the experimental studies carried out on the adsorption chiller located at the AGH UST Center of Energy. Therefore, this study aims at:

- Numerical modeling of the processes occurring in the evaporator, condenser, and bed of the adsorption chiller to understand better these processes;
- Analyzing the fields of temperature, pressure, and velocity;
- Indicating the locations with extremes of temperatures, pressures, and velocities;
- Determining the phenomena disrupting the operation of the main elements of the adsorption chiller;
- Determining potential changes in the structure of the main elements, which may improve the operating parameters of the individual elements of the chiller and increase its efficiency and reliability.

The direct value of this research is a possibility to determine the key elements and processes that impact the operation of the adsorption chiller. Consequently, this research indicates potential modifications of the chiller, which may be applied and improve its performance. Furthermore, the adsorption chiller located at the AGH UST Center of Energy has never been investigated using CFD analysis before. Thus, this research is a novelty in the context of getting to know the specifics of the chiller operation and the possibilities of its modification to improve the chiller efficiency and reliability as well as decrease its size and weight.

2. Materials and Methods

2.1. Empirical Research

The simulations were preceded by empirical tests carried out at the research station for adsorption systems at the AGH UST Center of Energy on a unique adsorption system. The system can operate in a two- or three-bed mode, generating chilled water and purified water. The beds were built on the basis of a plate-fin exchanger filled with adsorption material. The absorbent material used consists of elements with a size of 500–1000 μm. The scheme of the installation is presented in Figure 1, and the technical specification of the investigated chiller is given in Table 1.

For the purpose of this study, the chiller operated in a two-bed mode, and a detailed description of its operating principle can be found in [6]. The parameters that were measured during the experiments, and the measuring devices are listed in Table 2. The measured values were recorded every 5 s on a personal computer using special software, and then the results were exported to CSV files.

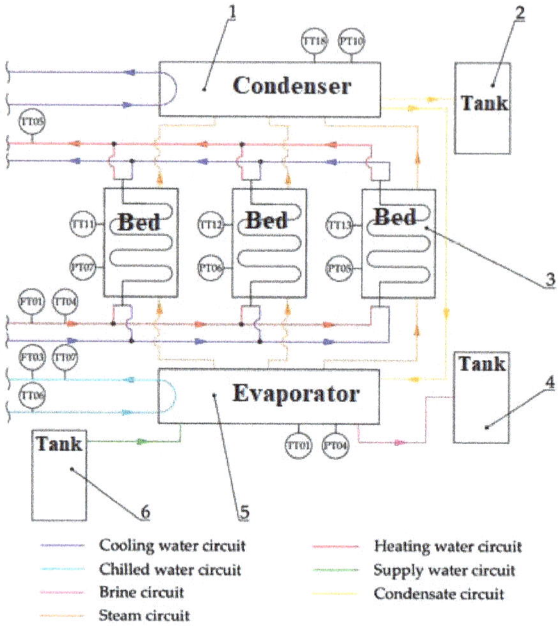

Figure 1. Scheme of the adsorption chiller with desalination test bench at the Center of Energy AGH [6,29]. 1—condenser; 2—distillate tank; 3—adsorbent bed; 4—brine tank; 5—evaporator; 6—deaerator; TT01—temperature in the evaporator; TT04—hot water inlet temperature; TT05—hot water outlet temperature; TT06—chilled water inlet temperature; TT07—chilled water outlet temperature; TT11—temperature in Bed 1; TT12—temperature in Bed 2; TT13—temperature in Bed 3; TT18—temperature in the condenser; PT04—pressure in the evaporator; PT07—pressure in Bed 1; PT06—pressure in Bed 2; PT05—pressure in Bed 3; PT10—pressure in the condenser; FT01—hot water flow; FT03—chilled water flow.

Table 1. Technical specification of a 3-bed adsorption chiller operating under nominal conditions. [6,29].

	Evaporator	
Cooling capacity	1.50	kW
Chilled water inlet temperature	32	°C
Chilled water outlet temperature	30	°C
Chilled water mass flow rate	0.184	kg/s
	Condenser	
Capacity	2.00	kW
Cooling water inlet temperature	30	°C
Cooling water outlet temperature	32	°C
Cooling water mass flow rate	0.25	kg/s
Daily distillate production	40	kg
	Beds	
Required cooling capacity (adsorption)	2.90	kW
Required heating capacity (desorption)	2.90	kW
Cooling water mass flow rate	0.25	kg/s
Heating water mass flow rate	0.25	kg/s
Cooling water inlet temperature	30	°C
Heating water inlet temperature	85	°C

Table 1. *Cont.*

Cooling Capacity		
Chilled Water: inlet/outlet temperature; mass flow rate: 32/30 °C; 0.184 kg/s	1.50	kW
Chilled Water: inlet/outlet temperature; mass flow rate: 16/11 °C; 0.0523 kg/s	1.32	kW
Chilled Water: inlet/outlet temperature; mass flow rate: 12/7 °C; 0.0523 kg/s	1.10	kW

Table 2. The parameters that were measured during the experiments and the measuring devices.

Temperature	Sensor	Range	Uncertainty
Heating water inlet to the beds Heating water outlet from the beds Free surface of the beds Inside the heat exchanger in the beds Cooling water outlet from the beds Cooling water outlet from the condenser Chilled water outlet from the evaporator Free space of the evaporator Free space of the condenser	Pt-1000	From −80 °C to 150 °C	±0.1 °C
Pressure	**Sensor**	**Range**	**Uncertainty**
Condenser Evaporator Beds	Pressure transducer	From 0 to 99 kPa	±0.5%

The results of the experimental studies are shown in Figures 2 and 3. Figure 2 depicts the temperature changes over time taking place in the main components of the investigated adsorption chiller. The pressure changes in the beds, evaporator, and condenser are shown in Figure 3. Some of the results shown in Figures 2 and 3 were used to define the boundary conditions and the rest to validate the simulation results. The average values between red lines in Figures 2 and 3 represent the values taken as the boundary conditions for the simulations. These values represent the operating parameters of the adsorption chiller after about 25 min from its start-up, as the operating conditions of the chiller stabilize after that time.

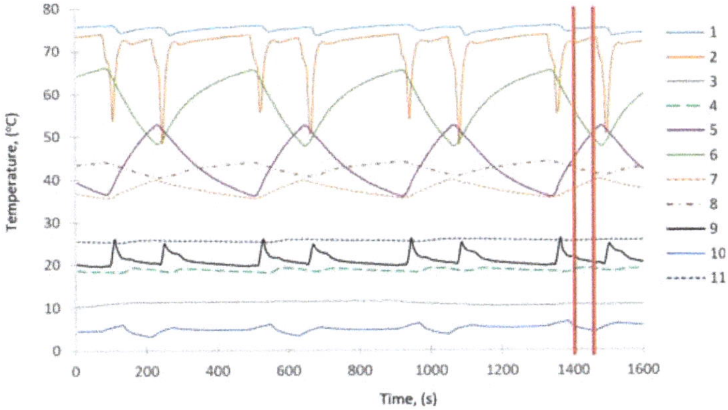

Figure 2. Temperature changes over time. 1—heating water inlet to the beds; 2—heating water outlet from the beds; 3—chilled water outlet from the evaporator; 4—cooling water outlet from the condenser; 5—heat exchanger in the first bed; 6—heat exchanger in the second bed; 7—free space of the first bed; 8—free space of the second bed; 9—cooling water outlet from the beds; 10—free space of the evaporator; 11—free space of the condenser.

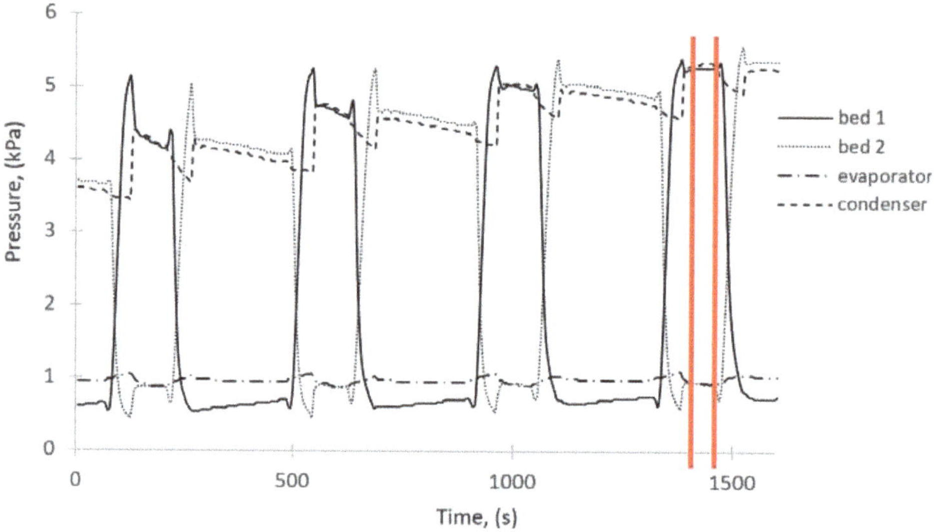

Figure 3. Pressure changes over time.

Figure 4 presents the destructive effect of the water vapor on the bed observed during the empirical research. Therefore, numerical simulations were applied to, among others, find the reasons for this destructive effect and to better understand the phenomena occurring in the system. Knowledge about the processes will allow modifying the components of the adsorption chiller to eliminate the destructive effect visible in Figure 4 and obtain a greater cooling capacity.

Figure 4. Destructive effect of the cumulative factor on the deposit.

2.2. *Construction of Spatial Geometry and a Computational Grid*

2.2.1. Generation of a Computational Mesh

During the construction of geometric models, the following simplifications were made:

- The housing of the elements was simplified to the form of a cylinder, without sight glasses and measuring connectors.

- Irregularly shaped elements such as heating and cooling junction boxes in the evaporator and condenser were simplified to a cylinder form.
- For the sorption chamber, the structural elements supporting the bed were omitted, and the bed itself was simplified to the form of a cuboid.

Then, after creating the structural elements, they were filled with the fluid domain in order to obtain the appropriate computational domains constituting the interior of the sorption chamber, evaporator, and condenser, respectively.

After the geometric model of the studied domain was prepared, it was exported to the ANSYS Meshing module. With it, a continuous domain was discretized in order to obtain a computational mesh.

In order to assess the mesh quality, an analysis of the cell quality parameters was carried out:

- Orthogonal quality: its value is in the range <0,1>, where the value 1 means the highest possible quality.
- Skewness: its value is in the range <0,1>, with the value 0 being the highest possible quality.

It was estimated that most of the elements for the orthogonal quality are in the range of 0.7–1.0 (approx. 2/3 elements). Likewise, most of the elements for skewness are in the range of 0–0.2 (approx. 2/3 elements). The obtained values of the orthogonal quality and skewness prove the good quality of the mesh.

The calculation grid for the sorption chamber consists of 472,686 nodes and 2,529,568 elements. The calculation grid for the evaporator consists of 2,165,228 nodes and 11,296,174 elements. The calculation grid for the condenser consists of 869,789 nodes and 4,562,582 elements. This enables performing calculations in a reasonably short time without using a supercomputer.

2.2.2. Boundary Conditions

The step preceding the calculations is to define the necessary solver settings. They were established on the basis of experimental studies (Figures 2 and 3) and studies in the literature described in the Introduction (Section 1).

- For the issues related to relatively low flow velocities (subsonic flow), flow solutions based on the pressure field ("pressure-based") were used.
- It was assumed that the simulation would be carried out in the "transient" mode, which enables the observation of changes in parameters over time.
- The influence of gravity on the fluid elements was taken into account by appropriately defining the acceleration vector.
- Due to the inclusion of gravity in the model and the occurrence of mass interactions, the scheme of coupling the velocity and pressure fields ("coupled") was applied.
- In the sorption chamber, reference conditions were defined in the entire domain as for sorption pressure of 1050 Pa and a temperature of 315.8 K, steam inlet at a temperature of 279.07 K, for desorption a pressure of 5250 Pa and a temperature of 301.98 K, and a steam outlet temperature of 312.82 K
- The sorption time was set at 100 s, and the desorption time 200 s
- Reference conditions for the evaporator was a pressure of 1050 Pa and a temperature of 279.09 K
- Evaporator water inlet temperature was 280.24 K, water outlet temperature was 280.29 K, and the steam outlet temperature was 279.09 K
- A second-order spatial discretization scheme was used for the governing equations (mass, momentum, and energy). However, for the dissipation of turbulence, a first-order scheme was used.
- Reference conditions for the condenser was a temperature of 301.98 K, and a pressure of 5250 Pa
- Steam inlet temperature to the condenser was 312.82 K, and the temperature of the cooling pipes was 292.82 K

- Condenser domain computation was set for a time equal to 500 s.

2.2.3. Computational Methods

For all phenomena, Ansys Fluent solves the equations of conservation of mass, momentum, and energy. The conservation of mass, momentum, and energy are shown in Equations (1)–(3), respectively [30].

$$\frac{\partial \rho}{\partial t} + \nabla \cdot (\rho \overline{v}) = S_m \tag{1}$$

$$\frac{\partial}{\partial t}(\rho \overline{v}) + \nabla \cdot (\rho \overline{v} \overline{v}) = -\nabla p + \nabla \cdot \left(\overline{\overline{\tau}}\right) + \rho \overline{g} + \overline{F} \tag{2}$$

$$\frac{\partial}{\partial t}(\rho E) + \nabla(\overline{v}(\rho E + p)) = \nabla \cdot (k_{eff} \nabla T - \sum_j h_j \overline{J}_j + \left(\overline{\overline{\tau}}_{eff} \cdot \overline{v}\right)) + S_h \tag{3}$$

where ρ, t, $\overline{v}$, S_m, p, $\overline{\overline{\tau}}$, $\rho\overline{g}$, $\overline{F}$, E, k_{eff}, h_j, $\overline{J}_j$, $\left(\overline{\overline{\tau}}_{eff} \cdot \overline{v}\right)$, and S_h stand for the density, time, velocity, source term, static pressure, stress tensor, gravitational body force, external mass forces acting on the fluid, energy, effective thermal conductivity coefficient, enthalpy of species j, diffusive flux of species j, viscous dissipation, and heat source, respectively.

To model turbulence occurring in the flows, the k-ε realizable turbulence model was used. In order to model the multiphase flow, the "Species Transport" model was selected for the calculations.

3. Results and Discussion

3.1. Results of Numerical Calculations

Numerical simulations are a tool that can explain many phenomena, the study of which with experimental methods is problematic. The simulation studies conducted in this work consider the interactions between the various components of the chiller at given phases of the work cycle. The simulation also allows for a time-dependent analysis of the temperature, pressure, and velocity variations during the operation of the unit. However, it should be remembered that they are based on assumptions and models. In order to assess the applicability and correctness of the tested model, it should be compared with the experimental results.

The results were developed using selected methods available as components of the Ansys CFD-Post post-processor module. To show the results of the simulations and the phenomena occurring in the main components of the investigated adsorption chiller, i.e., sorption chamber, evaporator, and condenser, the appropriate cross-sections of these components were selected, and they are shown in Figures 5–7.

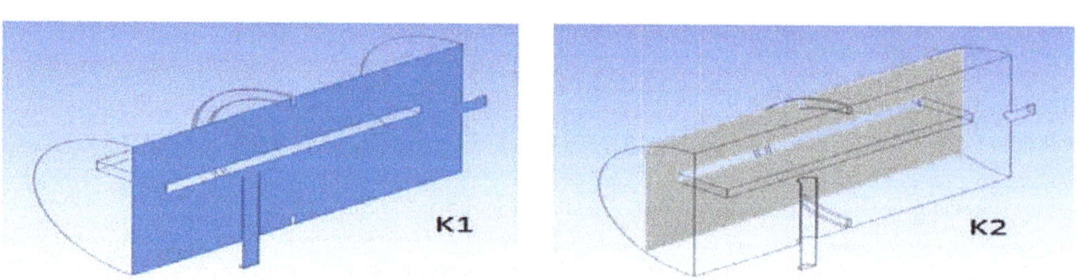

Figure 5. Visualization of the arrangement of cross-sections in the sorption chamber.

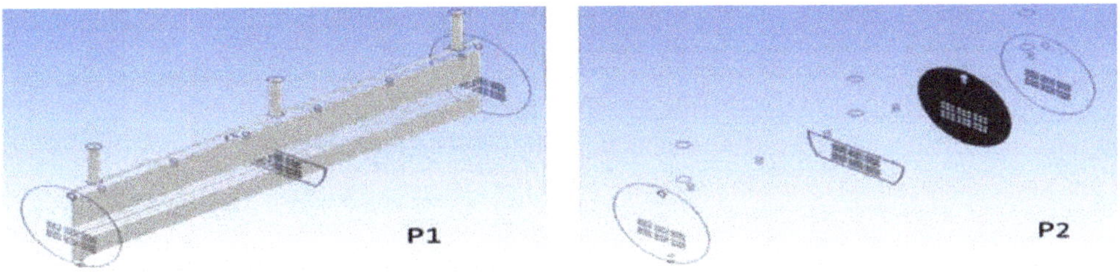

Figure 6. Visualization of the arrangement of cross-sections in the evaporator.

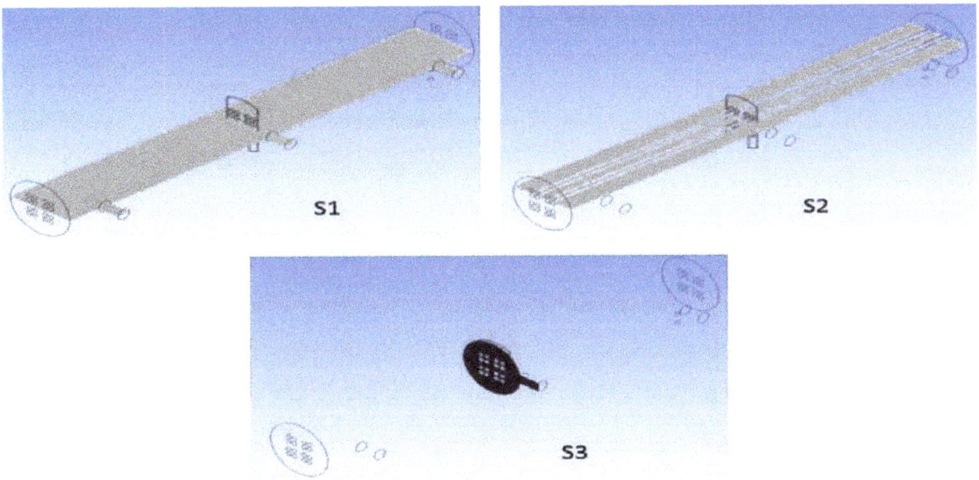

Figure 7. Visualization of the arrangement of cross-sections in the condenser.

3.2. *Simulation Results for the Sorption Chamber*

3.2.1. Sorption Process

In this study, the temperature and pressure distribution, as well as velocity field, during the sorption processes were determined. The presented results concern the state of the system at the time point t = 100 s.

Figure 8a, shows the temperature distribution in the computational domain in Section K1. Uniform temperature distribution was obtained close to 311 K, which proves that the dimensions of the sorption chamber are sufficient. If the sorption chamber was too small, there could be a non-uniform temperature distribution. The lowest temperature value 279 K occurs only in the vicinity of the deposit location. It results from the modeling of the heat flux leaving the system on these planes.

Figure 8b shows the pressure distribution in the computational domain in Section K1. Uniform pressure distribution in the range of 1043–1045 Pa was obtained. The area of increased pressure up to 1067 Pa occurs in the vicinity of the location of the bed on the extension of the steam inlet nozzle. This may be due to the close location of the stub pipe and the direct influence of the inlet stream. Likewise, high pressure is one of the reasons for the negative impact on the viability of the bed, as shown in Figure 4.

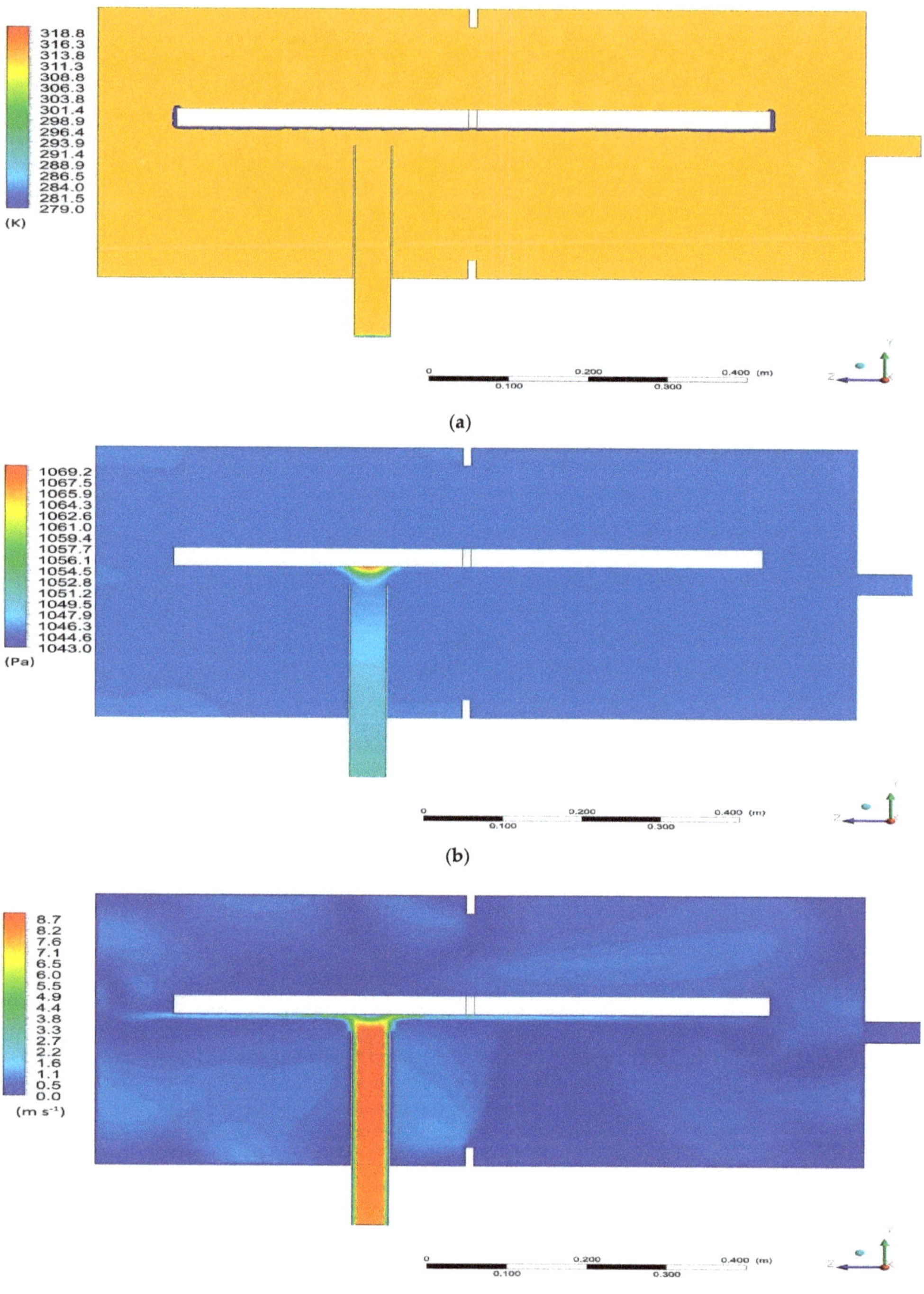

Figure 8. Cross-Section K1 for the sorption: (**a**) temperature distribution; (**b**) pressure distribution; (**c**) velocity distribution.

The visualization of the velocity field distribution is shown in Figure 8c. A large variation of the vapor velocity in the space around the bed is observed. The velocity of the vapor varies from 0.1 to 8.7 m/s. The greatest velocity, 8.7 m/s, is in the area of the outflow from the inlet connection, while lower velocity values, in the range of about 0–4 m/s, are present in the remainder of the domain. This is due to the small diameter of the inlet nozzle and the continuity equation, according to which a velocity of an incompressible fluid always decreases as the fluid flows from a channel with a small diameter (small cross-sectional area) to a channel with a larger diameter (larger cross-sectional area).

3.2.2. Desorption Process

The presented results concern the state of the system at the time point t = 200 s.

Figure 9 shows the pressure distribution in the computational domain in Section K1. Uniform pressure distribution of 5247 Pa on a global scale was obtained. The lower pressure values in the range of 5170–5230 Pa can be observed in the vicinity of the steam outlet from the system.

Figure 9. Pressure distribution in Section K1 for the desorption.

The visualization of the velocity field distribution for Section K2 is presented in Figure 10. On a global scale, the velocity level inside the domain is at a similar level. The highest velocities, up to 0.137 m/s, occur near the steam outlet, which result from the geometry of the system. Taking into account the local values for individual cross-sections, an increase in speed at the points of narrowings can be noticed. In addition, the speed also increases in the lower part of the tank relative to the upper part. This may be due to the geometrical imbalance of these parts—in the lower major part, the fluid must move faster in order to maintain a continuous flow. The results of the simulation also show the formation of steam vortices and their shapes and sizes, which are related to the adopted geometry of the bed.

3.3. Simulation Results for the Evaporator

The presented results concern the state of the system at the time point t = 100 s. The temperature distribution in the computational domain is presented in Figure 11a,b. The highest temperature, 284 K, occurs in the lower part of the evaporator tank, near the pipes through which the chilled water flows. When the fluid elements come into contact with the surface of the tubes at the higher temperature, the temperature of the fluid element increases until vaporization occurs. P1 (Figure 11a) shows brighter spots near the water entry into the system through the spray system. This is due to the lower temperature of the entering water compared to the surrounding fluid. In Cross-Section P2 (Figure 11b), one can see the increased temperature in the lower part to 284 K. Water in the system falls under the influence of gravity. At the same time, it heats up from other fluid elements and heating surfaces.

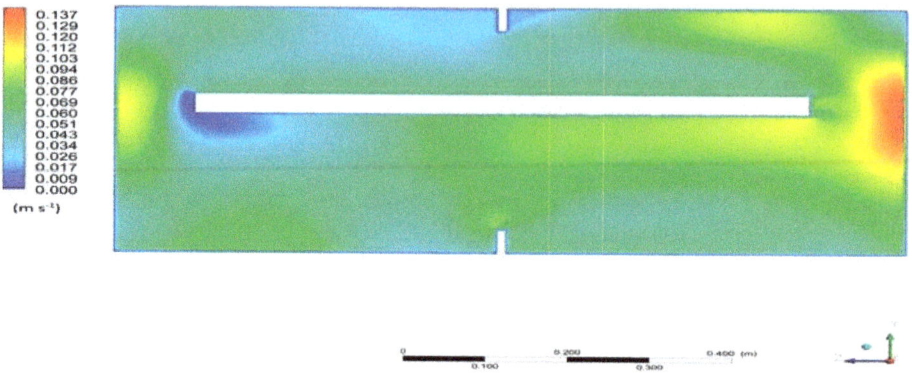

Figure 10. Velocity field distribution in Section K2 for the desorption.

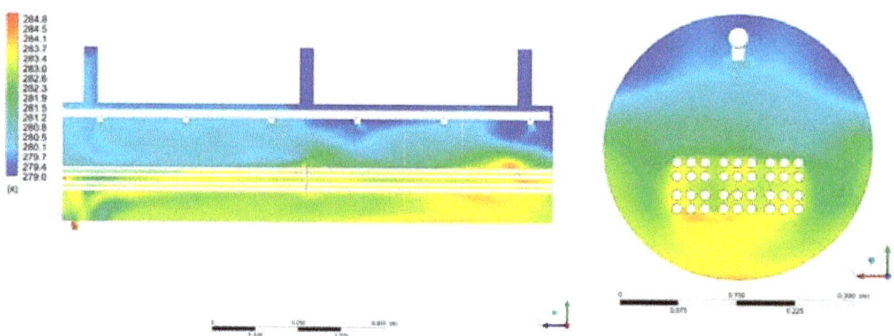

(a) (b)

Figure 11. Temperature distribution in: (**a**) Section P1; (**b**) Section P2.

Figure 12a,b shows the velocity field distributions in the Cross-Sections P1 and P2, respectively. The vapor velocity ranges from about 0.1 to 9.7 m/s. The highest speed of 9.7 m/s is at the steam outlet nozzles. It is caused by the fact that the steam flows from the tank to the nozzles that have a small diameter. As a result of the laws of flow continuity, speed increases.

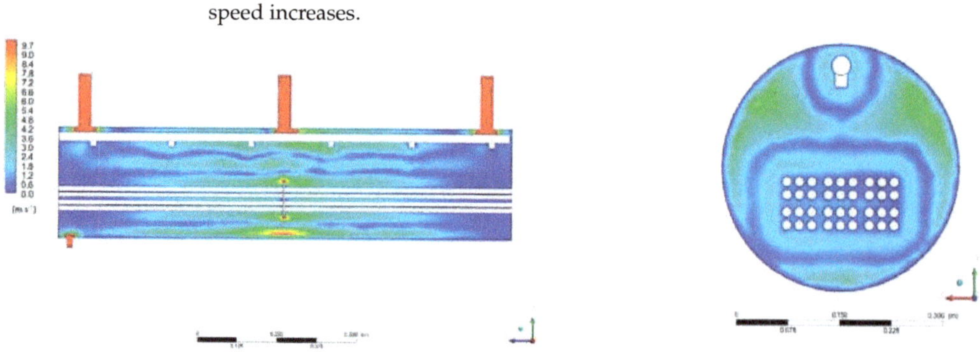

(a) (b)

Figure 12. Velocity field distribution: (**a**) Cross-Section P1; (**b**) Cross-Section P2.

The distribution of the velocity field on Section P1 indicates an increase in the velocity of the fluid elements in the vicinity of the obstacle constituted by the structural element inside the tank. When the fluid encounters this obstacle, there is less space available for it to move, so its velocity in this area must be increased.

The formation of closed zones with a similar value of the velocity of the fluid elements can be seen in Cross-Section P2. One of these zones was formed around the heat exchanger pipes, and the other around the pipe supplying water to the system. This may indicate that the flows between the layers of fluid in the reservoir are established.

Figure 13 shows the mass fraction of liquid water in Section P2. The greatest concentration of the liquid occurs in the lower part of the tank in accordance with the acting gravity. The greatest concentration occurs in the area of the outflow of non-evaporated water from the system and between the heat exchanger pipes. Above the area of the heat exchanger pipes, the water content is lower, which means that the water vapor content is predominant there. The distribution in this section also shows a greater liquid content on the left side compared to the right side.

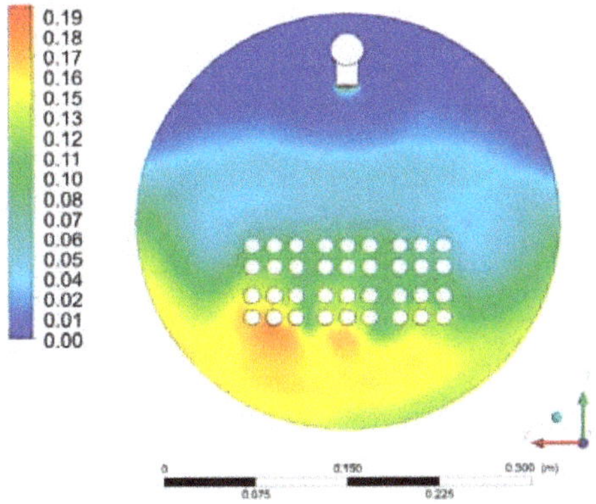

Figure 13. Distribution of liquid water content for Cross-Section P2 on the local scale.

3.4. Simulation Results for the Condenser

The presented results concern the state of the system at the time point t = 200 s.

The temperature distribution inside the condenser computational domain is shown in Figure 14a,b. The highest temperature of 312 K is visible near the steam input to the system. As the temperature moves away from the entry point, the temperature decreases, which indicates that the incoming vapor cools down in contact with other fluid components already present in the condenser vessel. The lowest temperature of 294 K is around the pipes with the cooling liquid. Additionally, as can be seen in Cross-Section S3, the temperature of fluid decreases when the fluid passes through the vicinity of the cooling surfaces. The fluid temperature is the lowest in the space opposite the steam inlet ports. On Cross-Sections S1 and S3, one can find the shift of isotherms in the direction opposite to the Z axis. This may indicate a local fluid flow in the direction opposite to the Z axis.

The distribution of velocity fields is presented in Figure 15. The vapor velocity varies from about 0.1 to 3.4 m/s. The highest velocity of 3.4 m/s prevails in the area of the steam inlet nozzles. This is understandable because the fluid entering the reservoir from the pipe slows down in velocity due to the increase of the space in which it can be held.

Additionally, it can be noticed that the velocity field distribution in the vicinity of the connection between the stub pipe and the tank is shifted in the direction opposite to the Z axis, similar to the temperature. There is also a greater value of velocity in the vicinity of the tank walls. This may indicate the establishment of solid current lines inside the condenser, along which the fluid elements move.

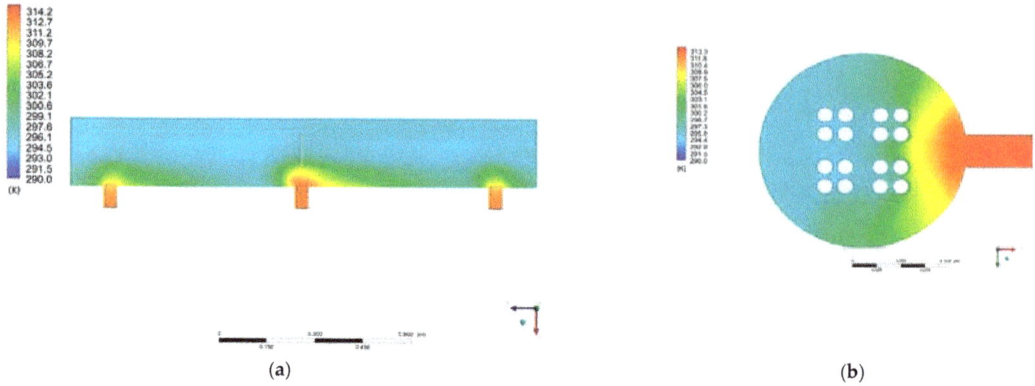

Figure 14. Temperature distribution for: (**a**) Cross-Section S1; (**b**) Cross-Section S3.

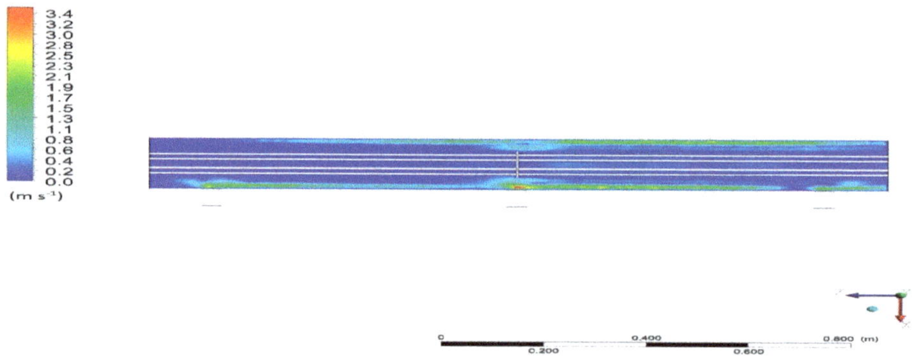

Figure 15. Distribution of the velocity field for Cross-Section S2.

3.5. Validation of Simulation

Numerical simulations are a tool that can explain many phenomena, the study of which with experimental methods is problematic. However, it should be remembered that they are based on assumptions and models. In order to assess the applicability and correctness of the tested model, it should be compared with the experimental results. The experimental type of measurements makes it impossible to measure local values such as velocity in cross-section. Validation is done by comparing global values. The most reliable parameter is the pressure for the sorption and desorption process.

Figures 16 and 17 show the pressure change over time for the sorption and desorption processes, respectively. The figures show a 2.5% error; however, the maximum discrepancy between the results of simulation and experiments is 1.5% and 1.8% for the sorption and desorption, respectively. The simulation results are within this range and can therefore be considered reliable.

Figure 16 shows the mean pressure inside the domain as a function of time for the sorption process. As can be seen, the pressure decreases almost linearly from the initial value of 1070 Pa to about 1050 Pa. The pressure drop is caused by the fact that during the sorption, water vapor is adsorbed by the sorbent.

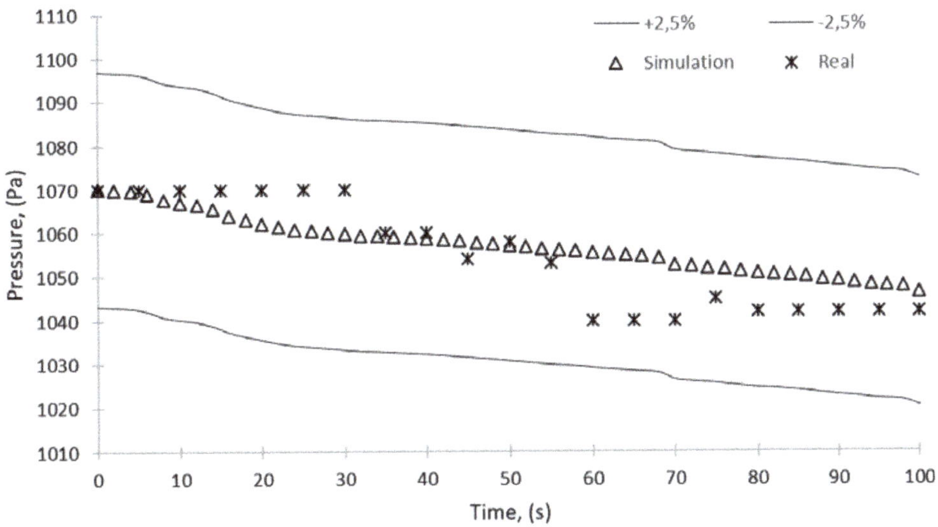

Figure 16. Comparison of the results of simulation and experimental studies—pressure over time for the sorption process.

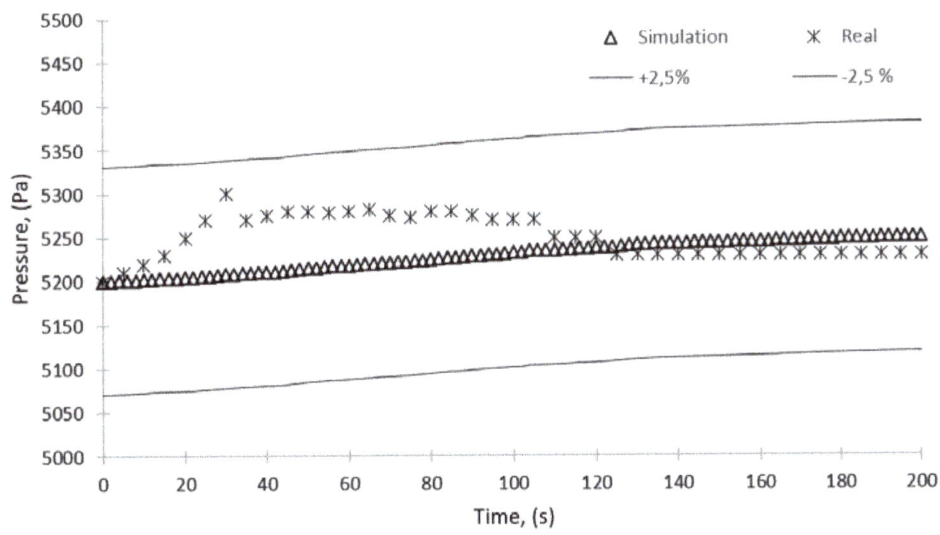

Figure 17. Comparison of the results of simulation and experimental studies—pressure over time for the desorption process.

As mentioned before, the discrepancy between the results of simulation and experimental studies is no more than 2.5%. These results of simulation reflect the quality of the model and the reliability of the performed calculations. The simulation will allow for an in-depth analysis of, firstly, the phenomena taking place during the adsorption and desorption and, secondly, any local changes in the operating parameters and the efficiency of the device caused by, e.g., the structure of the individual elements of the device.

Figure 16 presents the pressure in the bed during the sorption, and both the experimental and simulation results are shown. Pressure fluctuations obtained in the experimental studies can be attributed to the opening of the valve connecting the sorption chamber with the evaporator and pressure equalization between these two components, which was not included in the simulation. Additionally, from the experimental data, a pressure drop of about 30 Pa after 60 s from the start of operation can be observed. This pressure drop is caused by the sorption of water vapor in the bed, which results in the pressure drop in the entire unit. The sorption continues until the silica gel is saturated with water. When saturation is complete, the pressure equalizes, and the process stops.

For the desorption process (Figure 17), the mean domain pressure during the simulation showed an upward trend. The increase of the pressure is a result of the desorption process, during which the water vapor is released from the silica gel. Thus, more water vapor particles are in the given space, and according to the ideal gas law, its pressure must increase. However, this increase is below 1% from 5200 to 5250 Pa. It is also observed that the pressure in the bed obtained experimentally is about 100 Pa higher than the pressure from the simulation in the initial phase of the process. This is because the temperature in the bed increases as a result of supplying the heating water to the heat exchanger and opening the valve between the condenser and the bed, which could also cause pressure fluctuations. As a consequence of bed heating, the adsorbed water vapor is released from the silica gel, and this process continues until the absorbed water is fully desorbed and the pressure between the bed and the condenser equalizes.

Figures 18 and 19 show the temperature change over time for the sorption and desorption processes, respectively. The figures show a 2.5% error; however, the maximum discrepancy between the results of simulation and experiments is 0.5% and 0.4% for the sorption and desorption, respectively. The simulation results are within this range and can therefore be considered reliable.

Although the simulation model does not contain the description of all phenomena occurring in the device, the generated results of the simulation reflect the operation of the device with high accuracy and allow reliable simulation of the phenomena occurring in the device in different conditions, which can be used to improve the performance of the device.

The results of simulations show the differences in the velocity of water vapor in the main elements of the adsorption chiller. Detecting differences in the velocity of vapor in particular zones of a given element allows protecting the device from damage, e.g., erosion.

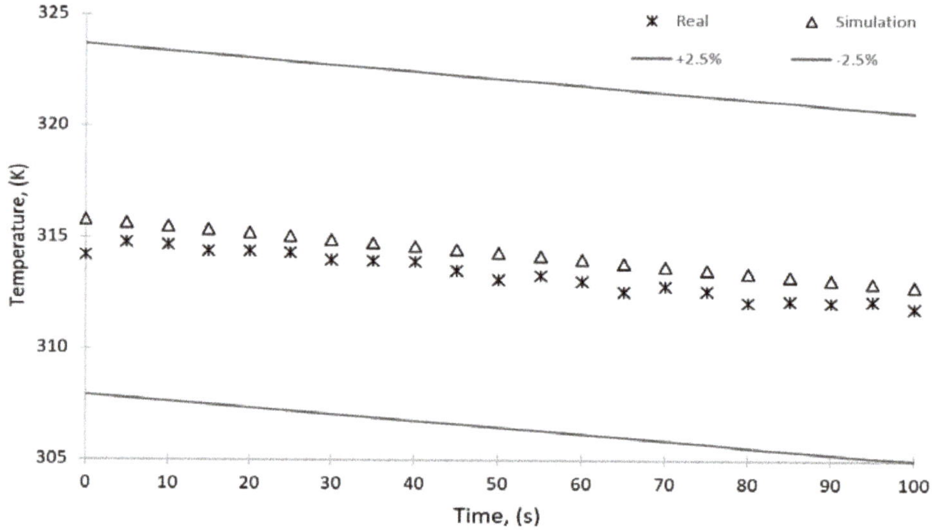

Figure 18. Comparison of the results of simulation and experimental studies—temperature over time for the sorption process.

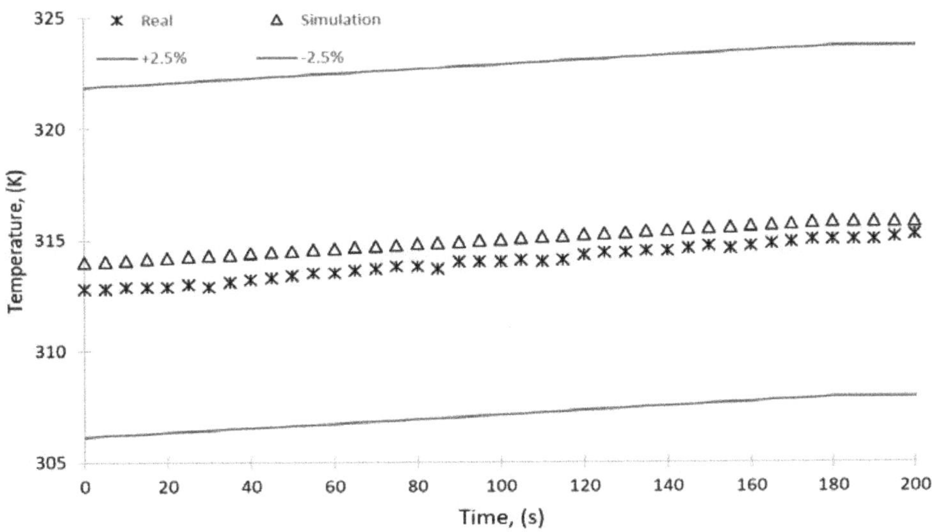

Figure 19. Comparison of the results of simulation and experimental studies—temperature over time for the desorption process.

The prepared numerical model is analyzed based on the selected variants, and the compatibility with the experimental data used for validation is achieved. The average values of velocity and pressure obtained in the simulation correspond to the values collected in the experimental research. Based on the results of the simulation, the probable reason for the destruction of the bed during experimental studies is also obtained. Taking into account all correctness criteria applied to the model, it can be concluded that the results obtained from the numerical calculations are sufficiently accurate and correct. The model created in this study can be further verified in the study of other configurations of adsorption chillers and also will be useful in the process of prototyping and testing new design solutions of individual elements of the adsorption chiller. In the next stages of the research, the model will be used to improve the design solutions of the device to improve its efficiency, optimize its operation cycle, and eliminate potential design faults that may damage the device. It should be noted that the individual components of the chiller are interdependent. Therefore, it is likely that modifying just one of them, such as the bed, without making changes to the other components may have a small impact on increasing the efficiency of the chiller.

4. Conclusions

The combination of equations used in the described simulation allows for the creation of a reliable model of a system that allows the analysis of the work cycle in time. The models used allow the results to be obtained in an acceptable time without access to a supercomputer unit. The results of the simulations are in good agreement with the results of the experiments. The maximum discrepancy between the pressure obtained experimentally and by the simulations is 1.8%, while the discrepancy between the temperatures obtained experimentally and by the simulations is no more than 0.5%. The promising results of the validation of the model make it possible to undertake further studies of this type of issue with the use of computational fluid dynamics (CFD). They can be used to test new configurations and design solutions without the need to create real test units. The results helped to identify the problem spots and formulate design recommendations to optimize the operation of the device as listed below.

- Changing the tube banks in the evaporator from an in-line arrangement to a staggered arrangement while simultaneously maintaining the same heat transfer surface area. This change can improve the cooling capacity of the evaporator and provide a more uniform temperature distribution;
- Using a turbulator inside the tubes in the evaporator, as the heat transfer rate is significantly greater for turbulent flow compared to laminar flow.
- Changing the arrangement of the tubes in the condenser, as the temperature distribution in the condenser is non-uniform. Another arrangement of the tubes could provide more uniform temperature distribution, which could result in a faster vapor condensation and thus a more efficient performance of the entire device;
- Reducing the length of the water vapor supply pipe or using a jet diffusion cone or a straight baffle at the stream outlet should be considered. The proposed solutions will lower the velocity of water vapor and improve its dispersion. As a result, the force acting on the sorbent will be reduced;
- Using a distribution manifold that distributes the vapor uniformly over the entire surface of the bed, which will accelerate the sorption. As shown in Figure 10, the water vapor diffusion in the bed is not uniform, which lengthens the adsorption time and decreases the overall efficiency of the adsorption chiller.

Author Contributions: Conceptualization, K.S. and T.S.; methodology, T.S. and K.S.; validation, K.S. and T.S.; formal analysis, T.S., K.S. and L.L.; investigation, T.S., K.S., W.K. and L.L.; resources, W.N. and L.M.; data curation, W.K., L.L., E.R. and L.M.; writing—original draft preparation, T.S., L.L. and K.S.; writing—review and editing, K.S., T.S., L.L. and L.M.; visualization, L.L., W.K. and E.R.; supervision, W.N. and L.M.; project administration, L.M., W.N. and K.S.; funding acquisition, L.M., W.N. and K.S. All authors have read and agreed to the published version of the manuscript.

Funding: This research was funded by the Ministry of Science and Higher Education, Poland, Grant AGH No. 16.16.210.476, and partly supported by the program "Excellence initiative—research university" for the AGH University of Science and Technology.

Institutional Review Board Statement: Not applicable.

Informed Consent Statement: Not applicable.

Data Availability Statement: Not applicable.

Conflicts of Interest: The authors declare no conflict of interest.

References

1. Lawrence Livermore National Laboratory, Energy Flow Charts. Available online: https://flowcharts.llnl.gov (accessed on 10 July 2021).
2. Kisielew, A.W.; Jaszin, J.I. *Adsorpcyjna Chromatografia Gazowa*; Państwowe Wydawnictwo Naukowe: Warszawa, Poland, 1969.
3. Ugale, V.D.; Pitale, A.D. A Review on Working Pair Used in Adsorption Cooling System. *Int. J. Air-Cond. Refrig.* **2015**, *23*, 1530001. [CrossRef]
4. Sztekler, K.; Kalawa, W.; Mika, Ł.; Mlonka-Medrala, A.; Sowa, M.; Nowak, W. Effect of Additives on the Sorption Kinetics of a Silica Gel Bed in Adsorption Chiller. *Energies* **2021**, *14*, 1083. [CrossRef]
5. Sztekler, K.; Kalawa, W.; Mlonka-Medrala, A.; Nowak, W.; Mika, Ł.; Krzywanski, J.; Grabowska, K.; Sosnowski, M.; Debniak, M. The Effect of Adhesive Additives on Silica Gel Water Sorption Properties. *Entropy* **2020**, *22*, 327. [CrossRef] [PubMed]
6. Sztekler, K.; Kalawa, W.; Nowak, W.; Mika, L.; Gradziel, S.; Krzywanski, J.; Radomska, E. Experimental study of three-bed adsorption chiller with desalination function. *Energies* **2020**, *13*, 5827. [CrossRef]
7. Szyc, M.; Nowak, W. Operation of an adsorption chiller in different cycle time conditions. *Chem. Process Eng. Inz. Chem. Process.* **2014**, *35*, 109–119. [CrossRef]
8. Uyun, A.S.; Miyazaki, T.; Ueda, Y.; Akisawa, A. Experimental Investigation of a Three-Bed Adsorption Refrigeration Chiller Employing an Advanced Mass Recovery. *Energies* **2009**, *2*, 531–544. [CrossRef]
9. Jribi, S.; Saha, B.B.; Koyama, S.; Bentaher, H. Modeling and simulation of an activated carbon–CO_2 four bed based adsorption cooling system. *Energy Convers. Manag.* **2014**, *78*, 985–991. [CrossRef]
10. Hong, S.W.; Ahn, S.H.; Kwon, O.K.; Chung, J.D. Optimization of a fin-tube type adsorption chiller by design of experiment. *Int. J. Refrig.* **2015**, *49*, 49–56. [CrossRef]

11. Cao, N.V.; Duong, X.Q.; Lee, W.S.; Park, M.Y.; Chung, J.D.; Hong, H. Effect of heat exchanger materials on the performance of adsorption chiller. *J. Mech. Sci. Technol.* **2020**, *34*, 2217–2223. [CrossRef]
12. Vodianitskaia, P.J.; Soares, J.J.; Melo, H.; Gurgel, J.M. Experimental chiller with silica gel: Adsorption kinetics analysis and performance evaluation. *Energy Convers. Manag.* **2017**, *132*, 172–179. [CrossRef]
13. Xua, S. Experiment on a New Adsorption Bed about Adsorption Refrigeration Driven by Solar Energy. *Energy Procedia* **2012**, *14*, 1542–1547. [CrossRef]
14. Seiler, J.; Lanzerath, F.; Jansen, C.; Bardow, A. Only a wet tube is a good tube: Understanding capillary-assisted thin-film evaporation of water for adsorption chillers. *Appl. Therm. Eng.* **2019**, *147*, 571–578. [CrossRef]
15. Thimmaiah, P.C.; Sharafian, A.; Rouhani, M.; Huttema, W.; Bahrami, M. Evaluation of low-pressure flooded evaporator performance for adsorption chillers. *Energy* **2017**, *122*, 144–158. [CrossRef]
16. Kleiner, T.; Rehfeldt, S.; Klein, H. CFD model and simulation of pure substance condensation on horizontal tubes using the volume of fluid method. *Int. J. Heat Mass Transf.* **2019**, *138*, 420–431. [CrossRef]
17. Zhang, L.; Zhang, G.; Tian, M.; Zhang, J.; Zhang, Y. Modeling of laminar filmwise condensation of methane with nitrogen on an isothermal vertical plate. *Int. Commun. Heat Mass Transf.* **2019**, *105*, 10–18. [CrossRef]
18. Tahir, F.; Mabrouk, A.; Koç, M. Impact of surface tension and viscosity on falling film thickness in multi-effect desalination (MED) horizontal tube evaporator. *Int. J. Therm. Sci.* **2020**, *150*, 106235. [CrossRef]
19. Mohammed, H.I.; Giddings, D.; Walker, G.S. CFD multiphase modelling of the acetone condensation and evaporation process in a horizontal circular tube. *Int. J. Heat Mass Transf.* **2019**, *134*, 1159–1170. [CrossRef]
20. Muttakin, M.; Pal, A.; Rupa, M.J.; Ito, K.; Saha, B.B. A critical overview of adsorption kinetics for cooling and refrigeration systems. *Adv. Colloid Interface Sci.* **2021**, *294*, 102468. [CrossRef]
21. Mitra, S.; Muttakin, M.; Thu, K.; Saha, B.B. Study on the influence of adsorbent particle size and heat exchanger aspect ratio on dynamic adsorption characteristics. *Appl. Therm. Eng.* **2018**, *133*, 764–773. [CrossRef]
22. Hong, S.W.; Kwon, O.K.; Chung, J.D. Application of an embossed plate heat exchanger to adsorption chiller. *Int. J. Refrig.* **2016**, *65*, 142–153. [CrossRef]
23. Solmuş, I.; Andrew, D.; Yamali, C.; Baker, D. A two-energy equation model for dynamic heat and mass transfer in an adsorbent bed using silica gel/water pair. *Int. J. Heat Mass Transf.* **2012**, *55*, 5275–5288. [CrossRef]
24. Mohammed, R.H.; Mesalhy, O.; Elsayed, M.L.; Chow, L.C. Novel compact bed design for adsorption cooling systems: Parametric numerical study. *Int. J. Refrig.* **2017**, *80*, 238–251. [CrossRef]
25. Mohammed, R.H.; Mesalhy, O.; Elsayed, M.L.; Chow, L.C. Performance enhancement of adsorption beds with silica-gel particles packed in aluminum foams. *Int. J. Refrig.* **2019**, *104*, 201–212. [CrossRef]
26. Solmuş, I.; Yamali, C.; Yildirim, C.; Bilen, K. Transient behavior of a cylindrical adsorbent bed during the adsorption process. *Appl. Energy* **2015**, *142*, 115–124. [CrossRef]
27. Yaïci, W.; Entchev, E. Coupled unsteady computational fluid dynamics with heat and mass transfer analysis of a solar/heat-powered adsorption cooling system for use in buildings. *Int. J. Heat Mass Transf.* **2019**, *144*, 1–17. [CrossRef]
28. Elsheniti, M.B.; Hassab, M.A.; Attia, A.E. Examination of effects of operating and geometric parameters on the performance of a two-bed adsorption chiller. *Appl. Therm. Eng.* **2019**, *146*, 674–687. [CrossRef]
29. Sztekler, K.; Kalawa, W.; Nowak, W.; Mika, L.; Grabowska, K.; Krzywanski, J.; Sosnowski, M.; Al-Harbi, A.A. Performance evaluation of a single-stage two-bed adsorption chiller with desalination function. *J. Energy Resour. Technol.* **2020**, *143*, 1–22. [CrossRef]
30. ANSYS FLUENT Documentation. Available online: https://www.afs.enea.it/project/neptunius/docs/fluent/index.htm (accessed on 1 March 2021).

energies

MDPI

Article

Possibility of Advanced Modified-Silica-Based Porous Materials Utilisation in Water Adsorption Processes—A Comparative Study

Karol Sztekler [1,*], Agata Mlonka-Mędrala [1], Nezar H. Khdary [2], Wojciech Kalawa [1], Wojciech Nowak [1] and Łukasz Mika [1]

1 Faculty of Energy and Fuels, AGH University of Science and Technology, Mickiewicza 30, 30-059 Krakow, Poland; amlonka@agh.edu.pl (A.M.-M.); kalawa@agh.edu.pl (W.K.); wnowak@agh.edu.pl (W.N.); lmika@agh.edu.pl (Ł.M.)

2 King Abdulaziz City for Science and Technology, Riyadh 11442, Saudi Arabia; nkhdary@kacst.edu.sa

* Correspondence: sztekler@agh.edu.pl

Abstract: Due to a high risk of power outages, a heat-driven adsorption chillers are gaining the attention. To increase the efficiency of the chiller, new adsorbents must be produced and examined. In this study, four newly developed silica–based porous materials were tested and compared with silica gel, an adsorber commonly paired with water. Extended sorption tests using mercury intrusion porosimetry, gas adsorption, and dynamic vapor sorption were performed. The morphology of the samples was determined using a scanning electron microscope. The thermal properties were defined using simultaneous thermal analysis and a laser flash method. Metal organic silica (MOS) nanocomposites analysed in this study had thermal properties similar to those of commonly used silica gel. MOS samples have a thermal diffusivity coefficient in the range of 0.17–0.25 mm^2/s, whereas silica gel of about 0.2 mm^2/s. The highest water adsorption capacity was measured for AFSMo-Cu and equal to 33–35%. For narrow porous silica gel mass uptake was equal about 25%. In the case of water adsorption, it was observed that the pore size of the sorbent is essential, and adsorbents with pore sizes higher than 5 nm, are most recommended in working pairs with water.

Keywords: metal organic silica; nanocomposites; sorption; thermal diffusivity; adsorption chiller

Citation: Sztekler, K.; Mlonka-Mędrala, A.; Khdary, N.H.; Kalawa, W.; Nowak, W.; Mika, Ł. Possibility of Advanced Modified-Silica-Based Porous Materials Utilisation in Water Adsorption Processes—A Comparative Study. *Energies* **2022**, *15*, 368. https://doi.org/10.3390/en15010368

Academic Editors: Adrián Mota Babiloni and Laura Savoldi

Received: 22 October 2021
Accepted: 29 December 2021
Published: 5 January 2022

1. Introduction

Adsorption chillers are cooling devices that use heat instead of electricity. The main advantages of this type of chillers are quiet and easy operation, high reliability, and the absence of moving mechanical parts (except valves), as well as the possibility of water desalination [1]. However, their disadvantages include: a relatively low coefficient of cooling performance COP, usually in the range of 0.5 to 0.6, low specific cooling power SCP, small mass and heat transfer in the adsorption bed, as well as considerable dimensions and large weight [2]. To increase the competitiveness of adsorption chillers in the cooling devices market, the performance parameters must be constantly improved. Companies manufacturing sorption cooling devices are looking for more advanced adsorbate-adsorbent working pairs and construction materials.

In this study, a group of new advanced materials is considered as effective adsorbents in adsorption chillers that work in pairs with water. Nanocomposites were synthesised through incorporation of metal nanoparticles into silica matrix.

Metal-organic silica (MOS) are dynamically developing materials with significant industrial potential, because of the wide range of useful properties such as high stability and resistibility to chemical changes [3]. The mesoporous structure of the silica matrix and extensive surface area of about 1500 m^2/g [4] enables to obtain material with high sorption capacity [5,6]. Modification of MOS properties is performed through introduction of additional atoms or functional groups [4,7]. The unique features of MOS make them candidates for a variety of applications such as wastewater treatment [8], CO_2 sequestration [9], catalysis [10,11], as heavy metal adsorbents [12,13] and other toxic contaminants detectors and adsorbents [14,15].

Metal-doped mesoporous silica obtained by spray drying was examined by Endo et al. to enhance its water adsorption capacity [16]. Very promising results were obtained also by Yanagihara et al. who was testing Zr-doped two-dimensional hexagonal mesoporous silica (Zr-MPS) in working pair with water, in this case a water uptake of about 70 g/g was noted [17].

Water used as an adsorbate in this study, is a readily available, cheap and nontoxic adsorbate, which is additionally characterised by high environmental friendliness. Another benefit is the possibility of desalination of water from seas and oceans in adsorption chiller and increase of global drinking water resources [18–20]. Moreover, water is the most thermally stable adsorbate, and its heat of vaporization is higher than in the case of other refrigerants [21]. However, since its evaporation in the evaporator takes place in a low pressure environment (below atmospheric pressure), it is necessary to create a vacuum inside the system, which will reduce the reliability of the entire device due to the additional risk of leakage. In combination with an extremely tight structure, it allows to eliminate the possibility of air leakage, which could result in incorrect adsorption, evaporation or condensation. Furthermore, water has a very low saturation pressure compared to the other types of adsorbates [22].

In this study, four novel adsorbents and one reference material were analysed in a working-pair with water. The presented results cover a knowledge gap regarding the possibilities and limitations of sorbents utilization in water vapour adsorption processes, with a special input on adsorption cooling devices. A minimal pore size diameter was determined for sorbents used in working pair with water. To properly address the problem of adsorbents used in sorption cooling devices, sorption properties and structure were examined, as well as thermal properties of all analysed materials were also tested.

2. Materials and Methods

The present study is a comparative analysis of new adsorbents of high potential to be used in adsorption chillers. In addition to water uptake capacity, thermal properties and morphology were examined to make a proper comparison of the materials.

2.1. Materials

Three samples of metal-organic silica (MOS) nanomaterials, and a sample of highly porous silica sorbent were delivered by a research group from King Abdulaziz City for Science and Technology. In this study also a commercial sample of silica gel was tested as a reference material. Narrow porous silica gel, of particle size 2–7 mm was manufactured by Chemland. The photographs of the analysed sorbents are presented in Figure 1.

The samples analysed in the study were as follows:

- Metal Organic Silica: AFSPd-Cu in the form and colour similar to the fine sand, AFSPd-Cu (NP) (MOS with metal nanoparticles (NP)) material was similar to the previous one, but the colour of the sample was grey, AFSMo-Cu in the form of small blue crystals,
- High-porous silica material MPSilica: which was a very fine white powder.
- Narrow porous silica gel of particle size 2–7 mm.

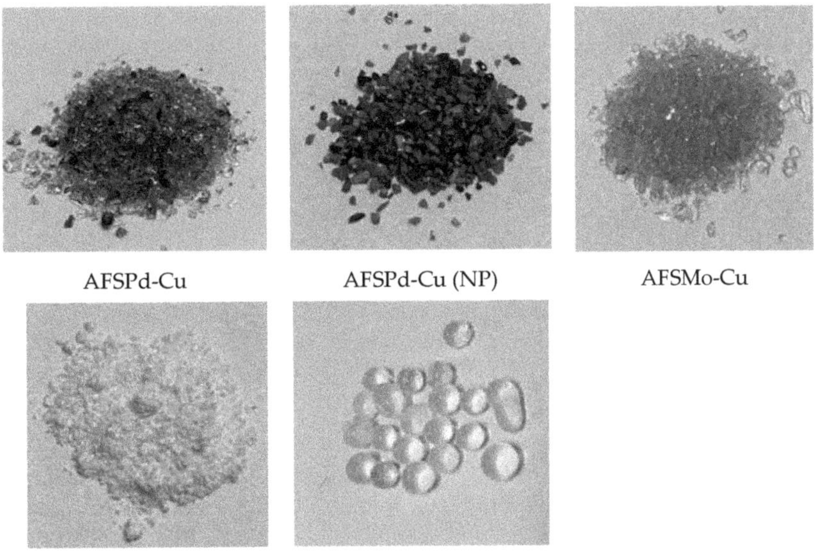

AFSPd-Cu AFSPd-Cu (NP) AFSMo-Cu

MPSilica Silica gel

Figure 1. Photographs of the samples analysed in the study.

2.2. Methods

Several analytical methods were used to determine sample morphology and structure, together with thermal and sorption properties. The STA method was used to analyse thermal decomposition of the materials in an adsorption chiller working temperature range. The experiments were carried out on a Mettler Toledo high-temperature thermogravimetric analyser. The sorbents were placed in aluminium oxide crucibles, the temperature range was from the ambient temperature to 300 °C at a constant heating rate of 10 °C/min in 50 mL/min air. TG, DSC, and DTG curves were prepared.

The morphology of the sorbents was analysed using a scanning electron microscope coupled with energy-dispersive X-ray (Nova NanoSEM 450) to identify the chemical composition of the materials. In this study, the morphology was tested at a beam acceleration voltage of 2 kV, whereas EDS tests were performed at 15 kV up to 30 kV.

Structural analyses of the materials were performed using two methods. First, the analysis was performed on an ASAP 2020 volumetric analyzer (Micromeritics) using low-pressure nitrogen physisorption at −196 °C. The specific surface area and average pore diameter of each sample were determined using the Brunauer–Emmet–Teller (BET) model based on the adsorption isotherms with P/P_0 ranging between 0.06 and 0.20. The total volume of the pores was determined by applying the Brunauer–Emmet–Teller (BET) and Barrett-Joyner-Halenda (BJH) methods to the adsorption and desorption curves. Average pore diameters were calculated on the basis of total pore volume and surface area.

Mercury intrusion porosimetry was used to determine the effective porosity, as it does not include closed porosity, which is inaccessible to the injected mercury. The method shows pore size distributions of mesopores and macropores. Quantachrome Poremaster 33 was used to define the pore size distribution in the range of ~7 nm to 1 mm.

The Laser Flash Method was used to analyse thermal properties of the sorbents in this work. The method allows to determine the thermal diffusivity coefficient of the given solid material. In this study, samples were tested using Netzsch LFA 457 MicroFlash. In this apparatus, the thermal diffusivity coefficient is defined automatically with a measurement accuracy equal to ±3%. The measurement conditions of thermal diffusivity were performed with argon flow rate of 50 mL/min at four temperatures: 30, 40, 50, and 60 °C. Before the measurement all samples were grounded into dust.

Water sorption properties of samples examined in this study were determined using Dynamic Gravimetric Vapor Sorption System DVS Vacuum. The sample mass during the adsorption and desorption processes is constantly measured by the equipment, the measurement is characterised by high sensitivity, equal to 0.1 μg. The stability of the temperature at 25 °C is equal to ±0.02 °C and the humidity conditions generated are typically in the range of ±0.1% with respect to the given value [2]. In this study, water was used as an adsorbate. Before the experiment, approximately 20 mg of sample was dried, the sample was placed in the apparatus at 100 °C for 60 min. After 60 min of stabilisation at a given process temperature, a series of 20 experimental stages of 20 min, each started. Each stage had a different setting of a relative pressure P/P_0, starting from 10% to 100%. On the basis of experimental results, adsorption and desorption isotherms were calculated. The water intake for all samples was obtained depending on its saturation pressure. The steam flow rate was set constant and equal to 15 sccm (standard cubic centimeters per minute). The experiment was carried out at four process temperatures: 30, 40, 50, and 60 °C.

3. Results and Discussion

The results of the performed experiments are presented and discussed in detail in this Section. Adsorbents were compared according to their thermal and sorption properties. An additional comparison was performed with previously published studies.

3.1. Structural Analysis and Morphology

Figure 2 shows SEM images of the sorbents analysed, the structure and morphology of the sorbents are an essential factor revealing how developed is the porous structure of the materials. More SEM images were included in Supplementary Materials in Figures S1–S5.

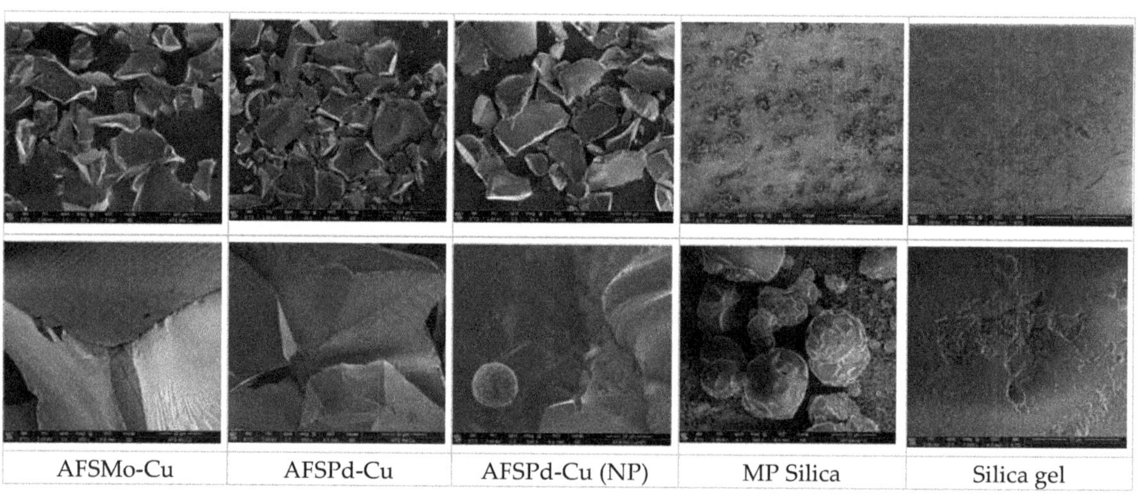

Figure 2. Morphology analysis of adsorbents analysed in this study.

MOS samples denoted as AFS group had a very similar morphology. They consist of irregular particles of various shapes and sizes. The structure of the sorbent particles was smooth and sharp. The average size of these particles was approximately 500 μm. In some images, areas where "grooves" appear on the surface or a "stepped" structure are present. Fine round pores on the surface of the particles might be observed at some points. In the case of samples doped with metal nanoparticles—AFSPd-Cu (NP), also sphere-shaped metal particles were present.

The MPSilica sorbent was a heterogeneous material. The particle size was not uniform, but most particles were characterised by an irregular spherical shape. In this sample, there are present large particles (up to approx. 80 μm), these particles had a developed surface, and smaller particles sticked or grown on them to form agglomerates. In addition, this sorbent consisted of a large number of very fine round particles, approx. 1 μm in size, which also formed agglomerates. These very small particles also stuck to larger particles. The pores were not visible in the photographs, what might be associated with the microporous structure of the material.

Silica gel is characterised by a very uniform structure, regular shape of the sorbent is recommended for sorption chiller adsorption bed. Small damages in adsorbent structure showed internal, porous morphology of the material.

EDS analysis was used to determine the chemical composition, but the results of the analysis are not clear because the peaks from some elements were poorly visible. Table 1 summarises the selected results of the quantitative EDS analysis for individual samples. The main constituent of all tested sorbents was SiO_2 with some additives which are determined in the samples names. It should be noted that the amounts of doped elements are rather small.

Table 1. Chemical composition of the analysed sorbents on the basis of EDS analysis, wt.%.

Sample	C	O	Si	F	S	Cl	Ni	Mo	Cu	Pd	Na
MP Silica (15)	-	49.2	50.8	-	-	-	-	-	-	-	-
AFS Mo-Cu	-	54.6	42.9	-	-	-	-	1.4	1.2	-	-
AFS Pd-Cu	-	56.3	40.6	-	-	1.0	-	-	0.6	1.5	-
AFS Pd-Cu (NP)	-	51.7	42.3	-	-	-	-	-	0.5	3.5	2.0
Silica gel	8.6	49	42.4	-	-	-	-	-	-	-	-

3.2. Porosity

In this study, two different metal-organic silica nanoparticles with high potential for water adsorption were selected, one of them, AFSMo-Cu, had a moderate BET surface area, not exceeding 300 m^2/g, while the AFSPd-Cu sample had a specific surface area greater than 600 m^2/g. Both samples differed in mean pore diameters, in the case of nanocomposite with molybdenum addition, the mean pore diameter was about 5 nm, whereas in the case of the Pd-doped sample, this value was two times lower. The addition of metal nanoparticles to the AFSPd-Cu sample significantly reduced the specific surface area of the material, over 10 times, and an increase in the mean pore diameter was observed, which also increased almost 10 times.

In general, metal-organic silica nanocomposites are characterised by noticeable active surface area, exceeding in some cases 1000 m^2/g [23], but in the literature MOS with lower BET surface area were also reported [3]. However, it was noted that for porous materials with developed surface area like e.g.,: MOFs, a BET surface area exceeding 7000 m^2/g is an indicator of the potential for the use of sorbents in adsorption chillers [24].

The material with the highest BET specific surface area was a highly porous silica-based material, MPSilica with a BET specific surface area exceeding 2000 m^2/g, this material was also characterised by narrow pores with a diameter of about 2.5 nm.

The analysis of summation curves as well as the pore size distribution during adsorption and desorption processes shows that, in most cases, a very wide unimodal pore distribution is observed. Table 2 summarizing the results of the gas sorption analysis shows the average values and the average values for a given maximum.

Mercury intrusion porosimetry is a technique frequently used to determine the volume of macropores and the size distribution of pores in adsorbents [25,26]. The results of mercury intrusion porosimetry—pore distribution and cumulative pore volume—are shown in Figures 3 and 4, respectively.

Table 2. Summary of BET + BJH analysis results.

Sample	BET Surface Area, m^2/g	Micropore Volume, m^2/g	External Surface Area, m^2/g	Adsorption Average Pore Diameter (4 V/A by BET), nm	Desorption Average Pore Diameter (4 V/A by BET), nm	BJH Adsorption Average Pore Diameter (4 V/A), nm	BJH Desorption Average Pore Diameter (4 V/A), nm
AFSMo-Cu	283.71	21.08	262.63	5.21	5.21	4.87	3.95
AFSPd-Cu	636.62	42.11	594.51	2.50	2.50	3.29	2.84
AFSPd-Cu (NP)	51.86	4.67	47.19	21.59	21.14	23.22	15.12
MPSilica	2144.68	0	3956.55	2.64	2.61	3.02	2.62
Silica gel	789.18	204.29	584.89	2.19	2.19	3.24	2.75

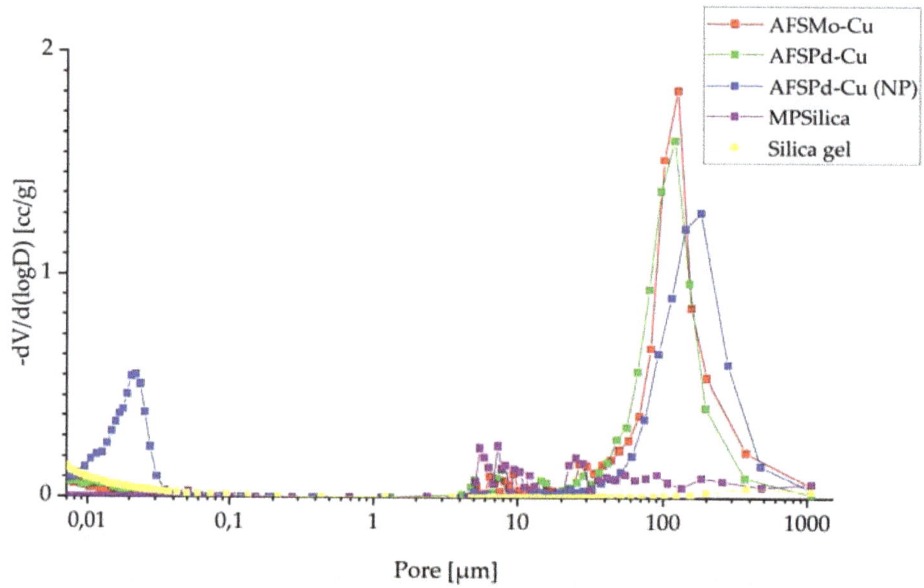

Figure 3. Pore distribution determined using mercury intrusion porosimetry.

The distribution of the pores in Figure 3 shows that for MOS materials, the highest recorded initial increase related to the presence of macropores was observed. For the range of pore diameter between 100 and 200 μm, the highest values of the total pore volume were observed, which will be influenced both by the fact that they are the pores with the largest volume, but in this case their number could also be significant. For MPSilica, the highest number of pores was in the size of 5 to 20 μm. The cumulative pore volume in Figure 4 was associated with the pore distribution and sorbents characterised by a smaller pore size had a noticeably smaller pore volume.

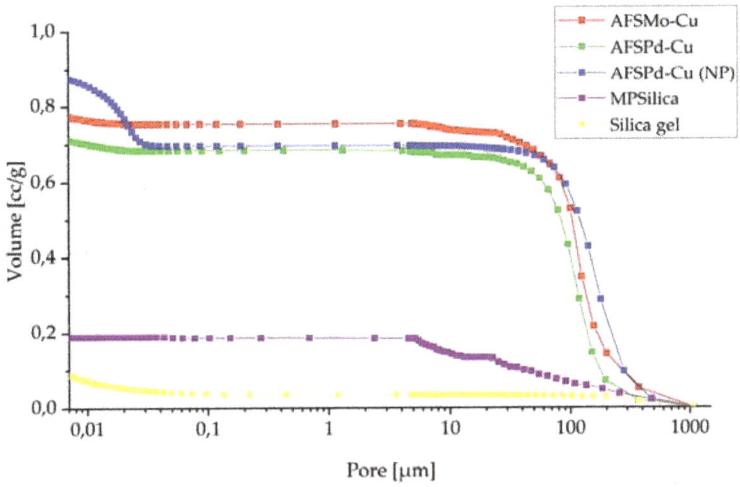

Figure 4. Cumulative pore volume determined using mercury intrusion porosimetry.

3.3. *Thermal Diffusivity Coefficient*

The results of thermal diffusivity coefficient, together with a measurement uncertainty equal to the standard deviation, are shown in Figure 5.

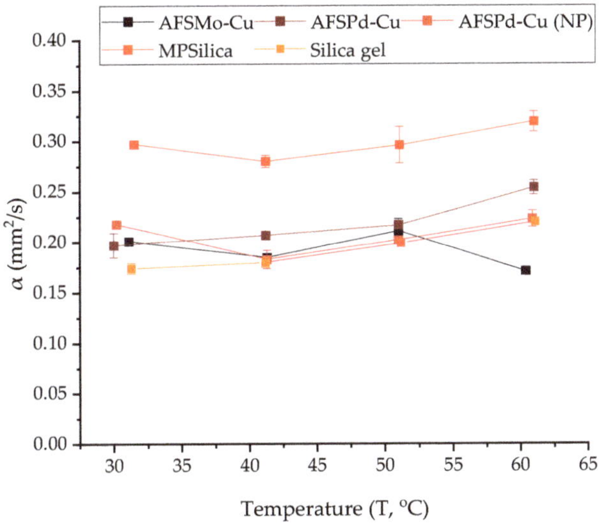

Figure 5. Thermal diffusivity coefficient of selected samples measured at four process temperatures (30, 40, 50, 60 °C).

The highest thermal diffusivity coefficient was measured for the MPSilica sample—about 0.3 mm^2/s, while the lowest was measured for silica gel—about 0.2 mm^2/s for both samples. The interesting thing is that the tested sorbents were characterised by a stable value of the thermal diffusivity coefficient. In the analysed temperature range the thermal diffusivity coefficient in all cases was constant or slightly increased with temperature. Compared to the most commonly used adsorbent in sorption cooling devices, silica gel, whose thermal diffusivity is 0.137 mm^2/s [20], thermophysical properties of the analysed sorbents are slightly higher.

3.4. Sorption Characteristics

The water intake was tested in the temperature range of 30–60 °C, the sorption isotherms are shown in Figures 6–10.

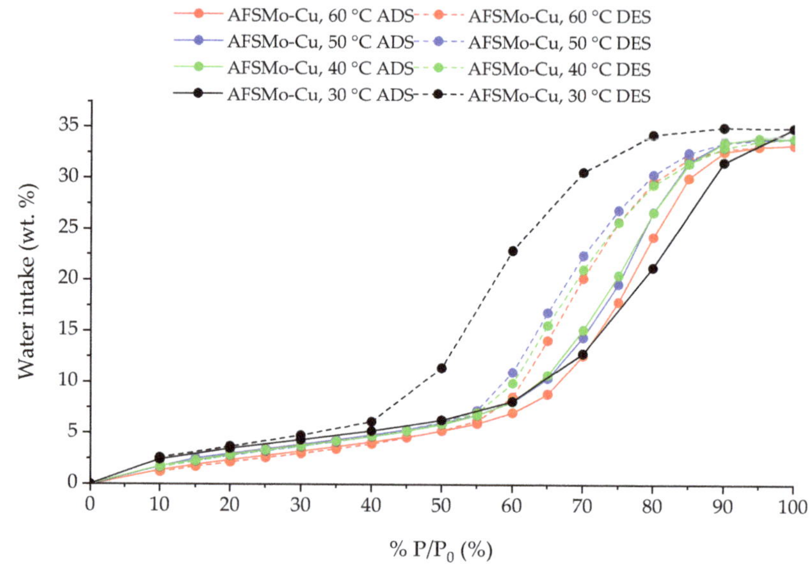

Figure 6. Adsorption and desorption isotherms for the AFSMo-Cu sample, at 30, 40, 50, and 60 °C.

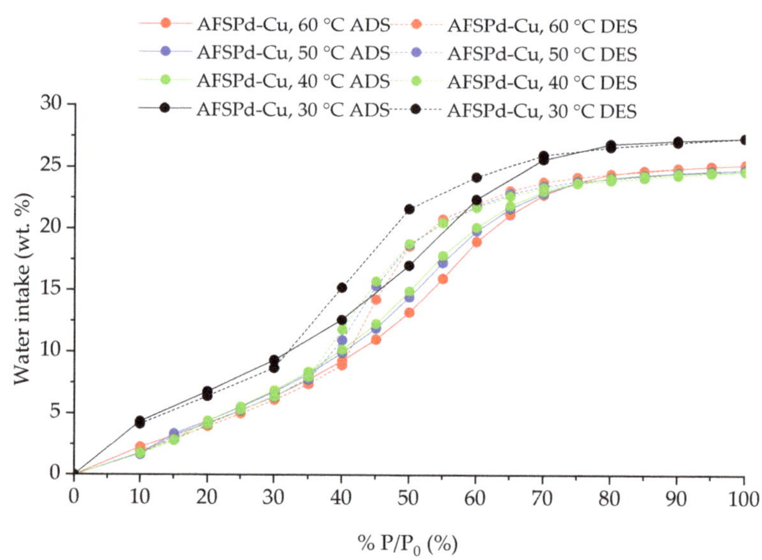

Figure 7. Adsorption and desorption isotherms for the AFSPd-Cu sample, at 30, 40, 50, and 60 °C.

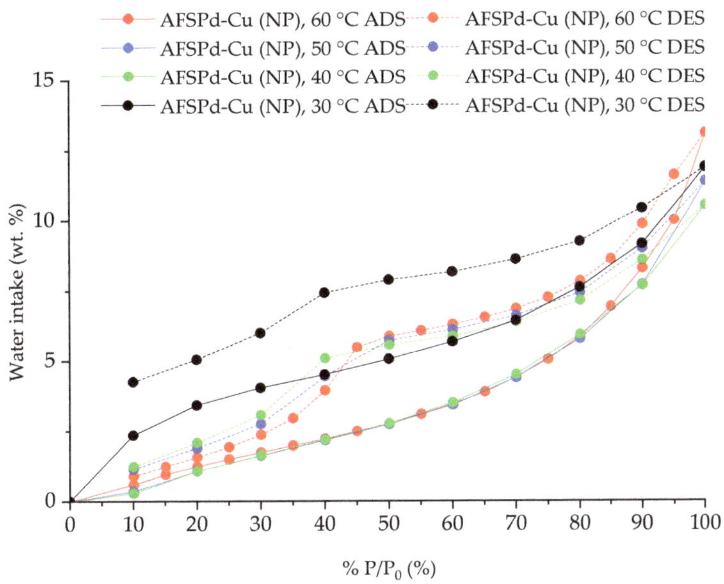

Figure 8. Adsorption and desorption isotherms for AFSPd-Cu (NP) sample, at 30, 40, 50 and 60 °C.

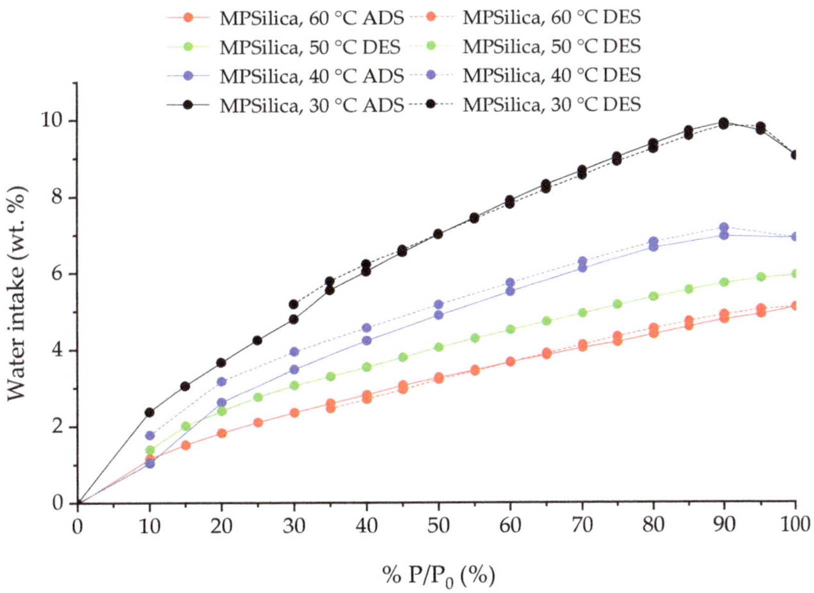

Figure 9. Adsorption and desorption isotherms for MPSilica samples, at 30, 40, 50 and 60 °C.

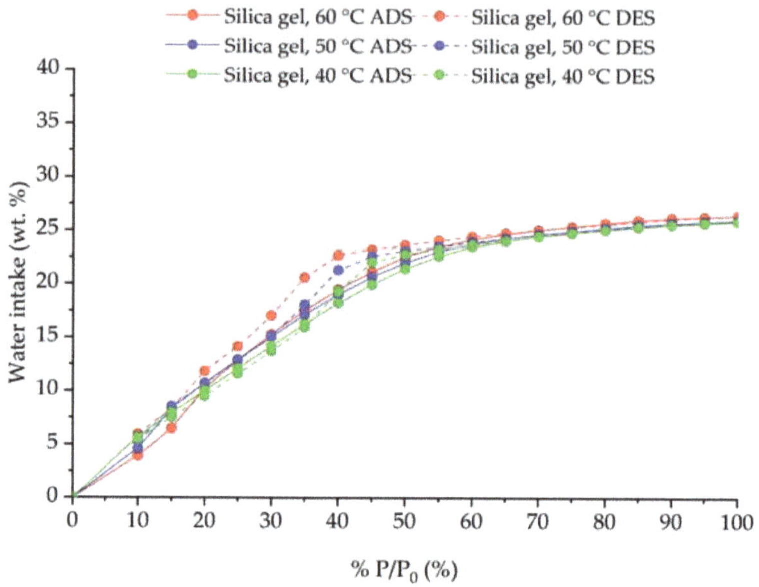

Figure 10. Adsorption and desorption isotherms for Silica gel samples, at 40, 50 and 60 °C.

The AFSMo-Cu sample was characterised by a high increase in mass in relation to the other tested sorbents. The type IV adsorption isotherm with capillary hysteresis characteristic for the H2 type was the highest at the lowest analysed temperature, equal to30 °C. The results of the BET analysis (presented in Table 2) showed that the specific surface area of this sorbent is not high, as it is less than 300 m^2/g, the pore size might be of key importance here, and it was approx. equal to 5 nm. This means that in the case of sorbents dedicated to water adsorption, the pore size (around 5 nm) and the high specific surface area are most important factors.

The highest amount of adsorbed water vapor for the AFSMo-Cu sorbent at 30 °C expressed as the percentage difference between the mass of the reference sample and the mass at a given pressure P/P_0 was 34.93%, while for 40 °C–33.90%, 50 °C–33.92% and 60 °C–33.28%.

The results shows that the sorbent can work at low process temperatures, its ability to adsorb water is not dependent on the temperature, but high hysteresis is observed at the lowest process temperature.

The AFSPd-Cu material has a structure similar to the AFSMo-Cu material, it has a much larger specific BET surface area (637 m^2/g), but a smaller pore size: approx. 2.5 nm. Therefore, a lower water vapor adsorption capacity of the adsorbent was observed.

Similarly to the previous sorbent, the adsorption isotherm has a type IV shape with a capillary hysteresis of type H2. The highest amount of adsorbed water vapor for the AFSPd-Cu sorbent at 30 °C expressed as a percentage difference between the reference mass of the sample and the mass at a given pressure P/P_0 was 27.39%, while for 40 °C–24.66%, 50 °C–24.81% and 60 °C–25.20%.

Also, in this case, it was observed that the sorbent can work at low process temperatures, its ability to adsorb water is not temperature dependent, and the highest sample mass change was obtained for the lowest process temperature.

The AFSPd-Cu (NP) sorbent is a mixture of AFSPd-Cu and metal nanoparticles (NP). For a P/P_0 of 40%, a characteristic breakdown of the adsorption isotherm was observed, most probably resulting from the presence of nanoparticles in the sample. Metal nanoparticles take the shape of the adsorption isotherm characteristic for type VI, while the AFSPd-Cu sorbent for type II, together, it can be assumed that the shape of the adsorption isotherm for the AFSPd-Cu (NP) mixture is similar to type III—very rarely present and characteristic for microporous adsorbents.

The highest amount of adsorbed water vapor for the AFSPd-Cu sorbent at 30 °C expressed as water intake at a given pressure P/P_0 was 11.91%, while for 40 °C–10.56%, 50 °C–11.41% and 60 °C–13.12%.

The tested sorbent showed poor water adsorption capacity, which resulted from the low active BET surface area, equal to 52 m^2/g.

MP Silica material was characterised by an adsorption isotherm of type I and there was practically no hysteresis, which confirms that the tested material has a microporous structure. The material was characterised by a low water adsorption capacity, although this capacity increased with the temperature decrease. The highest weight gain was observed for the process temperature of 30 °C, for the P/P_0 value of 90% and it was equal to 9.83%. For a temperature of 40 °C, the maximum for P/P_0 90% was 7.16%. On the other hand, for temperatures of 50 and 60 °C, the highest weight gain was observed for P/P_0 equal to 100% and they were 5.94% and 5.11%, respectively.

Despite the large active surface, as the BET surface area was over 2000 m^2/g, the material has a microporous structure and the pore size slightly exceeds 2.5 nm, which most likely significantly reduces its ability to adsorb water.

Based on the test results for narrow porous silica gel, it can be concluded that the adsorption isotherms for the temperatures of 40 °C, 50 °C and 60 °C at a water vapor saturation pressure from 10% to 100% P/P_0 according to the IUPAC classification take the shape characteristic for type IV isotherm. Both the shape of the adsorption isotherms and the maximum weight gain of the sorbent are similar in the range of the tested temperatures and equals to a maximum value of 26% on average. In the case of the desorption process, a slight hysteresis is observed, depending on the process temperature and it is characteristic for the type IV isotherm. The hysteresis takes a shape similar to the H2 type, which may indicate that spherical pores with numerous constrictions and open ends are present in the material.

The obtained water adsorption capacities results for analysed materials were compared with literature data and presented in Table 3.

The results of water adsorption for silica gel analysed in this study are slightly lower than in case of literature data, the reason might be a narrow porous structure of the analysed adsorbent. The sorption properties of analysed MOS materials are comparable to those achieved for zeolites, but they are definitely lower than in case of metal organic frameworks (MOFs). As the sorption properties are not the only parameter taken into consideration, but also economic factor is essential, it should be emphasised that the price of silica gels depends upon its purity and structure and in general it is less than 10 €/kg of adsorbent, whereas MOFs prices range is very wide, but starts at about 100 €/g of material. Further modifications of MOS materials might result in their enhanced water adsorption capacity and its value on adsorbent market will definitely increase.

3.5. Simultaneous Thermal Analysis

MOS and a highly porous sample of the silica-based material together with reference sample were tested up to a maximum temperature of 300 °C due to a lack of knowledge about the behaviour of the samples at higher temperatures. The results of thermal analysis are presented in Figure 11.

Table 3. Comparison of water adsorption capacities determined for sorbents analysed in this study with literature data regarding porous materials characterised by high water adsorption capacities.

No.	Material	Process Temperature	Maximal Water Loading, %	Reference
1	AFSMo-Cu	30 °C	35.01	Exp.
2	AFSMo-Cu	40 °C	33.90	Exp.
3	AFSMo-Cu	50 °C	33.92	Exp.
4	AFSMo-Cu	60 °C	33.28	Exp.
5	AFSPd-Cu	30 °C	27.39	Exp.
6	AFSPd-Cu	40 °C	24.66	Exp.
7	AFSPd-Cu	50 °C	24.81	Exp.
8	AFSPd-Cu	60 °C	25.20	Exp.
9	AFSPd-Cu (NP)	30 °C	11.91	Exp.
10	AFSPd-Cu (NP)	40 °C	10.56	Exp.
11	AFSPd-Cu (NP)	50 °C	11.41	Exp.
12	AFSPd-Cu (NP)	60 °C	13.12	Exp.
13	MPSilica	30 °C	9.83	Exp.
14	MPSilica	40 °C	7.16	Exp.
15	MPSilica	50 °C	5.93	Exp.
16	MPSilica	60 °C	5.11	Exp.
17	Silica gel	40 °C	26.39	Exp.
18	Silica gel	50 °C	25.94	Exp.
19	Silica gel	60 °C	25.86	Exp.
20	Silica gel	40 °C	30	[27]
21	TAPSO-34	40 °C	28	[27]
22	Silica gel	25 °C	34.35	[2]
23	Silica gel	40 °C	34.21	[2]
24	Silica gel	60 °C	33.79	[2]
25	MIL-100(Fe)	25 °C	90	[28]
26	MIL-100(Al)	25 °C	50	[28]
27	MOF-841	25 °C	64	[29]
28	MOF-806	25 °C	26	[29]
29	Zeolite 13X	25 °C	33	[29]
30	Zeolites	-	11–38.7	[30]
31	SMOF	-	42	[24]

AFSMo-Cu and AFSPd-Cu samples initially contained high amounts of moisture, which came from the storage of the samples. In both cases below 100 °C, a very fast evaporation of moisture from the surface of the material was observed. The samples were stable in the analysed temperatures range (up to 300 °C), however, a change in their colour was observed after the analysis, which may indicate that some changes in the structure or chemistry may have taken place under the influence of temperature. The sample with the addition of AFSPd-Cu metal nanoparticles showed much longer moisture release, practically, up to 300 °C a change in mass was observed.

For MP Silica sample, a significant loss was observed in the entire analysed temperature range. Almost half of the sample was evaporated up to the temperature of 300 °C, a major thermal decomposition process took place already at 150 °C. Silica gel sample water release was observed in a wider temperature range between 100–200 °C what might affect the kinetics of the adsorption process and the length of the cycle in an adsorption chiller.

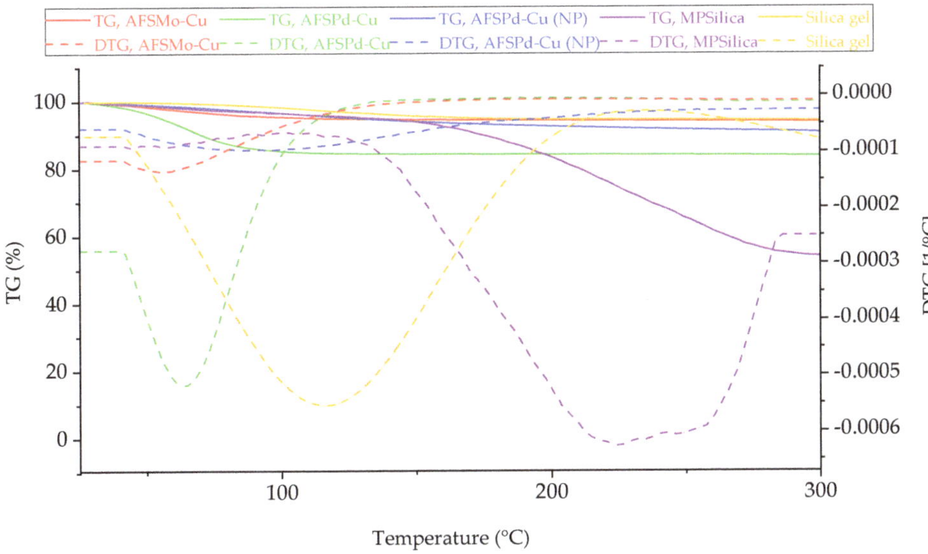

Figure 11. Thermal behaviour of analysed sorbents up to 300 °C in oxidising atmosphere.

4. Conclusions

In this study, four newly developed sorbents were analysed in terms of both, sorption and thermal properties. A possibility to use them in adsorption cooling device driven by low-temperature heat was analysed, and water intake was measured form 30 up to 60 °C. To sum up, MOS materials tested in this study presented enhanced sorption properties in comparison to narrow-porous silica gel. Further modifications of MOS pore size diameter might further increase water sorption properties of these materials and make them even more competitive on the porous materials market. However, addition of metal nanoparticles to MOS samples does not enhance sorption and thermal properties of the adsorbent.

It was noted that in water sorption processes not only active surface area is important, but also pore size defines the water intake properties of the sorbent. Pore diameters smaller than 5 nm are too narrow for water sorption process. Sample AFSMo-Cu was characterised by the smallest active surface area (283.71 m^2/g), but mean pore diameter was 5.21 nm, whereas for AFSPd-Cu, BET surface area was 636.62 m^2/g and mean pore diameter only 2.5 nm. Comparing water sorption capacities, it was noted that AFSMo-Cu sorption capacity was 33–35%, but for material with higher active surface area it was only 25–27%. Therefore, for adsorption chillers working with water as adsorbate we are looking for adsorbents of large active surface area, pores of diameter higher than 5 nm and noticeable thermal conductivity coefficient. Further modifications of AFSPd-Cu adsorbent pore size diameter might result in definitely larger sorption capacities of the material.

Supplementary Materials: The following supporting information can be downloaded at https://www.mdpi.com/article/10.3390/en15010368/s1, Figure S1: Morphology analysis of AFSMo-Cu, Figure S2: Morphology analysis of AFSPd-Cu, Figure S3: Morphology analysis of AFSPd-Cu (NP) Figure S4: Morphology Analysis of MP Silica, Figure S5: Morphology Analysis of Silica gel.

Author Contributions: Conceptualization, K.S. and A.M.-M.; methodology, A.M.-M., N.H.K., W.K. and K.S.; validation, K.S., W.N. and Ł.M.; formal analysis, K.S.; investigation, A.M.-M. and W.K.; resources, K.S., W.N. and Ł.M.; data curation, A.M.-M.; writing—original draft preparation, A.M.-M. and K.S.; writing—review and editing, K.S.; visualization, A.M.-M.; supervision, W.N. and Ł.M.; project administration, Ł.M., W.N. and K.S.; funding acquisition, Ł.M., W.N. and K.S. All authors have read and agreed to the published version of the manuscript.

Funding: This research was funded by Ministry of Science and Higher Education, Poland, grant AGH no 16.16.210.476 and partly supported by program "Excellence initiative—research university" for the AGH University of Science and Technology.

Conflicts of Interest: The authors declare no conflict of interest.

References

1. Sztekler, K.; Kalawa, W.; Mlonka-Medrala, A.; Nowak, W.; Mika, L.; Krzywanski, J.; Grabowska, K.; Sosnowski, M.; Debniak, M. The effect of adhesive additives on silica gel water sorption properties. *Entropy* **2020**, *22*, 327. [CrossRef]
2. Sztekler, K.; Kalawa, W.; Mika, Ł.; Mlonka-Mędrala, A.; Sowa, M.; Nowak, W. Effect of additives on the sorption kinetics of a silica gel bed in adsorption chiller. *Energies* **2021**, *14*, 83. [CrossRef]
3. Khdary, N.H.; Ghanem, M.A.; Merajuddine, M.G.; Bin Manie, F.M. Incorporation of Cu, Fe, Ag, and Au nanoparticles in mercapto-silica (MOS) and their CO_2 adsorption capacities. *J. CO_2 Util.* **2014**, *5*, 17–23. [CrossRef]
4. Kankala, R.K.; Zhang, H.; Liu, C.G.; Kanubaddi, K.R.; Lee, C.H.; Wang, S.B.; Cui, W.; Santos, H.A.; Lin, K.L.; Chen, A.Z. Metal species—Encapsulated mesoporous silica nanoparticles: Current advancements and latest breakthroughs. *Adv. Funct. Mater.* **2019**, *29*, 1902652. [CrossRef]
5. García-Fernández, A.; Sancenón, F.; Martínez-Máñez, R. Mesoporous silica nanoparticles for pulmonary drug delivery. *Adv. Drug Deliv. Rev.* **2021**, *177*, 113953. [CrossRef] [PubMed]
6. Chen, H.; Wang, W.; Wei, X.; Ding, J.; Yang, J. Experimental and numerical study on water sorption over modified mesoporous silica. *Adsorption* **2015**, *21*, 67–75. [CrossRef]
7. Sierosławska, A.; Borówka, A.; Rymuszka, A.; Żukociński, G.; Sobczak, K. Mesoporous silica nanoparticles containing copper or silver synthesized with a new metal source: Determination of their structure parameters and cytotoxic and irritating effects. *Toxicol. Appl. Pharmacol.* **2021**, *429*, 115685. [CrossRef]
8. Sachan, D.; Ramesh, A.; Das, G. Green synthesis of silica nanoparticles from leaf biomass and its application to remove heavy metals from synthetic wastewater: A comparative analysis. *Environ. Nanotechnol. Monit. Manag.* **2021**, *16*, 100467. [CrossRef]
9. Khdary, N.H.; Ghanem, M.A.; Abdesalam, M.E.; Al-Garadah, M.M. Sequestration of CO_2 using Cu nanoparticles supported on spherical and rod-shape mesoporous silica. *J. Saudi Chem. Soc.* **2018**, *22*, 343–351. [CrossRef]
10. Salazar Hoyos, L.A.; Faroldi, B.M.; Cornaglia, L.M. A coke-resistant catalyst for the dry reforming of methane based on Ni nanoparticles confined within rice husk-derived mesoporous materials. *Catal. Commun.* **2020**, *135*, 105898. [CrossRef]
11. Garbarino, G.; Cavattoni, T.; Riani, P.; Busca, G. Support effects in metal catalysis: A study of the behavior of unsupported and silica-supported cobalt catalysts in the hydrogenation of CO_2 at atmospheric pressure. *Catal. Today* **2020**, *345*, 213–219. [CrossRef]
12. Di Natale, F.; Gargiulo, V.; Alfè, M. Adsorption of heavy metals on silica-supported hydrophilic carbonaceous nanoparticles (SHNPs). *J. Hazard. Mater.* **2020**, *393*, 122374. [CrossRef] [PubMed]
13. El Kurdi, R.; Chebl, M.; Sillanpää, M.; El-Rassy, H.; Patra, D. Chitosan oligosaccharide/silica nanoparticles hybrid porous gel for mercury adsorption and detection. *Mater. Today Commun.* **2021**, *28*, 102707. [CrossRef]
14. Venkatesvaran, H.; Kannan, A.; Mayavan, A.; Dhanabal, A.; Gandhi, S. Metal nanoparticle ornated mesoporous silica: A potential nano-interface for uric acid detection. *Microporous Mesoporous Mater.* **2021**, *324*, 111313. [CrossRef]
15. Chirra, S.; Siliveri, S.; Gangalla, R.; Goskula, S.; Gujjula, S.R.; Adepu, A.K.; Anumula, R.; Sivasoorian, S.S.; Wang, L.F.; Narayanan, V. Synthesis of new multivalent metal ion functionalized mesoporous silica and studies of their enhanced antimicrobial and cytotoxicity activities. *J. Mater. Chem. B* **2019**, *7*, 7235–7245. [CrossRef] [PubMed]
16. Endo, A.; Inagi, Y.; Fujisaki, S.; Yamamoto, T.; Ohmori, T.; Nakaiwa, M. *Synthesis of Metal-Doped Mesoporous Silica by Spray Drying and Their Adsorption Properties of Water Vapor*; Elsevier: Amsterdam, The Netherlands, 2007; Volume 165, ISBN 044452178X.
17. Yanagihara, H.; Yamashita, K.; Endo, A.; Daiguji, H. Adsorption-desorption and transport of water in two-dimensional hexagonal mesoporous silica. *J. Phys. Chem. C* **2013**, *117*, 21795–21802. [CrossRef]
18. Oh, B.; Jabbari-hichri, A.; Bennici, S.; Auroux, A. Enhancing the heat storage density of silica—Alumina by addition. *Sol. Energy Mater. Sol. Cells* **2015**, *140*, 351–360. [CrossRef]
19. Hashimoto, K.; Kumagai, N.; Izumiya, K.; Takano, H.; Shinomiya, H.; Sasaki, Y.; Yoshida, T.; Kato, Z. The use of renewable energy in the form of methane via electrolytic hydrogen generation using carbon dioxide as the feedstock. *Appl. Surf. Sci.* **2016**, *388*, 608–615. [CrossRef]
20. Kulakowska, A.; Pajdak, A.; Krzywanski, J.; Grabowska, K.; Zylka, A.; Sosnowski, M.; Wesolowska, M.; Sztekler, K.; Nowak, W. Effect of metal and carbon nanotube additives on the thermal diffusivity of a silica-gel-based adsorption bed. *Energies* **2020**, *13*, 1391. [CrossRef]
21. Dammel, F.; Ochterbeck, J.M.; Stephan, P. Thermodynamics. In *Springer Handbook of Mechanical Engineering*; Grote, K., Antonsson, E., Eds.; Springer: Berlin/Heidelberg, Germany, 2009; pp. 223–294.
22. Wang, L.W.; Wang, R.Z.; Oliveira, R.G. A review on adsorption working pairs for refrigeration. *Renew. Sustain. Energy Rev.* **2009**, *13*, 518–534. [CrossRef]
23. Furtado, A.M.B.; Liu, J.; Wang, Y.; Levan, M.D. Mesoporous silica-metal organic composite: Synthesis, characterization, and ammonia adsorption. *J. Mater. Chem.* **2011**, *21*, 6698–6706. [CrossRef]

24. Palash, M.L.; Jahan, I.; Rupam, T.H.; Harish, S.; Saha, B.B. Novel technique for improving the water adsorption isotherms of metal-organic frameworks for performance enhancement of adsorption driven chillers. *Inorg. Chim. Acta* **2020**, *501*, 119313. [CrossRef]
25. Yang, Z.; Cao, L.; Li, J.; Lin, J.; Wang, J. Facile synthesis of Cu-BDC/Poly(*N*-methylol acrylamide) HIPE monoliths via CO_2-in-water emulsion stabilized by metal-organic framework. *Polymer* **2018**, *153*, 17–23. [CrossRef]
26. Feng, Y.; Hu, H.; Wang, Z.; Du, Y.; Zhong, L.; Zhang, C.; Jiang, Y.; Jia, S.; Cui, J. Three-dimensional ordered magnetic macroporous metal-organic frameworks for enzyme immobilization. *J. Colloid Interface Sci.* **2021**, *590*, 436–445. [CrossRef]
27. Hinze, M.; Ranft, F.; Drummer, D.; Schwieger, W. Reduction of the heat capacity in low-temperature adsorption chillers using thermally conductive polymers as heat exchangers material. *Energy Convers. Manag.* **2017**, *145*, 378–385. [CrossRef]
28. Karmakar, A.; Prabakaran, V.; Zhao, D.; Chua, K.J. A review of metal-organic frameworks (MOFs) as energy-efficient desiccants for adsorption driven heat-transformation applications. *Appl. Energy* **2020**, *269*, 115070. [CrossRef]
29. Furukawa, H.; Gándara, F.; Zhang, Y.B.; Jiang, J.; Queen, W.L.; Hudson, M.R.; Yaghi, O.M. Water adsorption in porous metal-organic frameworks and related materials. *J. Am. Chem. Soc.* **2014**, *136*, 4369–4381. [CrossRef]
30. Tatlier, M.; Munz, G.; Henninger, S.K. Relation of water adsorption capacities of zeolites with their structural properties. *Microporous Mesoporous Mater.* **2018**, *264*, 70–75. [CrossRef]

MDPI
St. Alban-Anlage 66
4052 Basel
Switzerland
Tel. +41 61 683 77 34
Fax +41 61 302 89 18
www.mdpi.com

Energies Editorial Office
E-mail: energies@mdpi.com
www.mdpi.com/journal/energies

www.ingramcontent.com/pod-product-compliance
Lightning Source LLC
Chambersburg PA
CBHW061847210926
51850CB00006B/993
9783036559131